Computer Analysis Methods for Power Systems

COMPUTER ANALYSIS METHODS FOR POWER SYSTEMS

G. T. HEYDT
Purdue University

Macmillan Publishing Company
New York
Collier Macmillan Publishers
London

Printed in the United States of America

Macmillan Publishing Company
866 Third Avenue, New York, New York 10022

Collier Macmillan Canada, Inc.

Library of Congress Cataloging in Publication Data

Heydt, G. T. (Gerald T.)
Computer analysis methods for power systems.

Includes bibliograhies and index.
1. Electric power systems--Data processing. I. Title.
TK1005.H49 1986 621.319'028'5 85-10520
ISBN 0-02-352860-5

Printing: 1 2 3 4 5 6 7 8 Year: 6 7 8 0 1 2 3 4 5

ISBN 0-02-352860-5

Preface

This textbook has been assembled from lecture notes for a course in digital computational methods in electric power engineering as offered at Purdue University. The course has also been offered in short course format at various electric utility companies in the United States and South America. The course is intended for graduate or senior Electrical Engineering students although anyone with an elementary knowledge of three phase circuits and computer programming experience should have no difficulty. Graduate students from Mechanical and Nuclear Engineering have taken the course with good success.

The objectives of this text are broadly stated as being twofold: graduate or senior level exposure to computer analysis of electric power systems and as a tutorial text for engineers in the power industry. The approaches taken in most topics presented in the text reflect this twofold objective: a student should be exposed to the theoretical basis for each method and technique, but the practicing engineer needs a clearly stated procedure unencumbered by theoretical developments. The reader will have to judge whether the balance was properly struck. Exercises are given at the end of each chapter which are intended to illustrate the theory and also practical pitfalls, "tricks", and numerical values. Generally, the exercises are divided into two types: one requires computer programming, the other does not. When this text is used in connection with a graduate or senior level class, it is expected that the students will have already had a course in power engineering using a text such as *Introduction to Electric Power Engineering* by R. Schultz and R. Smith or *Elements of Power System Analysis* by W. Stevenson. While there is overlap between these introductory texts and the present text, an attempt was made to minimize the overlap and deal with the subject matter in more depth and technical rigor. Also, the present text strongly emphasizes methods using digital computers, such as sparsity programming methods, *in situ* techniques, and other memory and processing time efficient algorithms.

It is required that the reader have a background in computer programming. The familiarity with Fortran is very desirable since most production programs in power engineering are written in this language. In power engineering, it is common for software to be "traded" or otherwise exchanged between companies and other agencies. It is not usual for software to be used on several diverse computer systems. If exotic, machine specific versions of Fortran were used, it would be difficult to retain enough software flexibility to run a given program on several different computers. The use of American National Standards Institute (ANSI) Fortran is desirable since the student is well advised to conform to usual power engineering practices and standards. Familiarity with the use of subscripted variables, complex arithmetic, alphameric quantities and subroutines is required.

Concerning the required knowledge in alternating current circuits, the student should be familiar with active and reactive power, the Kirchhoff laws, power factor, and complex (apparent) power. It is desirable that the student have some background in the solution of alternating current circuits, including an exposure to symmetrical components.

Chapter 2 contains the fundamental matrix quantities upon which much of the text is based. The principal topics presented are the technologies of the bus impedance matrix and bus admittance matrix. The important concept for the student to retain in view is that these matrices are nothing more than the Kirchhoff laws written in compact notation. A discussion of mutual coupling is included.

Chapter 3 is a "change of pace" from the preceding chapter in that attention turns to programming techniques. In some disciplines, programming techniques are not emphasized because the algorithms used do not have extensive time or memory requirements; in power engineering, on the other hand, the existence of programming techniques often make an algorithm practical. For example, the technique of triangular factorization to solve

$$Ax = b$$

renders the Newton Raphson method practical for the power flow study of large networks. The triangular factors of the Jacobian are sparse, and sparsity programming is used to solve the network. If the triangular factors of the Jacobian were not sparse, the Newton Raphson method would have an inordinate memory requirement, and most commercially available computers could not be used.

Chapters 4 and 5 deal with the power flow problem. The emphasis is on the Newton Raphson method in polar form. This is the method most often used in industry for power flow studies. Experience has shown that actual programming of the algorithms described is very motivational to the student. For this reason, the exercises are considered particularly important in this chapter. Chapter 5 gives two important modifications of the power flow study for fast solutions: Stott's decoupled power flow algorithm and approximate solutions (the method of distribution factors).

Chapter 6 deals with optimal dispatch methods. For reasons of comprehension and history, the equal incremental cost rule is described. Emphasis in this chapter, however, is on B-coefficients and gradient methods which employ a more exact formulation. These methods are essentially the minimization of a cost function subject to the constraint that the power flow equations must be satisfied. The topic of inequality constraints is also presented.

Chapter 7 concerns computer analysis of faults. The topic is presented for the single phase case in a simple way. The extension to the three phase case is presented in terms of symmetrical components. This topic is presented as a matrix transformation.

The concept of power system stability is discussed in Chapter VIII. The emphasis in this chapter is on the algorithms suitable for computer solution. Both the transient and small signal cases are considered. Numerical integration methods are organized into the predictor and predictor-corrector types. In the former case, the Adams predictor formulas and the trapezoidal rule are most commonly used. Several direct alternatives to numerical integration are also discussed: Liapunov methods, small signal analytical methods, and more exotic techniques. The Liapunov approach is based on calculation of a measure of system "energy", V, and the demonstration that this function is of constant sign but decreasing in magnitude. The technique is known as the second method of Liapunov or the "direct method."

The final chapter is an attempt at treating some stochastic phenomena in power engineering. Perhaps the best known methods in this domain are techniques for load forecasting. Topics in power system reliability and stochastic power flow studies are presented as an introduction to stochastic methods in power engineering. In the area of capacity outage tables, both the traditional and approximate methods are presented. The reader should have an elementary familiarity with probability theory and random variables in order to facilitate the study of the topics of this chapter. This familiarity includes the definitions, properties, and interrelationships between probability density and distribution functions. Also, a knowledge of Fourier transforms will help in the understanding of the development of the first and second characteristic functions.

This text has been used at Purdue University for several years. Also, it has been used in several short courses sponsored by the Purdue Electric Power Center. Thanks are due to many keen-eyed students who spotted errors and contributed directly to the text, particularly Buck Johnson, who is somewhat of an expert on the topic of pseudoinverses (some results of Mr. Johnson's Master's thesis are included in the text). Some of the exercises have been used at Purdue, and several have been inspired by sponsors of the Purdue Electric Power Center. Although most of the text was developed at Purdue, several sections were completed during a sabbatical leave spent at the University of Nevada, Reno. The author gratefully acknowledges his colleagues' help at the University of Nevada, particularly that of Dr. Mehdi Etezadi-Amoli, whose remarks on modifications to the Newton-Raphson algorithm were very useful.

I would like to thank Dr. A. H. El-Abiad, Professor Emeritus of Purdue University (presently at the University of Petroleum and Minerals, Dhahran, Saudi Arabia) for his comments and useful discussions. Dr. El-Abiad is credited with several algorithms now in common use in power flow studies and related areas. I consider Ahmed El-Abiad not only knowledgeable in power engineering, but, more important, in the philosophy of man seeking to use science to his benefit yet retaining a professional approach to problems of human engineering. Also, Professor Homer E. Brown of North Carolina State University provided very useful information particularly on the bus impedance matrix. Professor Peter W. Sauer of the University of Illinois provided valuable insight into fast power flow study formulation, and he was kind enough to permit reprinting of his details of the Jacobian matrix calculation for Newton-Raphson power flow studies. Dr. S. W. Anderson of the Commonwealth Edison Company, Chicago, provided the basic theory of power system distribution factors. Thanks also goes to Dr. B. Wollenberg for proofreading the text and Mr. R. L. Baum for special assistance on several problems. Finally, Mrs. Vicky Spence typed (and retyped) most of the manuscript. Thank you Vicky!

A well known investigator of vector–matrix analysis methods once compared this area of study to a gold field in which many sourdoughs are panning for gold. There are many aspects of vital problems that are left unsolved--and even a beginning sourdough might happen upon a very useful technique or algorithm. As the author of this text, I find it interesting to conjecture whether a reader might find his or her fortune in this area.

G. T. Heydt

Contents

Computer Analysis Methods for Power Systems

CHAPTER 1

Introduction

1.1

MODERN ANALYSIS METHODS AND ELECTRIC POWER ENGINEERING

The description and analysis of electric power systems has traditionally been given in scalar terms. The Kirchhoff laws and interrelationships between currents, voltage, and power are easily written and manipulated for power systems; but when system size exceeds a few buses, circuit data and equations become so voluminous that scalar notation is not convenient, and analysis without a computer is tedious. It is natural to turn to vector notation of nodal voltages and mesh currents. In this notation, the Kirchhoff laws are written by "building" the matrix coefficients occurring in these equations – and the term vector–matrix is used to descriptively denote the notation. This book focuses generally on vector–matrix methods of power system analysis and, in particular, on techniques suitable for digital computer analysis.

In North America prior to 1940 few (if any) electric power circuits of great complexity existed. Electric power systems were characterized by generation consisting of a few machines geographically close to each other and connected through transmission circuits of substantial capacity to subtransmission and distribution circuits. Interconnection of circuits was not widespread. Design and analysis problems centered about devices such as generators, transformers, phase shifters, insulators, and transmission lines. The electronic digital computer was in its infancy and was largely unused in electric power system analysis and design.

In the period 1940–1950, and shortly thereafter, an average growth rate of 8 to 15% in electric power demand per annum created an ever–increasing need for more transmission circuits and more sources of generation. Also, the advantages of increased reliability realized through interconnection of several power systems gave rise to systems of more extensive size. The concept of exchange of electric power to assist a neighboring company when the need occurs resulted in the formulation of power pools. In subsequent years, interconnection and growth continued. A direct result of the decrease in the operational margin of transmission circuits and generation levels coupled with the increased interconnection of transmission circuits was the requirement for rapid, accurate analysis methods for large electrical networks. This mode of analysis is not relegated to highly developed parts of the world; in underdeveloped parts of the world, special problems associated with long transmission circuits, complex exciters for synchronous generators, and demanding economic problems also require solution methods that can be implemented on a digital computer.

Systems of more than a dozen circuit nodes or buses become difficult to handle in scalar notation. Computation by hand is impractical. Vector-matrix notation offers a versatile notation that is amenable to digital computation and, in many ways, insensitive to system size. The latter advantage is particularly attractive since, for example, analysis of one 80–bus system is much more difficult than the analyses of two 40–bus systems. Modern power networks consist of thousands of buses, are loaded in the gigavoltampere range, and consist of transmission voltages up to 765 kV. Direct–current transmission systems may be embedded in the AC network, and generation is derived from a variety of types of generation from nuclear (characterized by thousands of MVA output) through low–level hydro sources. The only practical way to describe such a complex electric circuit is through computer–stored, vector–matrix data. It is no exaggeration to claim that the modern interconnected power system is one of the largest, if not *the* largest, nonlinear, dynamic, multivariate, high–order systems in existence.

The first algorithms for power system analysis suitable for computer implementation appeared in the 1940s. In 1961, a study of transmission voltages and currents of the entire United States (simultaneously) was performed. This type of study is known as a power flow study; studies for systems of over 10,000 buses have been performed. Future trends focus on high–speed power flow studies, approximate methods, microprocessor applications, techniques to improve operating economy, reliability, stability enhancement, and methods to include phenomena previously not considered. Although electrification of cities has been with us for more than a century and interconnection of power systems has been a common practice for nearly half as long, contemporary requirements for low operating and construction costs have resulted in increased interest in automated, optimal, and accurate methods of power system analysis, operating, and design.

TERMINOLOGY AND NOTATION

The following terms are used throughout this book.

Matrix. A matrix is a rectangular array of data. Usually,these data are numeric. The array is denoted symbolically without the use of bold type or other special printing format since the context of the use of the notation will always specify the dimensions of the matrix. If Z is an m by n matrix, then $(Z)_{ij}$ denotes the row i, column j entry of Z. Other notation used includes

$$det\ (A) = \text{determinant of square matrix } A$$

$$tr\ (A) = \sum_{i=1}^{n} (A)_{ii} = \text{trace of square, } n \text{ by } n \text{ matrix } A$$

$$row_i\ (A) = \text{row } i \text{ of matrix } A$$

$$col_j\ (A) = \text{column } j \text{ of matrix } A.$$

The upper right triangle of a square matrix A is that portion for which $(A)_{ij}$ lies in $i \leq j$. The lower left triangle is $i \geq j$. These terms are abbreviated URT and LLT, respectively.

Axis of a matrix. The l axis of matrix M is the entries in the ℓ row and ℓ column of M. Element $(M)_{\ell\ell}$ is called the pivot element of the l axis.

Diagonal matrix. A diagonal matrix is a square matrix with all zero entries off the principal diagonal.

Bus. A bus is a circuit node or equipotential region in space. Practically, a power system bus is a metallic bar or other form of circuit node. The term *busbar* is occasionally found in the literature. Bus voltages are measured with respect to (i.e., "referred to") a reference bus. The ground bus is a node corresponding to earth ground. Often, but not always, the ground bus is chosen as reference. When the reference bus is not specified, the ground bus is usually assumed to be the reference for all voltages. This is the case since it is particularly convenient to measure voltages with respect to ground. A bus at which the net active power injected into the power system is positive is usually called a *generator bus* or an *intertie bus* (depending on the source of the power). A bus at which the net active power injected is negative is usually called a *load bus* or an *intertie bus* (depending on the destination of that power).

Swing bus. A swing bus is a circuit node whose voltage magnitude and angle are fixed. The voltage source at the swing bus is a perfect source known as the *swing machine* or *slack machine.* Note that the swing bus is voltage regulated (i.e., the voltage amplitude is held fixed). The phase angle at the swing bus is chosen as the system reference phase and hence this angle is fixed. The term *swing* originates from the concept that a generator at a swing bus has a torque angle and excitation which vary or "swing" as the demand changes. The variation is such as to produce fixed bus voltage.

Eigenvalue and eigenvector of a matrix. Consider an n by n matrix, A. The n eigenvalues of A are denoted as λ and are implicitly defined by

$$det\ (A - \lambda I) = 0,$$

where I is the identity matrix. This determinant is an nth–order polynomial in λ. The equation is known as the ***characteristic equation*** and it has n roots, $\lambda_1, \lambda_2, \cdots, \lambda_n$. For the case of nonrepeated eigenvalues ($\lambda_i \neq \lambda_j$, all i, j), the following is an implicit definition for eigenvector e_i corresponding to eigenvalue λ_i

$$(A - \lambda_i I)e_i = 0 \qquad ||e_i|| \neq 0.$$

Spectral radius. The spectral radius of a square matrix is the radius of the smallest circle centered at the origin in the complex plane such that all eigenvalues are enclosed. Denoted ρ, the spectral radius is

$$\rho = \max_i (|\lambda_i|).$$

Hermitian operation. The notation $(\cdot)^H$ denotes the Hermitian operation on a matrix. The operation is defined on matrix M by

$$(M^H)_{ij} = (M)_{ji}^*,$$

where $(\cdot)^*$ denotes complex conjugation. The Hermitian operation is also known as complex conjugate transposition. When matrix M consists of all real entries, M^H is equivalent to transposition, M^t,

$$(M^t)_{ij} = (M)_{ji}.$$

Note that a matrix for which $M^H = M^{-1}$ is termed *unitary*.

Symmetric matrix. A symmetric matrix is a matrix, M, such that

$$M^t = M.$$

The term *skew symmetric* implies that $M^t = -M$.

Hermitian matrix. A Hermitian matrix is a matrix, M, such that

$$M^H = M.$$

Real matrix. A real matrix is a matrix, M, such that

$$M^* = M.$$

Sparse matrix. A matrix is sparse if the majority of its entries are zero. The percent sparsity is

$$100\ \frac{number\ of\ zero\ entries}{total\ number\ of\ entries} = percent\ sparsity.$$

Matrix inverse. The right inverse of matrix A is A^{-1} if

$$AA^{-1} = I$$

where I is the identity. Similarly, the left inverse of B is B^{-1} if

$$B^{-1}B = I.$$

If either the right or left inverse of a matrix exists, it is equal to the other inverse. Also, a matrix for which $A^t = A^{-1}$ is termed *orthonormal.*

If a matrix inverse does not exist, the matrix is said to be *singular.* Otherwise, it is *nonsingular.* An operation closely related to the matrix inverse is the pseudoinverse or Moore–Penrose pseudoinverse. The pseudoinverse of matrix A is denoted A^+ and is described in Appendix B.

1.3

POWER SYSTEM DATA IN A DIGITAL COMPUTER

Power system data fall into two broad types: numeric and alphameric. *Numeric data* include line impedances, transformer reactances, load and generation levels, and other system data. These data are generally complex quantities and most computer languages require that the labels for these quantities be "declared" as complex. In Fortran, the quantity Z is declared complex using

```
COMPLEX Z
```

Alphameric data include such elements as bus names and component labels. Alphameric (or Hollerith) quantities usually do not have to be declared in computer languages; they are often handled in the same way as one might handle integers (except that input/output usually requires specialized alphameric formats). Alphameric data are usually read and assembled into a "dictionary" which effectively assigns a number to the input label. For example, if input bus names are given as

ADAVEN
S. ADAVEN
KINGSTON
UNION

the bus dictionary appears as

External Name	Internal Name
ADAVEN	1
S. ADAVEN	2
KINGSTON	3
UNION	4

Usually, the happenstance appearance of names in the external input data results in the order of the dictionary. Thus transmission line data given as

ADAVEN	S. ADAVEN	0.01	0.10
S. ADAVEN	KINGSTON	0.05	0.43
S. ADAVEN	UNION	0.05	0.05
S. ADAVEN	WILLOUGHBY	0.05	0.07
WILLOUGHBY	LIBERTY	0.01	0.09

result in the bus name directory

ADAVEN 1
S. ADAVEN 2
KINGSTON 3

UNION 4
WILLOUGHBY 5
LIBERTY 6

This is the case since ADAVEN appears in the line list first, S. ADAVEN second (S. ADAVEN is then repeated), followed by KINGSTON, and so on. Note that the two floating point quantities in the transmission line list represent line resistance and reactance.

External bus names, lines, and other power system components observed in the field will often appear in the digital computer in list form. A *bus list* is a list of system nodes or buses, and the list contains each system bus in a unique entry. Typical bus lists are shown above. A *line list* consists of unique entries keyed to the bus list; this array is a list of lines in the system. By way of example, consider a system of 10 lines:

AJAX to KING	$0.01 + j0.1$
AJAX to LINCOLN	$0.01 + j0.1$
AJAX to UNION	$0.02 + j0.2$
UNION to OX BOW	0.001
OX BOW to WILL	0.001
WILL to KING	0.001
KING to EDISON	$0.01 + j0.1$
EDISON to UNION	$0.05 + j0.5$
EDISON to ROCK	0.001
UNION to ROCK	0.0005

The bus list is found by scanning each line and searching for a bus name not yet in the bus list:

AJAX 1
KING 2
LINCOLN 3
UNION 4
OX BOW 5
WILL 6
EDISON 7
ROCK 8

As the bus list is being constructed, the line list is also formed. The first line, AJAX to KING, is read and "AJAX" is passed to a subroutine for lookup in the bus list. Since it is the first bus processed, it is not found and therefore the subroutine must add it to the list as the first entry and assign a number. Thus AJAX becomes bus 1. Similarly, KING becomes bus 2. The second line processed, AJAX to LINCOLN, is different in that AJAX has already been seen. The lookup subroutine simply returns a 1 for AJAX; the subroutine returns a 3 for LINCOLN, however, and this name is added to the bus list. A Fortran subroutine suitable for the lookup process is as follows:

```
      SUBROUTINE LOOKUP (NAME, I)
      DIMENSION LIST (100)
      DATA NBLANK/10H          /
      DO 1 J=1, 100
      I=J
      IF(LIST(J).EQ.NAME) GO TO 2
      IF(LIST(J).EQ.NBLANK) GO TO 3
1     CONTINUE
      STOP 700
```

```
2       RETURN
3       LIST(I)=NAME
        RETURN
        END
```

To use this subroutine, LIST must be "cleared" to all blanks before use; that is, LIST must be set equal to the Hollerith designation for 10 blanks, 10H   , before the first call to LOOKUP. Also, if STOP 700 is executed, more than 100 buses have been encountered and the dimension of LIST must be increased. As a result of the use of this subroutine for the line list given above, a bus list is formed which is a dictionary between bus names and internal code number. The line list is readily formed for this example as each line is read. The result is as follows:

LINEA	LINEB	LINE NUMBER
1	2	1
1	3	2
1	4	3
4	5	4
5	6	5
6	2	6
2	7	7
7	4	8
7	8	9
4	8	10

The arrays LINEA and LINEB form the line list. Thus LINEA(1) and LINEB(1) is line 1, LINEA(2) and LINEB(2) is line 2, and so on. It is not necessary to store the line numbers, since these are sequential and simply the subscript of LINEA and LINEB. The arrays LINEA and LINEB are *parallel vectors* since they are synchronized to each other: the ith entry in LINEA corresponds to the ith entry in LINEB. The line impedances are stored in a *primitive impedance list* or primitive impedance matrix. These line data are keyed to the line number and are stored in a complex vector that is in parallel with the line list:

LINE NUMBER	IMPEDANCE	
1	0.01	0.1
2	0.01	0.1
3	0.02	0.2
4	0.001	0.0
5	0.001	0.0
6	0.001	0.0
7	0.01	0.1
8	0.05	0.5
9	0.001	0.0
10	0.0005	0.0

Other possible features of the line and bus lists include:

i. Keying transmission components to the line list so that transformers, circuit breakers, and switches may be identified as being in a particular line.

ii. Outage information describing which lines are out of service (outages).

iii. Statistical information.

iv. In the case of a bus list, the per unit voltage base for each bus may be stored in a vector parallel to the bus dictionary.

This section concludes with a caution and remark to potential programmers of power engineering algorithms. At the time of writing, the principal computer language used in power engineering is Fortran. In Europe, Algol is occasionally used, and there are some computer-specific codes written in other languages. Some students have used Pascal for applications that do not involve complex numbers, but the great majority of implementations are in Fortran. Commercial codes are usually ***not*** computer or compiler specific; in these codes, the following are avoided in order to render the Fortran "noncontroversial": assigned GO TO, EQUIVALENCE, negative subscripts, ENTRY statements, and calculations in indices of DO loops.

1.4 FUNDAMENTALS OF ALTERNATING CURRENT CIRCUITS

Power system analysis is generally done in the sinusoidal steady state in which voltages and current are sinusoids of fixed frequency and phase angle. Let V denote a bus voltage phasor and I the bus injection current phasor; then the active power, P, injected into that bus is

$$P = |V||I| \cos \phi,$$

where ϕ is the angle between phasors V and I and $|\cdot|$ denotes the magnitude of the phasor. The injected reactive voltamperes is

$$Q = |V||I| \sin \phi$$

$$\phi = arg(V) - arg(I)$$

where arg $(\cdot)$ denotes the phase angle of a phasor. The complex voltamperes (apparent voltamperes) injected is

$$S = VI^*$$

$$= |V||I| \underline{/arg\,(V) - arg\,(I)}$$

$$= P + jQ.$$

The quantity $\cos \phi$ is called the *power factor* and is said to be "leading" when I leads V, "lagging" when I lags V. In three phase circuits, V and I are often expressed in *per unit*. In such a case, P, Q, and S are also in per unit on a consistent base.

For a power system of n buses plus one reference bus, the current injection vector, I_{bus},

$$I_{bus} = \begin{pmatrix} I_1 \\ I_2 \\ \vdots \\ I_n \end{pmatrix},$$

represents the n injection currents which find a return path through the reference bus. Also, the bus voltage vector, V_{bus},

$$V_{bus} = \begin{bmatrix} V_1 \\ V_2 \\ \vdots \\ V_n \end{bmatrix},$$

consists of the n bus voltages measured with respect to the reference bus. The total apparent power injected into the system is S,

$$S = I_{bus}^H V_{bus},$$

S is a scalar that is the sum of the complex quantities $I_1^* V_1$, $I_{12}^* V_{12}$, and so on.

In power system engineering, the bus current vector, I_{bus}, is usually written as an injection current. Despite the feeling that a load represents an outward flow of active power, loads are written in this injection convention. For the case of an authentic load, the real part of $I_i^* V_i$ will be negative, since active power injected from a load into the system is negative.

The fundamental laws describing the relationships between voltages and currents in a power system are the following Kirchhoff laws:

1. *Kirchhoff voltage law:* The phasor sum of voltages around a closed loop in a power system is zero.

2. *Kirchhoff current law:* The phasor sum of currents entering a bus in a power system is zero.

There is a considerable literature of power engineering. References [1–10] are cited at the end of this chapter as modern textbooks on power engineering fundamentals. Reference [11] is now out of print, but is given as a classic text on computer analysis methods for power systems. References [12–18] are newer and more advanced textbooks on specific topics cited in the bibliography. Reference [19] contains many survey papers on a range of topics in systems engineering for power system analysis. In 1983, a text [20] on many of the topics covered in this book became available – it is exceptionally strong in HVDC systems and it is indispensible in this area.

Bibliography

[1] O. Elgerd, *Electric Energy Systems Theory: An Introduction,* McGraw–Hill, New York, 1978.

[2] R. Shultz and R. Smith, *Introduction to Electric Power Engineering,* Harper & Row, New York, 1984.

[3] W. Stevenson, *Elements of Power System Analysis,* 4th ed., McGraw–Hill, New York, 1982.

[4] H. E. Brown, *Solution of Large Networks by Matrix Methods,* Wiley, New York, 1975.

[5] J. Neuenswander, *Modern Power Systems,* International Press, Scranton, Pa, 1971.

[6] B. Weedy, *Electric Power Systems,* Wiley, New York, 1979.

[7] C. Gross, *Power System Analysis,* Wiley, New York, 1979.

[8] O. Elgerd, *Electric Energy Systems Theory,* McGraw-Hill, New York, 1982.

[9] A. Knable, *Electrical Power Systems Engineering,* McGraw-Hill, New York, 1967.

[10] R. Sullivan, *Power System Planning,* McGraw-Hill, New York, 1977.

[11] G. Stagg and A. El-Abiad, *Computer Methods in Power System Analysis,* McGraw-Hill, New York, 1968.

[12] H. Happ, *Diakoptics and Networks,* Academic Press, New York, 1971.

[13] H. Happ, *Piecewise Methods and Applications to Power Systems,* Wiley, New York, 1980.

[14] J. Endrenyi, *Reliability Modeling in Electric Power Systems,* Wiley, New York, 1978.

[15] R. Stein and W. Hunt, *Electric Power System Components,* Van Nostrand Reinhold, New York, 1979.

[16] M. El-Hawary, *Electrical Power Systems Design and Analysis,* Reston, Reston, VA, 1983.

[17] E. Kimbark, *Power System Stability,* Synchronous Machines, Dover, New York, 1956.

[18] P. Anderson, and A. Fouad, *Power System Control and Stability,* Iowa State University Press, Ames, Iowa, 1977.

[19] United Engineering Foundation, *Proc., Systems Engineering for Power,* Henniker, N.H., September 1975.

[20] J. Arrillaga, C. Arnold, and B. Harker, *Computer Modeling of Electric Power Systems,* Wiley (Interscience Publication), Chichester, UK, 1983.

CHAPTER 2

Power System Matrices

2.1

INCIDENCE AND CONNECTION MATRICES

Perhaps the most elementary task facing the engineer in the application of computers to power system studies is the representation of the system configuration in the machine. To the human being, this task is solved by visual inspection of a power system drawing; in the computer, the task is less evident. The general terms *incidence matrix* and *connection matrix* are used to characterize data arrays that describe such attributes as incidence of lines to system buses, incidence of loop circuits to sections of the power system that lie in circuit meshes, and connections between buses. Much of the information in such arrays is binary in nature (i.e., a bus is connected to another bus or it is not; a mesh has a common branch with another mesh or it does not). The notion of a binary valued matrix is introduced in order to characterize this information. A *binary valued matrix* is a matrix whose entries are binary (Boolean) variables. Note that these incidence matrices do not give the impedance information that is necessary for electrical analysis; only connection information is stored. Matrices that give impedance information are discussed in subsequent sections.

There are several types of incidence and connection matrices. The binary bus connection matrix, B, is introduced as an example. This square binary matrix is defined for an n bus power system by

$$(B)_{ij} = \begin{cases} 1 \text{ bus } i \text{ connected to bus } j \text{ by a line} \\ 1 \; i = j \\ 0 \text{ otherwise} \end{cases} \qquad i, j = 1, 2, ..., n.$$

Note that "1" denotes Boolean "TRUE" and "0" denotes Boolean "FALSE" and these integers should not be confused with actual numerical quantities. Evidently, B is symmetric. Before proceeding to determine other properties of B, it is necessary to introduce a few operations for binary valued matrices.

Negation. $\bar{A}$ is the negation of binary matrix A. All TRUE elements of A are replaced by FALSE, and all FALSE entries by TRUE.

Boolean AND. $A \circ B$ is the "AND" of binary matrices A and B, where

$$(A \circ B)_{ij} = (A)_{i1}(B)_{1j} + (A)_{i2}(B)_{2j} + \cdots + (A)_{im}(B)_{mj}$$

$$i = 1, 2, ..., r; \qquad j = 1, 2, ..., c$$

where $(A)_{il}(B)_{lj}$ denotes the Boolean AND of these binary scalar variables, the notation $(+)$ denotes the Boolean OR, and m is both the number of columns of A and the number of rows of B. Note that $A \circ B$ has dimension r by c, where A is r by m and B is m by c. Although similar to the conventional matrix multiply operation, the Boolean AND is *not* identical to numerical multiplication.

Boolean OR. $A + B$ denotes the Boolean OR operation between binary valued matrices A and B where

$$(A + B)_{ij} = (A)_{ij} + (B)_{ij}$$

for all i and j. It is necessary that A and B have the same dimensions. Although similar to the matrix addition operation, the Boolean OR operation is *not* identical to numerical addition.

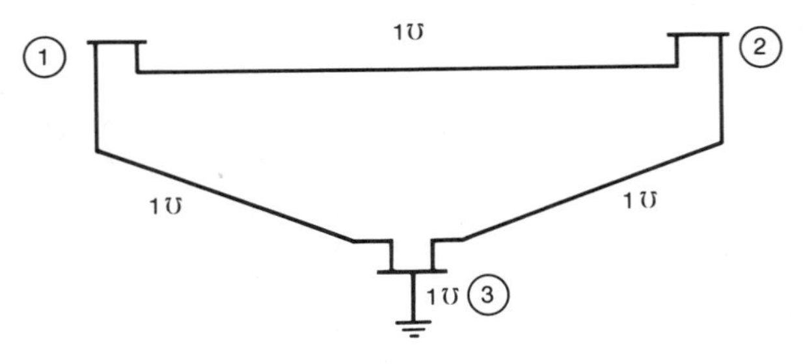

FIGURE 2.1 Three-bus system used in Example 2.1.

The properties of the binary bus connection matrix are found by observation of the steps in forming Boolean AND and OR operations on the B matrix. The operation $B \circ B$, for example, produces a square matrix of dimension n for an n−bus system where $(B \circ B)_{ij}$ is 1 when buses i and j are joined by a line *or* are joined through an intervening bus by a line. Otherwise, $(B \circ B)_{ij}$ is 0. Figure 2.1 provides an example:

$$B = \begin{array}{c} \\ 1 \\ 2 \\ 3 \\ 4 \\ 5 \end{array} \begin{array}{c} \begin{array}{ccccc} 1 & 2 & 3 & 4 & 5 \end{array} \\ \begin{bmatrix} 1 & 1 & 0 & 1 & 0 \\ 1 & 1 & 1 & 1 & 0 \\ 0 & 1 & 1 & 0 & 1 \\ 1 & 1 & 0 & 1 & 0 \\ 0 & 0 & 1 & 0 & 1 \end{bmatrix} \end{array}$$

$$B \circ B = \begin{array}{c} \\ 1 \\ 2 \\ 3 \\ 4 \\ 5 \end{array} \begin{array}{c} \begin{array}{ccccc} 1 & 2 & 3 & 4 & 5 \end{array} \\ \begin{bmatrix} 1 & 1 & 1 & 1 & 0 \\ 1 & 1 & 1 & 1 & 1 \\ 1 & 1 & 1 & 1 & 1 \\ 1 & 1 & 1 & 1 & 0 \\ 0 & 1 & 1 & 0 & 1 \end{bmatrix} \end{array}.$$

Note that B and $B \circ B$ are written above with row and column identifiers at the left and above the matrix so that the buses corresponding to the row and columns are readily identified. In this example, the rows and columns of B and $B \circ B$ correspond to the bus name in simple numerical order. Such is not always the case, unfortunately. Observe that $(B \circ B)_{13}$ is 1 since buses 1 and 3 are joined via an intervening bus (in fact, bus 2). The entry $(B \circ B)_{15}$ is 0 since there does not exist a way to go from bus 1 to bus 5 via two lines and one intervening bus.

This property of the binary bus connection matrix B will be generalized momentarily, but first note that matrix $B \circ B \circ B$ is also square and contains 1 in positions corresponding to buses joined by three or fewer lines and two or less intervening buses. The diagonal entries of B, $B \circ B$, $B \circ B \circ B$, and so on, are all 1. The property is generalized as follows:

Boolean AND operations on the binary bus connection matrix. If B is the binary bus connection matrix for an n–bus power system, and the notation $(B)^{(m)}$,

$$(B)^{(m)} = \underbrace{B \circ B \circ B \circ \cdots \circ B}_{\leftarrow m \text{ times} \rightarrow},$$

is used to denote repeated AND operations, and $(B)^{(1)} = B$, then $(B)^{(m)}$ consists of all 0s except in the diagonal position, where 1s appear. Also, the position ij of $(B)^{(m)}$ contains 1s if and only if buses i and j are joined via m lines or less (hence giving $m - 1$ intervening buses). Furthermore, for an n–bus system, $(B)^{(n-1)}$ consists of all 1s when all system buses are connected to the system.

This property of B is demonstrated by induction with $m = 1$ as an obvious case. For $m = 2$,

$$(B)_{(2)}^{ij} = (B)_{i1}(B)_{1j} + (B)_{i2}(B)_{2j} + \cdots + (B)_{in}(B)_{nj}. \qquad (2.1)$$

If for any k, $1 \leq k \leq n$, $(B)_{ik}(B)_{kj}$ happens to be 1, we immediately infer that both $(B)_{ik}$ and $(B)_{kj}$ are 1. This means that bus i is joined to bus k and bus k is joined to bus j. Hence buses i and j are joined via two lines (in fact the lines are ik and kj) and there exists on intervening bus (bus k). When $i = j$, $((B)^{(2)})_{ij}$ is 1 since in Eq. (2.1), the jth entry in the Boolean OR is $(B)_{jj}\ (B)_{jj}$ and $(B)_{jj} = 1$. When all entries in the Boolean OR in (2.1) are 0, then $(B)_{ij} = 0$ and no two lines exist that join buses i and j (and i and j are not joined by one line). The property is therefore demonstrated for $m = 2$.

Assume that the property holds for $m = \ell$ and complete the demonstration by induction. It is necessary to show that the property holds for $m = \ell + 1$. If the property holds for $m = \ell$, $((B)^{(\ell)})_{ij}$ is 1 when $i = j$ or is connected to j via ℓ fewer lines. Then

$$(B)_{ij}^{(\ell+1)} = (B)_{i1}^{(\ell)}\ (B)_{1j} + \cdots + (B)_{in}^{(\ell)}\ (B)_{nj} \ .$$

If one or more terms on the right hand side are 1, it is concluded that there exists some k such that $B_{ik}^{(\ell)}\ B_{kj}$ is 1. Therefore,

$$B_{ik}^{(\ell)} = 1 \qquad\qquad B_{kj} = 1$$

and ℓ intervening lines join buses i and j. The reasoning is similar to the $m = 2$ case. This completes the demonstration by induction.

One application of the binary bus connection matrix is to determine whether a given bus is close to another bus. The buses i and j may be evaluated for "proximity" by examining $B_{ij}^{(m)}$; if this entry is a 1, buses i and j are connected to each other through m or fewer lines. If m is set at 6 (for example), the degree to which buses i and j are near to each other is readily assessed. A potential application for the evaluation of bus pairs for proximities occurs in bus fault studies: If a fault study is to be done for a short circuit at bus k, only buses "nearby" bus k need to be examined.

Another application of the B matrix is to determine whether a given system is connected. The matrix $(B)^{(n-1)}$ must be all 1s when an n-bus system is connected since the farthest possible configuration between bus 1 and bus n occurs when 1 and n are on opposite ends of a radial string of buses. In such a radial configuration, there are $n-1$ intervening lines and $n-2$ intervening buses between buses 1 and n. If $(B)^{(n-1)}$ contains a 0, the system is disconnected.

The $\overline{B}$ matrix is the bus disconnection matrix and the matrices $\overline{B} \circ \overline{B}$, $\overline{B} \circ \overline{B} \circ \overline{B}$, and so on, have properties similar to $B^{(2)}$, $B^{(3)}$, and so on [1]. Allowing D to denote the bus disconnection matrix, it is easy to show that

$$B^{(k)}[D^{(\ell)}] = B^{(k-\ell)}.$$

An analog to the bus connection matrix may be drawn for circuit loops. The *loop incidence matrix* is defined similarly to B except that 1s occur in positions corresponding to loops with common branches. The loop incidence matrix has properties similar to B in that proximity of loops may be assessed by raising the loop incidence matrix to higher powers.

There are many other connection and incidence matrices that reflect how loops, lines, and buses are interconnected. Interest in these matrices is generally relegated to circuit theory. The *line incidence matrix, L,* has some application in power engineering and for this reason it is selected as a final example of these matrices. The line incidence matrix is defined by

$$(L)_{ij} = \begin{cases} 1 & \text{line } i \text{ starts at bus } j \\ -1 & \text{line } i \text{ ends at bus } j. \\ 0 & \text{otherwise} \end{cases}$$

For a system of ℓ lines and n buses, L is ℓ by n. This matrix is used to calculate the voltage difference between the buses at the terminals of each system line. Let V_{line} be a ℓ-vector of "line voltages" (i.e., voltage drops across each system line); then

$$V_{\text{line}} = L V_{bus}.$$

For the purpose of calculating of line voltage drops, each line must be considered to be directed (i.e., having a start or higher voltage bus and an end or lower voltage bus). This convention is reflected in the definition of line start and end in matrix L. While it is unimportant which bus is selected as the start of a line and which as the end, once the convention is established, it must be consistent in the definition of elements of V_{line}.

2.2

BUS ADMITTANCE MATRIX

The connection and incidence matrices described above do not give impedance information and cannot be used to model the Ohm's law. It is reasonable to turn attention to both Ohm's law and the Kirchhoff laws themselves to find a matrix formulation that may be used to solve for electrical quantities. This approach yields two matrices when the Kirchhoff current law is examined; these are called ***bus*** matrices since they result from the fact that the sum of currents entering a ***bus*** is zero. The Kirchhoff voltage law results in ***loop*** matrices since it deals with voltages in a circuit ***loop***. The bus matrix written in terms of admittances comes directly from the Kirchhoff current law applied to each circuit bus other than the reference bus. Let bus j be an arbitrary circuit bus other than the reference bus (which is denoted as bus 0). Then

$$\sum_{i=1}^{n} \bar{I}_{ji} = I_j, \tag{2.2}$$

where I_j is the current injected from an external source into the system at bus j, and $\bar{I}_{ji}$ is the current leaving bus j to bus i via a line. The bar above $\bar{I}_{ji}$ denotes a line current rather than a bus injection current. Evidently,

$$\bar{I}_{ji} = \bar{y}_{ji}(V_j - V_i) \tag{2.3}$$

where $\bar{y}_{ji}$ is the line admittance of a line from j to i. The bar notation in $\bar{y}_{ji}$ indicates that this admittance is a line admittance. Certain elements of (2.2) are zero: the current $\bar{I}_{jj}$ is zero since no line joins bus j with bus j [Eq. (2.3) gives zero for $\bar{I}_{jj}$]. Also, $\bar{I}_{ji}$ will be zero if no line exists between j and i, the "line impedance" is infinite, and $\bar{y}_{ji}$ is zero. Equation (2.3) is substituted into (2.2) to yield

$$\sum_{i=1}^{n} \bar{y}_{ji}(V_j - V_i) = I_j .$$

Note that V_j and V_i are the bus voltages with respect to bus 0, the reference bus. This equation is written for $i = 1, 2, ..., n$ and the coefficient of voltage V_j in the Kirchhoff equation at bus j is $\sum_{i=1}^{n} \bar{y}_{ji}$, while the coefficient of V_i $(i \neq j)$ in the jth equation is $-\bar{y}_{ji}$. If all n equations are written,

$$I_1 = (\sum_{i=1}^{n} \bar{y}_{1i})V_1 + (-\bar{y}_{12})V_2 + \cdots + (-\bar{y}_{1n})V_n$$

$$I_2 = (-\bar{y}_{21})V_1 + (\sum_{i=1}^{n} \bar{y}_{2i})V_2 + \cdots + (-\bar{y}_{2n})V_n \tag{2.4}$$

$$\vdots$$

$$I_n = (-\bar{y}_{n1})V_1 + (-\bar{y}_{n2})V_2 + \cdots + (\sum_{i=1}^{n} \bar{y}_{ni})V_n .$$

Observe the pattern of the coefficients of the bus voltages on the rigth–hand side of (2.4). This pattern will be generalized as the rule for the formation of a matrix of coefficients of V. These n equations are compactly written as

$$I_{bus} = Y_{bus} V_{bus},$$

where I_{bus} is the bus injection current vector representing the n phasor injection currents, Y_{bus} is an n by n matrix of coefficients formed as indicated above, and V_{bus} is the bus voltage vector of n phasor voltages referenced to bus 0. Note that injection currents find a return path through bus 0. Also note that the Kirchhoff current law is not applied to bus 0 in (2.4); however, the total current entering bus 0 is the negative sum of the currents $I_1, I_2, ..., I_n$. Therefore, the Kirchhoff current law at bus 0 is

$$-\sum_{j=1}^{n} I_j = -[(-\bar{y}_{21} - \bar{y}_{31} - \cdots - \bar{y}_{n1} + \sum_{i=1}^{n} \bar{y}_{1i})V_1$$

$$+ (-\bar{y}_{12} - \bar{y}_{32} - \cdots - \bar{y}_{n2} + \sum_{i=1}^{n} \bar{y}_{2i})V_2 + \cdots]$$

which is readily simplified as

$$I_0 = -\bar{y}_{01}V_1 - \bar{y}_{02}V_2 - \cdots - \bar{y}_{0n}V_n, \tag{2.5}$$

where $\bar{y}_{0k}$ is the primitive admittance joining buses 0 and k. Since (2.5) is a linear combination of the previous n equations, it *cannot* be used to obtain additional information beyond that given in (2.4). Discussion below on the indefinite bus admittance matrix further relates to inclusion of (2.5) to the set of n simultaneous equations.

The foregoing discussion should be familiar to persons who have studied elementary circuit theory. The bus admittance matrix is, in fact, the familiar nodal admittance used in circuit analysis. The terminology of power engineering is used here to introduce the notions and notation of this field. Although the nodal analysis methodology of circuit theory is certainly applicable here, it is worthwhile to point out that line impedances in power systems are generally small and, as a consequence, coefficients of V_{bus} in (2.4) are rather large numbers. In nodal admittance matrices used in circuit analysis, this is ***not*** generally the case. ***Line impedances*** are termed ***branch impedances*** in circuit theory terminology. There is an additional point to be made here concerning the very elementary transmission line model used above. A transmission line is a distributed parameter system which exhibits nonlinearities (primarily due to the skin effect present in the ac conduction process and the variation of resistance with conductor temperature) and effects due to the shunt displacement current. The latter is due to "charging current" and this effect is usually included by considering a shunt tie to ground of $j\frac{B}{2}$ mhos at the two terminals of the transmission line. In this approach, B is the lumped shunt capacitive susceptance which is an approximate model of the distributed parameter phenomenon. This modeling consideration and others are discussed in more detail in Section 4.6.

Inspection of the coefficients of the V's in (2.4) reveals a rule for the formation of Y_{bus}. The following rules are used to "build" the bus admittance matrix referenced to bus 0:

The Y_{bus} Building Algorithm

$(Y)_{jj}$: The diagonal entries of Y_{bus} are found by summing the primitive admittance of lines and ties to the reference at bus j.

$(Y)_{ij}$: The off diagonal entries are the negatives of the admittances of lines between buses i and j. If there is no line between i and j, this term is zero.

Usually, bus 0 is the ground bus, but other buses may be chosen as the reference for specialized application. Since bus 0 is commonly chosen as the reference, the notation Y_{bus} is used to denote the bus admittance matrix referenced to ground. If some other bus is used as the reference, for example bus k, the notation $Y_{bus}^{(k)}$ is used.

Inspection of the Y_{bus} building algorithm reveals several properties of the matrix:

1. The matrix is complex and symmetric.

2. The matrix is sparse since each bus is connected to only a few nearby buses (this causes many $\bar{y}_{ij} = 0$ terms). Note that the percent sparsity of Y_{bus} is identical to that of the binary bus connection matrix, B. The percent sparsity generally increases with the matrix dimension.

3. Provided that there is a net nonzero admittance tie to the reference bus, Y_{bus} may be inverted to find V_{bus} from $I_{bus} = Y_{bus} V_{bus}$. Hence, when there is a net nonzero admittance tie to the reference bus, Y_{bus} is nonsingular. By a similar argument, when there are no ties to the reference bus, Y_{bus} is singular.

4. When I_{bus} is augmented to include I_0, the injection current into the reference bus, and V_{bus} is augmented with a zero element denoted as V_0 and interpreted as the bus voltage at the reference with respect to itself. The matrix Y_{bus} is augmented with a row and column (i.e., an axis) corresponding to the reference bus. This $n+1$ by $n+1$ admittance matrix is called the ***indefinite bus admittance matrix*** and it is always singular.

5. The indefinite bus admittance matrix entries are formed with exactly the same building algorithm as that of the conventional bus admittance matrix.

Example 2.1

Find the bus admittance matrix for the system shown in Figure 2.1. Use the ground bus as the reference.

Solution

The admittance sum at bus 1 is 1 + 1 mho. Therefore, $Y_{11} = 2$ mhos. The sums at buses 2 and 3 are

$$Y_{22} = 1 + 1 = 2 \text{ mhos}$$
$$Y_{33} = 1 + 1 + 1 = 3 \text{ mhos.}$$

Note that at bus 3, the 1 mho load admittance must be summed into Y_{33}. The off diagonals are negatives of the admittance ties:

$Y_{12} = Y_{21}$ = the negative of the line admittance between buses 1 and 2

$$Y_{12} = Y_{21} = -1$$
$$Y_{31} = Y_{13} = -1$$
$$Y_{32} = Y_{23} = -1.$$

Therefore,

$$Y_{bus} = \begin{bmatrix} 2 & -1 & -1 \\ -1 & 2 & -1 \\ -1 & -1 & 3 \end{bmatrix}.$$

In Example 2.1 it is possible to augment Y_{bus} with one additional axis corresponding to the reference bus. The additional axis corresponds to the reference bus and is formed in the same way as any other axis. If the $(n+1)$st axis is axis 0, entries Y_{00} and Y_{0i} are found by

$$Y_{00} = \sum \bar{y} \text{ connected to bus } 0$$

$$Y_{0i} = -\bar{y} \text{ between buses 0 and } i, \quad i \neq 0.$$

The row sum of every row of the augmented matrix, Y_{bus}^{aug}, is zero. The column sums are also zero. The 0 axis is the sum of all other n axes and therefore

$$det\ (Y_{bus}^{aug}) = 0.$$

As observed earlier, Y_{bus}^{aug} is singular. In the case of the previous example, the Y_{bus}^{aug} matrix is

$$Y_{bus}^{aug} = \begin{bmatrix} 2 & -1 & -1 & 0 \\ -1 & 2 & -1 & 0 \\ -1 & -1 & 3 & -1 \\ 0 & 0 & -1 & 1 \end{bmatrix}.$$

CHANGES IN THE BUS ADMITTANCE MATRIX TO REFLECT SYSTEM CHANGES

The formation of Y_{bus} is rapid using the Y_{bus} building algorithm. A line is read from the line list, terminals are identified using a lookup subroutine, and $\bar{y}$ is added to the ii and jj positions of Y_{bus} and subtracted from the ij and ji positions. The computer Y_{bus} array must be cleared to zero before using this algorithm. The number of computer operations required to form Y_{bus} is proportional to the number of entries in the line list. Having formed Y_{bus}, it is reasonable to ask how Y_{bus} may be modified to reflect changes in the system or changes in the choice of a reference bus. These areas are examined in this section.

Line Outages

Let Y_{bus} be given for a system in which a line is to be outaged. To find Y_{bus} after the line has been outaged, note that mathematically, the line outage is equivalent to adding a new line of admittance $-\bar{y}_{out}$ in parallel with the line to be outaged, where $\bar{y}_{out}$ is the admittance of the line to be outaged. This is the case since the parallel combination of $\bar{y}_{out}$ and $-\bar{y}_{out}$ is a zero admittance or an open circuit. Since the outage of a line is equivalent to addition of a line of admittance $-\bar{y}_{out}$, the procedure to modify Y_{bus} is

Diagonal entries: Add $-\bar{y}_{out}$ to the ii and jj entries.

Off diagonal entries: Add $\bar{y}_{out}$ to the ij and ji entries.

Deletion of a Bus

Consider a system of n buses in which m buses ($m < n$) are considered to be "uninteresting" and about which no electrical information is sought. Partition I_{bus} and V_{bus} such that the m buses to be ignored in the study are represented in the lower m rows of these vectors:

$$I_{bus} = \left[\begin{array}{c} I_a \\ \hline I_b \end{array}\right] \begin{array}{c} \uparrow \\ n-m \\ \downarrow \\ \uparrow \\ m \\ \downarrow \end{array} \qquad V_{bus} = \left[\begin{array}{c} V_a \\ \hline V_b \end{array}\right] \begin{array}{c} \uparrow \\ n-m \\ \downarrow \\ \uparrow \\ m \\ \downarrow \end{array}$$

Consider the case in which no injection currents are received at the m "uninteresting" buses,

$$I_b = 0.$$

It is desirable to find Y_{bus}^{eq}, an admittance matrix for the power system equivalent circuit with the m "uninteresting" buses deleted. The matrix Y_{bus}^{eq} must be $(n - m)$ by $(n - m)$. To find Y_{bus}^{eq}, note that

$$I_{bus} = Y_{bus} V_{bus}$$

$$\left[\begin{array}{c} I_a \\ \hline I_b \end{array}\right] = \left[\begin{array}{c|c} Y_{aa} & Y_{ab} \\ \hline Y_{ab}^t & Y_{bb} \end{array}\right] \left[\begin{array}{c} V_a \\ \hline V_b \end{array}\right]. \tag{2.6}$$

In (2.6), Y_{bus} has been partitioned to correspond to the partitioning of I_{bus} and V_{bus}. To find Y_{bus}^{eq}, it is necessary to eliminate I_b and V_b. The I_b term is simply zero. V_b is eliminated from (2.6) by expanding the bottom m rows and solving for V_b,

$$I_b = 0 = Y_{ab}^t V_a + Y_{bb} V_b$$

$$V_b = -Y_{bb}^{-1} Y_{ab}^t V_a .$$

Then V_b is substituted into the top $n - m$ rows,

$$\begin{aligned} I_a &= Y_{aa} V_a + Y_{ab} V_b \\ &= Y_{aa} V_a - Y_{ab} Y_{bb}^{-1} Y_{ab}^t V_a \\ &= [Y_{aa} - Y_{ab} Y_{bb}^{-1} Y_{ab}^t] V_a . \end{aligned} \tag{2.7}$$

In (2.7), the total coefficient of V_a on the right hand side is Y_{bus}^{eq},

$$Y_{bus}^{eq} = Y_{aa} - Y_{ab} Y_{bb}^{-1} Y_{ab}^t . \tag{2.8}$$

Equation (2.8) is known as the *Kron reduction formula* and it is essentially the elimination of m variables from a set of n simultaneous equations.

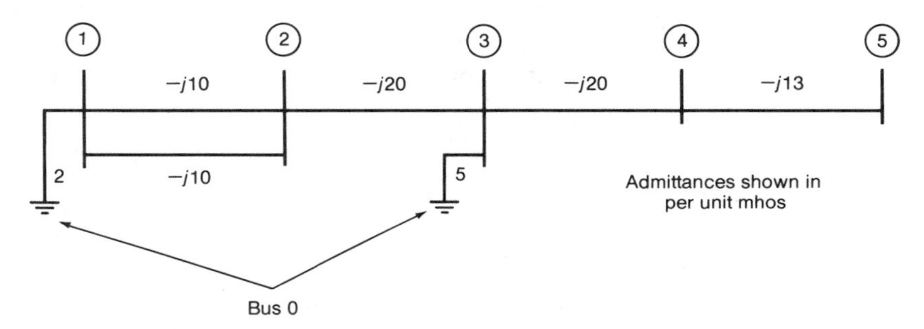

FIGURE 2.2 System used in Examples 2.2, 2.3, and 2.4.

Example 2.2

In this example, a single bus is eliminated from Y_{bus}. The reason for eliminating the bus is that no electrical information is sought at this bus and it is desired to work with a smaller Y_{bus} matrix. The system used is shown in Figure 2.2. The Y_{bus} matrix for the entire system is

$$Y_{bus} = \begin{bmatrix} 2-j20 & j20 & 0 & 0 & 0 \\ & -j40 & j20 & 0 & 0 \\ & & 5-j40 & j20 & 0 \\ & & & -j33 & j13 \\ \text{symmetric} & & & & -j13 \end{bmatrix}$$

where the axes are arranged numerically. In this example, bus 4 is to be deleted from consideration. It is assumed that I_4 is identically zero.

Solution

Rearrange the axes of Y_{bus} such that bus 4 is last (the order of the axes is now 1, 2, 3, 5, 4)

$$Y_{bus} = \begin{bmatrix} 2-j20 & j20 & 0 & 0 & 0 \\ j20 & -j40 & j20 & 0 & 0 \\ 0 & j20 & 5-j40 & 0 & j20 \\ 0 & 0 & 0 & -j13 & j13 \\ 0 & 0 & j20 & j13 & -j33 \end{bmatrix}$$

Partition Y_{bus} so that the first four axes are to be retained and the last deleted. Using the notation in (2.8),

$$Y_{aa} = \begin{bmatrix} 2-j20 & j20 & 0 & 0 \\ j20 & -j40 & j20 & 0 \\ 0 & j20 & 5-j40 & 0 \\ 0 & 0 & 0 & -j13 \end{bmatrix}$$

$$Y_{ab} = \begin{bmatrix} 0 \\ 0 \\ j20 \\ j13 \end{bmatrix} \qquad Y_{bb} = [-j33] \; .$$

Since only one bus is to be eliminated, Y_{bb} is a scalar. Apply the Kron reduction formula

$$Y_{bus}^{eq} = \begin{bmatrix} 2-j20 & j20 & 0 & 0 \\ j20 & -j40 & j20 & 0 \\ 0 & j20 & 5-j40 & 0 \\ 0 & 0 & 0 & -j13 \end{bmatrix} - \begin{bmatrix} 0 \\ 0 \\ j20 \\ j13 \end{bmatrix} \left[\frac{1}{-j33}\right] [0 \;\; 0 \;\; j20 \;\; j13]$$

$$= \begin{bmatrix} 2-j20 & j20 & 0 & 0 \\ j20 & -j40 & j20 & 0 \\ 0 & j20 & 5-j\dfrac{920}{33} & j\dfrac{260}{33} \\ 0 & 0 & j\dfrac{260}{33} & -j\dfrac{260}{33} \end{bmatrix} .$$

The order of the axes in Y_{bus}^{eq} is 1, 2, 3, 5. The validity of this result is checked very simply by noting that if bus 4 is eliminated, line 3–4 appears in series with line 4–5. The equivalent line 3–5 has admittance $-j260/33$ mhos. If Y_{bus} were built with this line admittance, Y_{bus}^{eq} would result.

Note that in the construction of Y_{bus} in Example 2.2, the parallel lines 1–2 result in a total admittance which is the sum of the individual admittances. The building algorithm applied to cases of parallel lines requires the negative sum of line admittances joining buses i and j in the off diagonal positions, and the positive sum of all admittances to bus i in the diagonal position. Thus the building algorithm is consistent with the rule for combining admittances in parallel.

Example 2.3

In this example, two buses are eliminated from the system in Figure 2.2: buses 2 and 4.

Solution

In an actual application, it may be desirable to eliminate hundreds of buses so that the equivalent Y_{bus} matrix is much smaller than the matrix for the entire system. It is assumed that the injection currents at the buses to be eliminated are zero. For this illustration, the axes of Y_{bus} are reordered to put buses 2 and 4 last. In a digital analysis, this would not be necessary since with some programming care, axes 2 and 4 may be considered as being Y_{ab} and Y_{bb}. The reordered Y_{bus} is

$$Y_{bus} = \begin{vmatrix} 2-j20 & 0 & 0 & j20 & 0 \\ & -j40 & 0 & j20 & j20 \\ & & -j13 & 0 & j13 \\ & & & 5-j40 & 0 \\ \text{symmetric} & & & & -j33 \end{vmatrix}.$$

This matrix is easily verified by checking term–by–term with the original Y_{bus} in Example 2.2. For example, row 2, column 4 of the reordered matrix corresponds to bus 3, bus 2 (the axes are reordered as 1, 3, 5, 2, 4). $Y_{row2\ col4}$ is, in fact, $Y_{bus3\ bus2}$. In both the original and reordered matrices, $Y_{bus3\ bus2}$ is $j20$ mhos. Proceeding to apply the Kron reduction formula yields

$$Y_{aa} = \begin{bmatrix} 2-j20 & 0 & 0 \\ 0 & 5-j40 & 0 \\ 0 & 0 & -j13 \end{bmatrix} \qquad Y_{ab} = \begin{bmatrix} j20 & 0 \\ j20 & j20 \\ 0 & j13 \end{bmatrix}$$

$$Y_{bb} = \begin{bmatrix} -j40 & 0 \\ 0 & -j33 \end{bmatrix}.$$

The inverse of Y_{bb} is evaluated and substituted into (2.8)

$$Y_{bb}^{-1} = \begin{bmatrix} \dfrac{1}{-j40} & 0 \\ 0 & \dfrac{1}{-j33} \end{bmatrix}.$$

$$Y_{bus}^{eq} = \begin{bmatrix} 2-j20 & 0 & 0 \\ 0 & 5-j40 & 0 \\ 0 & 0 & -j13 \end{bmatrix} - \begin{bmatrix} j20 & 0 \\ j20 & j20 \\ 0 & j13 \end{bmatrix} \begin{bmatrix} \dfrac{1}{-j40} & 0 \\ 0 & \dfrac{1}{-j33} \end{bmatrix} \begin{bmatrix} j20 & j20 & 0 \\ 0 & j20 & j13 \end{bmatrix}.$$

The result is

$$Y_{bus}^{eq} = \begin{bmatrix} 2-j10 & j10 & 0 \\ j10 & 5-j17\dfrac{29}{33} & j7\dfrac{29}{33} \\ 0 & j7\dfrac{29}{33} & -j7\dfrac{29}{33} \end{bmatrix}$$

and the order of the axes is 1, 3, 5.

When m buses are eliminated simultaneously from Y_{bus} ($m = 2$ in the previous example), it is necessary to invert an m by m matrix (Y_{bb}). Because the time required for inversion is high, it is more efficient to eliminate the m buses one at a time; the precise reason for this is that special advantage may be taken of positions in Y_{aa} which are zero. This special advantage, through a technique known as sparsity programming, is considered in Chapter 3. The essence of the advantage is that only nonzero elements of Y_{aa} need to be processed (except when a nonzero element of $Y_{ab} Y_{bb}^{-1} Y_{ab}^{t}$ occurs).

Change of Reference in Y_{bus}

Consideration of the change of reference bus for Y_{bus} studies is different from the modification topics considered above because the change of reference involves no alteration or modification of the system configuration. The change in reference bus refers to the change in the way voltages are measured and injection currents find return paths. The circuit lines and buses remain in the same configuration. Figure 2.3 shows a few buses in a large power system. The notation $V_i^{(0)}$ is used to denote the voltage at bus i with respect to bus 0, and the notation $I_i^{(0)}$ denotes the injection current at bus i with return path through bus 0. The superscript denotes the reference bus for the given measurement; the subscript denotes the bus at which the measurement or injection is made.

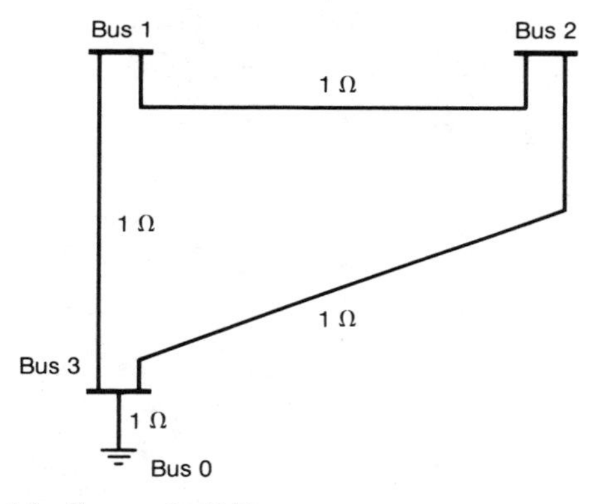

FIGURE 2.3 System used in Example 2.5.

The diagonal and off diagonal elements of Y_{bus} may be viewed as appropriate partial derivatives of the Kirchhoff current law equation at bus j (row j of $I_{bus} = Y_{bus} V_{bus}$)

$$I_j = \sum_{k=1}^{n} (Y_{bus})_{jk} V_k$$

$$(Y_{bus})_{jj} = \frac{\partial I_j}{\partial V_j} \qquad (Y_{bus})_{ji} = \frac{\partial I_j}{\partial V_i} \ .$$

The implication of these partial derivatives is that all other voltages are held fixed. Therefore, $\partial I_j / \partial V_j$ implies that the voltages at bus 0 and bus i are fixed. If bus 0 is used as a reference bus, $(Y_{bus}^{(0)})_{ji}$ is $\partial I_j^{(0)} / \partial V_j^{(0)}$. When the reference bus is shifted to bus i, physical considerations indicate that

$$\frac{\partial I_j^{(0)}}{\partial V_j^{(0)}} = \frac{\partial I_j^{(i)}}{\partial V_j^{(i)}}$$

since a voltage change at bus j holding all other bus voltages fixed results in the same injection current into bus j no matter what reference bus is used. Therefore,

$$(Y_{bus}^{(0)})_{jj} = (Y_{bus}^{(i)})_{jj} \ .$$

A similar argument applies to $(Y_{bus})_{ji}$. This reasoning does not imply that $Y_{bus}^{(0)}$ is identical to $Y_{bus}^{(i)}$ because different axes appear in these two matrices: in $Y_{bus}^{(0)}$, the axis for bus 0 is missing. Similarly, in $Y_{bus}^{(i)}$, no axis exists for bus i. These considerations lead to the following rule to change the reference bus in Y_{bus}:

To find $Y_{bus}^{(b)}$ *using* $Y_{bus}^{(a)}$ (i.e., to convert the reference bus from a to b in a given Y_{bus}), form the indefinite bus admittance matrix by augmenting $Y_{bus}^{(a)}$ with an a axis formed using the Y_{bus} building algorithm. This a axis is simply the negative of the sum of the previous n axes. Then delete the b axis. The resulting n by n matrix is $Y_{bus}^{(b)}$.

Example 2.4

The system used in Example 2.2 was used previously to form $Y_{bus}^{(0)}$. This matrix is modified to illustrate a change of reference bus to bus 3.

The indefinite admittance matrix is formed by adding a bus zero axis,

$$Y_{bus}^{indef} = \begin{bmatrix} 2-j20 & j20 & 0 & 0 & 0 & -2 \\ j20 & -j40 & j20 & 0 & 0 & 0 \\ 0 & j20 & 5-j40 & j20 & 0 & -5 \\ 0 & 0 & j20 & -j33 & j13 & 0 \\ 0 & 0 & 0 & j13 & -j13 & 0 \\ -2 & 0 & -5 & 0 & 0 & 7 \end{bmatrix} .$$

The last axis in this matrix corresponds to bus 0. Note that the entry in row 6, column 1 is the negative sum of the previous five entries in column 1. The other entries of row 6 are found similarly, as are the entries of column 6. Finally, the desired $Y_{bus}^{(3)}$ matrix is found by deleting axis 3

$$Y_{bus}^{(3)} = \begin{bmatrix} 2-j20 & j20 & 0 & 0 & -2 \\ j20 & -j40 & 0 & 0 & 0 \\ 0 & 0 & -j33 & j13 & 0 \\ 0 & 0 & j13 & -j13 & 0 \\ -2 & 0 & 0 & 0 & 7 \end{bmatrix} .$$

In this matrix, the bus ordering is 1, 2, 4, 5, 0.

2.4 BUS IMPEDANCE MATRIX

The simultaneous set of equations

$$I_{bus} = Y_{bus} V_{bus}$$

may be solved for the n bus voltages when there is a net connection to the reference bus (Y_{bus} is nonsingular)

$$V_{bus} = Y_{bus}^{-1} I_{bus} . \tag{2.9}$$

The inverse of the admittance matrix is known as the bus *bus impedance matrix* Z_{bus}. Sometimes it is more convenient to define Z_{bus} not in terms of Y_{bus} but rather, as the coefficients of the injection currents in the Kirchhoff current law solved for bus voltages

$$(Z_{bus})_{ij} = \frac{\partial V_i}{\partial I_j} \qquad i = 1, 2, ..., n \qquad j = 1, 2, ..., n. \tag{2.10}$$

For (2.10) to apply, there must exist a path for injection current I_j to flow.

It is possible for Z_{bus} to be singular if a nonzero voltage at some system bus can not exist. Such is the case if a bus is shorted to the reference. If bus k is shorted to the reference,

$$V_k = 0$$

and the $(Z_{bus})_{ik}$ entries are zero for all i. This results in an axis of zeros in Z_{bus} (axis k) and the matrix is singular. Under such circumstances, the Y_{bus} matrix does not exist. In simplified terminology, when all system buses are open circuited from the reference, Y_{bus} exists but is singular (Z_{bus} does not exist), and when any bus is shorted to the reference, Z_{bus} exists but is singular (Y_{bus} does not exist).

A few additional fundamental properties of Z_{bus} should be noted:

1. For linear systems, the reciprocity theorem applies and $\partial V_i / \partial I_j = \partial V_j / \partial I_i$. This means that $Z_{ij} = Z_{ji}$ and Z_{bus} is symmetric. The same conclusion could be obtained by noting that Z_{bus} is the inverse of a symmetric matrix.

2. When current is injected into bus k, bus voltages will rise throughout the system – but the greatest rise is usually expected at bus k. When bus k is tightly tied to another bus, ℓ, the voltage increase at ℓ will be, at most, equal to the increase at k. The reasoning leads to

   $$\frac{\partial V_\ell}{\partial I_k} \le \frac{\partial V_k}{\partial I_k}.$$

   Therefore, usually one finds that

   $$z_{\ell k} \le z_{kk}. \tag{2.11}$$

   The diagonal of Z_{bus} generally predominates. Note that in ac circuits, resonance effects may occur and it is possible that (2.11) will not hold.

3. Since injection current at k will in general produce a voltage rise at all other buses, $z_{k\ell} = z_{\ell k} \neq 0$, in general. Thus, Z_{bus} is not sparse.

4. The behavior of a system as viewed from just m buses of an n bus system, $m < n$, can be obtained by discarding unused rows and columns of Z_{bus}. This property makes the impedance matrix most useful for large system studies, and Z_{bus} is often used to describe a system rather than Y_{bus}. This is a rather remarkable property since a very large system may be studied, under some conditions, very conveniently by discarding the unneeded axes of Z_{bus}.

5. The bus impedance matrix is identical to the nodal impedance matrix used in circuit theory. The diagonal entries of Z_{bus} are the driving point impedances at that bus and the off diagonal entries are the transfer impedances.

The Indefinite Bus Impedance Matrix

The indefinite bus impedance matrix, Z_{bus}^{ind}, is defined as Z_{bus} augmented by an axis corresponding to the reference bus. This axis consists entirely of zeros since $\partial V/\partial I$ must be zero if V is the voltage of the reference bus with respect to itself. The axis of zeros causes Z_{bus}^{ind} to be singular. The indefinite bus impedance matrix is related to the indefinite bus admittance matrix through the Moore–Penrose pseudoinverse. This pseudoinverse, singular values, and singular vectors are defined and discussed in Appendix B. Let A^+ denote the pseudoinverse of matrix A. Note that by virtue of the fact that the pseudoinverse and conventional inverse are identical for nonsingular matrices,

$$Z_{bus}^+ = Y_{bus} \qquad Y_{bus}^+ = Z_{bus}$$

provided that Z_{bus} and Y_{bus} exist. The eigenvalues of these matrices cannot include a zero since they are nonsingular. When these matrices are augmented to form the indefinite matrix, a zero eigenvalue is added but the positive singular values are unchanged. Let U and V denote the $n+1$ by n and n by $n+1$ matrices, respectively, of singular vectors of Y_{bus}^{ind}. Then the definition of the pseudoinverse is used to obtain $(Y_{bus}^{ind})^+$,

$$(Y_{bus}^{ind})^+ = U\Sigma^+V,$$

where

$$\Sigma = diag\ (\sigma_1, \sigma_2, ..., \sigma_n, 0)$$

and the σ_i are the positive singular values of Y_{bus}. Since

$$Z_{bus} = Y_{bus}^{-1} = Y_{bus}^+ = U'(\Sigma')^+V',$$

where U' and V' are the singular vectors of Y_{bus} and Σ' is diag $(\sigma_1, \sigma_2, \ldots, \sigma_n)$, it is readily shown that $(Y_{bus}^{ind})^+$ is (Z_{bus}^{ind}) by noting that the matrices of singular vectors, U and V, are

$$U = \begin{bmatrix} U' \\ 0 \end{bmatrix} \qquad V = \begin{bmatrix} V' & 0 \end{bmatrix}$$

(the notation 0 in U is a row of zeros; the notation 0 in V is a column of zeros). Thus

$$\begin{bmatrix} Z_{bus} & 0 \\ 0 & 0 \end{bmatrix} = U\,\Sigma^+V = (Y_{bus}^{ind})^+ \qquad Z_{bus}^{ind} = (Y_{bus}^{ind})^+.$$

By a property of pseudoinverses,

$$Y_{bus}^{ind} = (Z_{bus}^{ind})^+.$$

Also, it is not difficult to show that

$$Z_{bus}^{ind}Y_{bus}^{ind} = \begin{bmatrix} & & & 0 \\ & I_{n,n} & & 0 \\ & & & 0 \\ 0 & 0 & 0 & 0 \end{bmatrix},$$

where $I_{n,n}$ is the n by n identity matrix.

Methods of Calculating Z_{bus}

Turning attention to methods of calculating Z_{bus}, the reader may well question why use of Z_{bus} is even considered in view of the lengthy computation requirements of inversion and the unfortunate fact that the inverse of a sparse matrix is, in general, full. The answer to this fundamental inquiry lies in the value of Z_{bus} in certain applications in power engineering – particularly in fault studies. Complete descriptions of this and other applications are relegated to a subsequent chapter. It is worthwhile to emphasize that Y_{bus} offers distinct advantages in formation and storage. There are three broad classes of methods for the calculation of Z_{bus}:

1. Inversion of Y_{bus}.
2. Individual calculation of terms using $\partial V/\partial I$
3. Z_{bus} building algorithms.

Inversion of the admittance matrix requires extensive computation time. There are inversion techniques that may be used to exploit the symmetry of Y_{bus} and Z_{bus}, but execution time is still proportional approximately to the *cube* of the matrix size for large matrices. A method particularly useful for inversion of Y_{bus} by computer is presented in Chapter 3, but actual industrial applications usually employ the Z_{bus} building algorithm rather than inversion. This algorithm is described later in this section.

For small systems, direct calculation of each entry in Z_{bus} is possible by elementary considerations. This method also yields insight into the properties of the impedance matrix. Direct calculation of Z_{bus} is based on

$$(Z_{bus})_{ij} = \frac{\partial V_i}{\partial I_j} .$$

Since the system is linear, the partial derivative is exactly evaluated using

$$(Z_{bus})_{ij} = \frac{\Delta V_i}{\Delta I_j},$$

where other ΔI_k $(k \neq j)$ are zero. The method is usually applied by assuming that a 1 per unit current is injected into bus j $(\Delta I_j = 1)$. The unit current injection at bus j causes voltage rises throughout the system which are numerically equal to $(Z_{bus})_{ij}$ and $(Z_{bus})_{ji}$. The following example serves to illustrate this method.

Example 2.5

Find Z_{bus} for the system in Figure 2.3.

Solution

To find z_{11}, z_{12}, and z_{13}, place a 1 Ampere injection at bus 1 and solve for v_1, v_2, v_3. These voltages are numerically equal to z_{11}, $z_{12} = z_{21}$, and $z_{13} = z_{31}$. Figure 2.4 shows this process. The 1 Ampere source undergoes current division at bus 1 as shown, resulting in

$$v_1 = 5/3\ V \qquad v_2 = 4/3\ V \qquad v_3 = 1\ V.$$

Therefore,

$$z_{11} = 5/3\ \Omega \quad z_{12} = 4/3\ \Omega \quad z_{21} = 4/3\ \Omega \quad z_{13} = 1\ \Omega \quad z_{31} = 1\ \Omega.$$

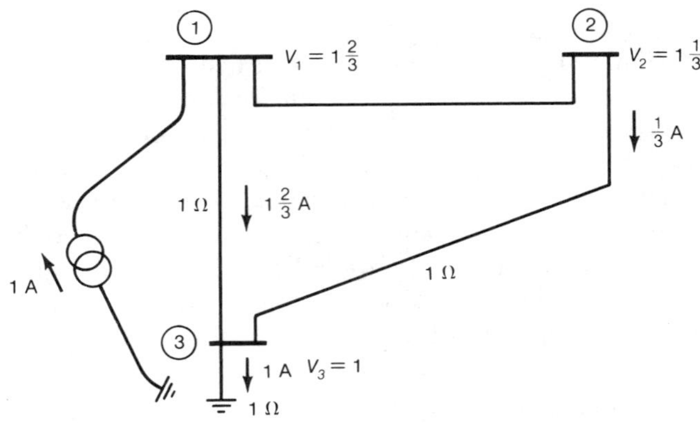

FIGURE 2.4 Current injection into bus 1.

To find z_{22} and $z_{23} = z_{32}$, locate the current source at bus 2 (Figure 2.5). Using current division at bus 2 results in

$$v_1 = 4/3\ V \quad v_2 = 5/3\ V \quad v_3 = 1\ V.$$

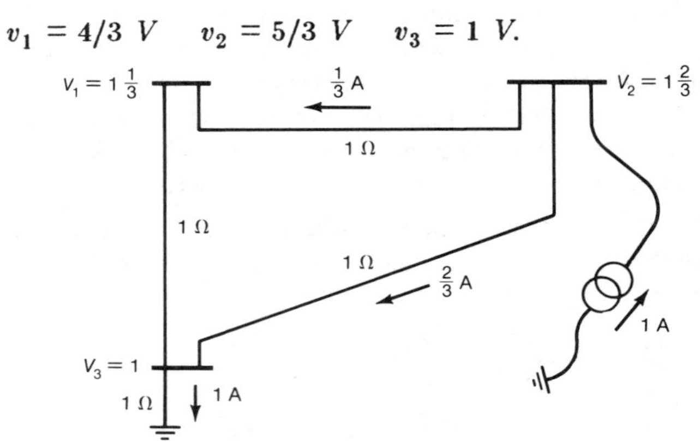

FIGURE 2.5 Current injection into bus 2.

Therefore,

$$z_{21} = z_{12} = 4/3\ \Omega \quad z_{22} = 5/3 \quad z_{23} = z_{32} = 1\ .$$

Similarly, z_{33} is found to be 1 Ω. Therefore,

$$Z_{bus} = \begin{bmatrix} 5/3 & 4/3 & 1 \\ 4/3 & 5/3 & 1 \\ 1 & 1 & 1 \end{bmatrix}.$$

The result is checked by inverting Y_{bus}

$$Z_{bus} = Y_{bus}^{-1} = \begin{bmatrix} 2 & -1 & -1 \\ -1 & 2 & -1 \\ -1 & -1 & 3 \end{bmatrix}^{-1}$$

$$= \frac{\begin{bmatrix} 5 & 4 & 3 \\ 4 & 5 & 3 \\ 3 & 3 & 3 \end{bmatrix}}{det\ (Y_{bus})} = \begin{bmatrix} 5/3 & 4/3 & 1 \\ 4/3 & 5/3 & 1 \\ 1 & 1 & 1 \end{bmatrix}.$$

The most common method to calculate Z_{bus} in practical applications is the bus impedance matrix building algorithm. This algorithm is now presented followed by an abbreviated derivation.

The Z_{bus} Building Algorithm

The Z_{bus} building algorithm is a method by which line data are sequentially processed to obtain the impedance matrix. The validity of the Z_{bus} building algorithm is, unfortunately, not as that of evident as the topics previously described; for this reason, the algorithm is first described as a six step procedure and the validity of these steps is discussed later. To use the building algorithm, a line list is used and one line at a time is "processed." Also, the line list must be ordered such that the first "line" processed is a tie to the reference bus. This is not an unreasonable requirement since there must be at least one tie to the reference bus (otherwise, the Y_{bus} matrix is singular). Also, at least one bus terminal of *all* subsequent lines in the line list must be a bus that has been previously "seen" (i.e., processed). This requirement is also reasonable since isolated lines, totally disconnected from the system, result in singular Y_{bus}. These two requirements imply that the Z_{bus} building algorithm requires line list preprocessing (i.e., line ordering). There is a third reason for line ordering, which is related to the speed of processing; this is considered briefly at the end of this section. In the description below, bus 0 is the reference bus.

1. Read a line. If the last line has been processed, stop.
2. Determine whether the line is a radial line from the system processed earlier, or a loop closure between two buses seen earlier in the list. A radial line is identified as a line from a bus seen earlier to a bus not yet seen. The first line processed is considered to be a radial from bus 0 to bus 1.
3. If the line read is the first line processed, simply establish Z_{bus} as a 1 by 1 "matrix" consisting of the line impedance 0–1, $Z_{bus} = [\bar{z}_{01}]$. The notation of a bar above z denotes a line impedance. The first line processed must be a tie to bus 0.
4. If the line read is a radial line, the Z_{bus} matrix after this line is added to the system is

$$Z_{bus}^{new} = \begin{bmatrix} Z_{bus}^{old} & col_i Z_{bus}^{old} \\ row_i Z_{bus}^{old} & (Z_{bus}^{old})_{ii} + \bar{z} \end{bmatrix}$$

   where Z_{bus}^{old} is the matrix prior to consideration of the present line, i is the bus already seen, and the new axis in the matrix corresponds to the new, yet unseen bus. The line impedance is $\bar{z}$. Note that when a radial line is processed, the matrix size increases by one axis.

5. If the line read is a loop closure between buses i and j, both of which have been seen earlier in the line list, form an intermediate matrix denoted as Z^{loop}

$$Z^{loop} = \begin{bmatrix} Z_{bus}^{old} & col_i Z_{bus}^{old} - col_j Z_{bus}^{old} \\ row_i Z_{bus}^{old} - row_j Z_{bus}^{old} & (Z_{bus}^{old})_{ii} + (Z_{bus}^{old})_{jj} - 2(Z_{bus}^{old})_{ij} + \bar{z} \end{bmatrix}.$$

The line impedance is $\bar{z}$. Note that the dimension of Z^{loop} is one axis greater than Z_{bus}^{old}. The added axis is termed the loop axis

$$Z^{loop} = \begin{bmatrix} Z_{bus}^{old} & col_{loop}(Z^{loop}) \\ row_{loop}(Z^{loop}) & (Z^{loop})_{loop\ loop} \end{bmatrix}.$$

Next, perform a Kron reduction of Z_{bus}^{old} using the loop-loop position of Z^{loop} as the pivot element. The result is Z_{bus}^{new}

$$Z_{bus}^{new} = Z_{bus}^{old} - [col_{loop}(Z^{loop})]\ [(Z^{loop})_{loop\ loop}]^{-1}\ [row_{loop}(Z^{loop})]. \quad (2.12)$$

Since Kron reduction is effectively the elimination of a variable, this process may be viewed as elimination of the "loop" variable. The dimension of Z_{bus}^{new} is the same as Z_{bus}^{old}.

6. If the radial or loop closure procedure indicated above involves the reference bus (bus 0), no special theoretical provision need be made except that $(Z_{bus})_{0j}$ is taken to be zero for all j. Since the reference axis is not stored in Z_{bus}, a computer implementation of the building algorithm requires logic to detect a radial or loop closure involving the reference bus. In such a case, appropriate impedances are assumed to be zero

$$(Z_{bus})_{0j} = 0.$$

The six steps above are repeated as necessary until all lines are processed. It is necessary to order the input line list so that the first entry is a tie to the reference bus, and subsequent lines are either "radials" or "loop closures." A line from a previously unseen bus to another previously unseen bus is not permitted.

Validity of the Z_{bus} Building Algorithm

The demonstration of the validity of the Z_{bus} building algorithm is divided into three parts, corresponding to processing the first line, a radial line, and a loop closure. The first line processed is a line from bus 0 to bus i. The matrix consists of only one axis at this point and the single entry is $(Z_{bus})_{ii}$. This is the driving point impedance at bus i, which is simply the line impedance,

$$Z_{bus} = [\bar{z}].$$

Radial lines are considered by inspection of Figure 2.6. The radial addition is a line of impedance $\bar{z}$ from bus i to bus $n+1$. If a unit current is injected into any bus in the system already processed, the voltages generated at buses 1 through n are unaffected by the new radial addition. Thus Z_{bus}^{new} will consist of Z_{bus}^{old} in the upper left n rows and columns.

The added column $(col_{n+1}(Z_{bus}^{new}))$ is evaluated by injecting a unit current into bus j, $j = 1, 2, ..., n$, and observing the voltage at bus $n+1$. Since bus $n+1$ is at the end of a radial line, no current flows in this line and

$$V_{n+1} = V_i\ .$$

Thus

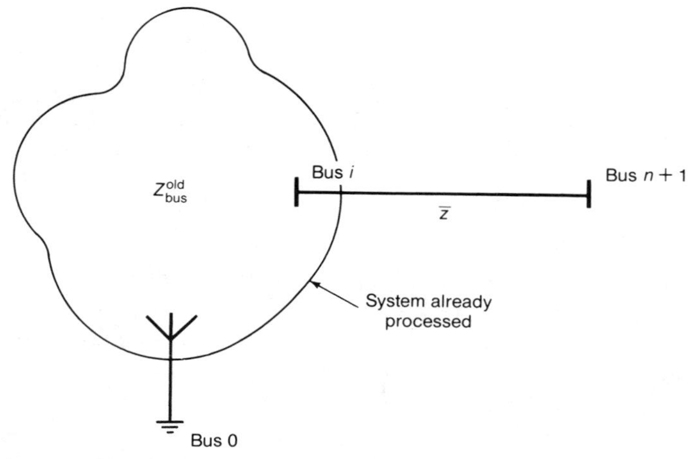

FIGURE 2.6 Radial line addition in the Z_{bus} building algorithm.

$$\frac{\partial V_{n+1}}{\partial I_j} = \frac{\partial V_i}{\partial I_j} = (Z_{bus}^{old})_{ij} \; .$$

Therefore, $col_{n+1}(Z_{bus}^{new})$ is simply $col_i(Z_{bus}^{old})$. A similar argument applies to row $n+1$ of Z_{bus}^{new}. The row $n+1$, column $n+1$ element is evaluated by injecting a unit current into bus $n+1$. The voltage drop in the new line (between buses i and $n+1$) is numerically equal to $\bar{z}$. The voltage rise at bus i is numerically equal to $(Z_{bus}^{old})_{ii}$. Therefore, the voltage at bus $n+1$ is $\bar{z} + (Z_{bus}^{old})_{ii}$. It is concluded that

$$(Z_{bus}^{new})_{n+1\; n+1} = \bar{z} + (Z_{bus}^{old})_{ii} .$$

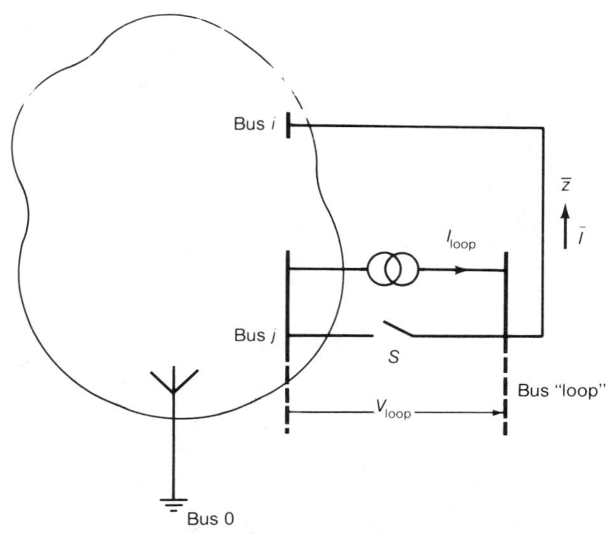

FIGURE 2.7 Verification of the loop closure procedure in the Z_{bus} building algorithm.

Verification of the loop closure procedure is slightly more involved. Consider Figure 2.7. Let switch S be open momentarily, and consider the voltage V_{loop},

$$V_{loop} = V_i - V_j + \bar{I}\bar{z}.$$

A new axis is now inserted into Z_{bus}^{old} corresponding to the voltage V_{loop}. Considering the definition of this voltage, this added axis is the difference between the i and j axes of Z_{bus}^{old}. This is the case since for line current $\bar{I}$ equal to zero, V_{loop} is simply V_i minus V_j. Let Z^{loop} designate the augmented $n+1$ by $n+1$ matrix. The loop–loop position of Z^{loop} is

$$\frac{\partial V_{loop}}{\partial I_{loop}},$$

where I_{loop} is the current injection by a source in parallel with the open switch S (see Figure 2.7). Note that I_{loop} equals $\bar{I}$ and this line current causes the $(Z^{loop})_{loop\ loop}$ entry to be the difference between the i and j entries of the loop axis [i.e., the difference between $(Z_{bus}^{old})_{ii} - (Z_{bus}^{old})_{ij}$ and $(Z_{bus}^{old})_{ij} - (Z_{bus}^{old})_{jj}$] plus the impedance of the line. The latter term is due to the fact that V_{loop} is not simply $V_i - V_j$. It also contains the drop in the line. The loop–loop diagonal entry is therefore

$$\begin{aligned}(Z^{loop})_{loop\ loop} &= (Z_{bus}^{old})_{ii} - (Z_{bus}^{old})_{ij} - [(Z_{bus}^{old})_{ij} - (Z_{bus}^{old})_{jj}] + \bar{z} \\ &= (Z_{bus}^{old})_{ii} + (Z_{bus}^{old})_{jj} - 2(Z_{bus}^{old})_{ij} + \bar{z}.\end{aligned}$$

When the switch S is closed, V_{loop} is zero and in the expression

$$\begin{bmatrix} V_{bus} \\ \hline V_{loop} \end{bmatrix} = Z^{loop} \begin{bmatrix} I_{bus} \\ \hline I_{loop} \end{bmatrix}$$

the scalar current I_{loop} must be solved and eliminated using

$$V_{loop} = 0.$$

As in the case of elimination of a bus in Y_{bus} where an injection current was zero and a bus voltage is eliminated from $I_{bus} = Y_{bus} V_{bus}$, Kron reduction applies. The Kron reduction in this case uses the loop–loop position of Z^{loop} as a pivot element. The Kron reduction causes I_{loop} to be solved and eliminated. This gives (2.12), which concludes the demonstration.

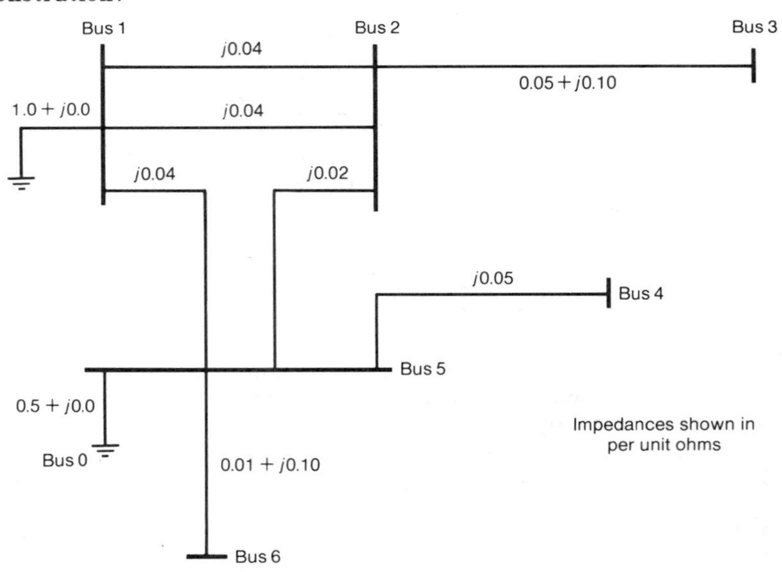

FIGURE 2.8 System used in Example 2.6.

Example 2.6

The power system of Figure 2.8 has the following line list:

From	To	$\bar{z}$
0	1	$1.00 + j0.00$
1	2	$j0.04$
1	2	$j0.04$
2	5	$j0.02$
5	0	$0.50 + j0.00$
1	5	$j0.04$
5	4	$j0.05$
5	6	$0.01 + j0.10$
2	3	$0.05 + j0.10$

Using the Z_{bus} building algorithm, form the bus impedance matrix.

Solution

The line list is already properly ordered: the first entry is a ground tie and all subsequent lines are either radials or loop closures. The first entry in the line list results in

$$Z_{bus} = \begin{matrix} & bus\ 1 \\ bus\ 1 & [1.0 + j0.0] \end{matrix}.$$

The second entry in the line list is identified as a radial (bus 1 to bus 2) since bus 1 has already been seen earlier but bus 2 has not. The bus 1 axis is repeated to find the bus 2 axis,

$$Z_{bus} = \begin{matrix} bus\ 1 \\ bus\ 2 \end{matrix} \begin{bmatrix} 1.0 + j0.0 & 1.0 + j0.0 \\ 1.0 + j0.0 & 1.0 + j0.04 \end{bmatrix}.$$

Note that the 2, 2 position is found using

$$(Z_{bus}^{new})_{22} = (Z_{bus}^{old})_{11} + \bar{z}.$$

The third entry of the line list is a loop closure since buses 1 and 2 have both been seen earlier in the line list. The Z^{loop} matrix is formed by augmenting Z_{bus} with an axis corresponding to axis 1 minus axis 2. The order of the subtraction is immaterial – the subsequent Kron reduction will yield the same results independent of the order of subtraction. Matrix Z^{loop} is found to be

$$Z^{loop} = \begin{bmatrix} 1.0 + j0.0 & 1.0 + j0.00 & 0 \\ 1.0 + j0.0 & 1.0 + j0.04 & -j0.04 \\ 0 & -j0.04 & j0.08 \end{bmatrix}. \quad \text{(last column: loop axis)}$$

The matrix Z^{loop} is Kron reduced using the loop–loop position as the pivot element,

$$Z_{bus} = \begin{bmatrix} 1.0 + j0.0 & 1.0 + j0.00 \\ 1.0 + j0.0 & 1.0 + j0.02 \end{bmatrix}.$$

Line 4, bus 2 to bus 5, is a radial line. The bus 2 axis must be repeated:

$$Z_{bus} = \begin{bmatrix} 1.0 + j0.0 & 1.0 + j0.00 & 1.0 + j0.00 \\ 1.0 + j0.0 & 1.0 + j0.02 & 1.0 + j0.02 \\ 1.0 + j0.0 & 1.0 + j0.02 & 1.0 + j0.04 \end{bmatrix}.$$

The axes of this matrix are 1, 2, 5. Line 5 in the line list is from bus 5 to bus 0, a loop closure. The bus 0 axis is not written in Z_{bus} but is assumed to be all zeros. The Z^{loop} matrix contains a loop axis which is the bus 5 axis minus the bus 0 axis

$$Z^{loop} = \begin{bmatrix} 1.0 + j0.0 & 1.0 + j0.00 & 1.0 + j0.00 & 1.0 + j0.00 \\ 1.0 + j0.0 & 1.0 + j0.02 & 1.0 + j0.02 & 1.0 + j0.02 \\ 1.0 + j0.0 & 1.0 + j0.02 & 1.0 + j0.04 & 1.0 + j0.04 \\ 1.0 + j0.0 & 1.0 + j0.02 & 1.0 + j0.04 & 1.5 + j0.04 \end{bmatrix}.$$

After Kron reduction and elimination of the loop axis,

$$Z_{bus} = \begin{bmatrix} 0.33387 + j0.01777 & 0.33345 + j0.00444 & 0.33310 - j0.00888 \\ 0.33345 + j0.00444 & 0.33336 + j0.01111 & 0.33327 - j0.00222 \\ 0.33310 - j0.00888 & 0.33327 - j0.00222 & 0.33345 + j0.00444 \end{bmatrix}.$$

This matrix was obtained using the Kron reduction formula on the nine elements outside the loop axis in Z^{loop}. One of these nine reductions is illustrated by the calculation of the row 1, column 2 entry of Z_{bus},

$$(Z_{bus})_{12} = (Z_{bus}^{loop})_{12} - \frac{(Z_{bus}^{loop})_{1\ loop}(Z_{bus}^{loop})_{loop\ 2}}{(Z_{bus}^{loop})_{loop\ loop}}$$

$$= (1 + j0) - \frac{(1 + j0)(1 + j0.02)}{1.5 + j0.04}$$

$$= 0.33345 + j0.00444.$$

Before proceeding to the remaining four lines, it is instructive to reflect on this Z_{bus} matrix and check it in a cursory fashion. Note that the highest impedance lines in the line list are lines 0–1 and 5–0. If these lines were the only lines to be processed and the other lines were of zero impedance, buses 1 through 5 would be joined together through zero impedance ties and all entries in Z_{bus} would be identical. Line 0–1 and line 5–0 would be in parallel to give 1/3 Ω (1 and 0.5 in parallel). Hence all entries in Z_{bus} would be 0.33333. Inspection of the present Z_{bus} matrix verifies this. Also note that the diagonal dominates in this matrix (the magnitude on the diagonal is the larger than the magnitudes off the diagonal).

The next line is from bus 1 to bus 5. This is a loop closure. The intermediate Z^{loop} and final Z_{bus} matrices (after Kron reduction) are

$$\begin{bmatrix} 0.33381 + j0.01777 & 0.33345 + j0.00444 & 0.33310 - j0.00888 & 0.00071 + j0.02665 \\ 0.33345 + j0.00444 & 0.33336 + j0.01111 & 0.33327 - j0.00222 & 0.00018 + j0.00666 \\ 0.33310 - j0.00888 & 0.33327 - j0.00222 & 0.33345 + j0.00444 & -0.00035 - j0.01332 \\ 0.00071 + j0.02665 & 0.00018 + j0.00666 & -0.00035 - j0.01332 & 0.00106 + j0.07997 \end{bmatrix}$$

$$Z_{bus} = \begin{bmatrix} 0.33345 + j0.00889 & 0.33336 + j0.00222 & 0.33328 - j0.00444 \\ 0.33336 + j0.00222 & 0.33334 + j0.01056 & 0.33331 - j0.00111 \\ 0.33328 - j0.00444 & 0.33331 + j0.00111 & 0.33336 + j0.00222 \end{bmatrix}.$$

The axes of this matrix correspond to buses 1, 2, 5 as before. The next three lines, 5–4, 5–6, and 2–3, are all radials since buses 4, 6, and 3 had not yet been encountered. To add line 5–4, the present Z_{bus} matrix is augmented with an axis corresponding to bus 4. The new axis is a copy of the bus 5 axis. The bus 4–bus4 position is

$$(Z_{bus})_{44} = (Z_{bus})_{55} + j0.05.$$

The result of the three radial additions is the following matrix

$$\begin{bmatrix} 0.33345+j0.00889 & 0.33336+j0.00222 & 0.33328-j0.00444 & 0.33328-j0.00444 & 0.33328-j0.00444 & 0.33336+j0.00222 \\ & 0.33334+j0.01056 & 0.33331-j0.00111 & 0.33331-j0.00111 & 0.33331-j0.00111 & 0.33334+j0.01056 \\ & & 0.33336+j0.00222 & 0.33336+j0.00222 & 0.3336+j0.00222 & 0.33331-j0.00111 \\ & & & 0.33336+j0.05222 & 0.33336+j0.00222 & 0.33331-j0.00111 \\ & \text{matrix is symmetric} & & & 0.34336+j0.10222 & 0.33331-j0.00111 \\ & & & & & 0.38334+j0.11056 \end{bmatrix}.$$

The axes of this final matrix correspond to buses 1, 2, 5, 4, 6, 3.

This concludes this example, but it is appropriate to add a few final remarks and observations:

1. The final Z_{bus} matrix has dominant diagonal.
2. The most time–consuming steps in the construction of Z_{bus} are the Kron reductions, which require n^2 modifications of existing entries (where n is the order of the matrix at that stage). If loops are closed as early as possible, n is small, and the Kron reductions require less time.
3. The axes of the final matrix do not correspond to the buses with the buses in numerical order. Row 1 corresponds to bus 1, row 2 corresponds to bus 2, but row 3 corresponds to bus 5. Further, row 4 corresponds to bus 4 but row 5 corresponds to bus 6. The last axis corresponds to bus 3.
4. When the system ground ties are not of low impedance, the Z_{bus} entries will all be a similar value. This is the case in this example. The transmission system is of low impedance in Example 2.6, but the ground ties (two of them) are comparatively high impedance. Therefore, the Z_{bus} entries are all approximately $0.33333 + j0.0$.
5. If the ties to the reference bus were generally less than or of a similar order of magnitude to the transmission line impedances, the Z_{bus} entries would not be of similar order of magnitude to each other. This is the case when the swing bus is used as a reference bus.
6. In Example 2.6, some negative signs appear in the final Z_{bus} matrix. This may, at first, be surprising since the system to be modelled has no capacitors or negative resistances. However, an off diagonal element of Z_{bus} cannot be related to physical components in an elementary way. The negative reactive part of the row 1, column 3 entry does not suggest that a capacitor is present in the circuit. The Z_{bus} matrix must satisfy certain sign constraints, however, which are related to the passivity of the network. Since the network consists

of passive elements, the real part of the total complex power injected into all buses of the network (see Figure 2.9) must be greater than or equal to zero

$$Re\ (V_1 I_1^* + V_2 I_2^* + \cdots) \geq 0$$

$$Re\ (I_{bus}^H V_{bus}) \geq 0.$$

This requirement is written in terms of Z_{bus} as follows:

$$Re\ (I_{bus}^H Z_{bus} I_{bus}) \geq 0. \tag{2.13}$$

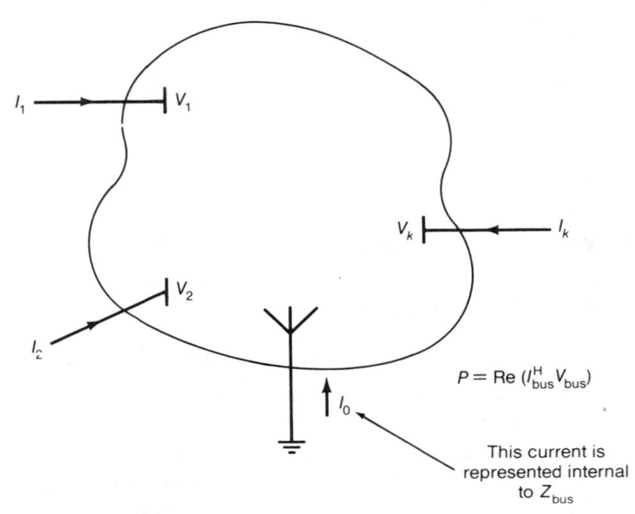

FIGURE 2.9 Total active power injected into a system.

A matrix, Z_{bus}, which satisfies (2.13) for all possible complex vectors I_{bus} is called a positive real matrix. Hence passive linear systems are characterized by positive real bus impedance matrices. The consequences of being a positive real matrix are numerous and too involved to consider here; one consequence, however, is that the real part of the diagonal entries must be greater than or equal to zero. If A is Hermitian (i.e., $A = A^H$), a convenient test for the positive real property is known as *Sylvester's test*. This test requires that the row 1, column 1 entry be positive, and the determinant

$$det \begin{bmatrix} a_{11} & a_{12} \\ a_{21} & a_{22} \end{bmatrix} > 0$$

and all successively larger determinants along the principal diagonal must also be positive,

$$det \begin{bmatrix} a_{11} a_{12} \ldots \\ a_{21} a_{22} \ldots \\ \ldots \quad a_{nn} \end{bmatrix} > 0 .$$

This test may be applied to Z_{bus} only if Z_{bus} is all real since Sylvester's test applies to Hermitian matrices only, and the only symmetric Hermitian matrices are real matrices.

It is possible to extend Sylvester's test to symmetric, complex matrices such as Z_{bus}. This is done by defining

$$A_h = \frac{1}{2}(Z_{bus} + Z_{bus}^H)$$

$$A_{sh} = \frac{1}{2}(Z_{bus} - Z_{bus}^H) .$$

Matrix A_h is Hermitian, and A_{sh} is skew Hermitian ($A_{sh}^H = -A_{sh}$). Also,

$$Z_{bus} = A_h + A_{sh} .$$

Note that (2.13) implies that

$$Re\ (I_{bus}^H A_h I_{bus}) + Re\ (I_{bus}^H A_{sh} I_{bus}) \geq 0.$$

Since the $I_{bus}^H A_{sh} I_{bus}$ term is zero, only A_h need be checked for the positive real property. Sylvester's test applies to A_h since it is Hermitian. This rather complicated procedure may be used to check the validity of Z_{bus} with regard to passivity; there are other alternatives that generally relate to calculating or otherwise locating the eigenvalues of Z_{bus}. By the power invariance property of similarity transformations, the eigenvalues of Z_{bus} must have positive real parts. It is easy to show that the inverse of a positive real matrix is also positive real; thus Y_{bus} is positive real. There is no simple requirement, however, on the signs of the elements of Z_{bus} except that the real parts of the diagonal entries must be positive.

7. The line list order affects the number of matrix elements that must be Kron reduced, and hence the speed of the algorithm. The optimal (fastest) line ordering is that which "closes loops" as early in the list as possible so that the least number of entries need to be modified.
8. A nearly optimal, easy to program line ordering algorithm is one that places certain lines early in the list. These early lines have terminal buses which are encountered very frequently in the entire list.
9. If no tie to the reference exists, one must be introduced to render Z_{bus} well defined. If the ground bus is used as the reference, it is convenient to introduce a ground tie at the swing bus. This will not alter the bus voltage solution or line power flows since the swing bus is a fixed voltage bus and introduction of the artificial tie will not affect the system.

Note that items 7 through 9 give the essence of the requirements of an optimal line ordering algorithm. Such an algorithm requires scanning the line list to count the number of line connections to each bus. Heavily interconnected buses require early processing and therefore lines with these terminal buses are reordered to near the top of the list.

References [2, 3] provide historical documentation of the Z_{bus} matrix. El-Abiad and Stagg [3,4] give extensive examples of the building algorithm, including mutually coupled cases. Brown [5] is often considered a definitive reference in Z_{bus} technology.

CHANGES IN THE BUS IMPEDANCE MATRIX TO REFLECT SYSTEM CHANGES

The Z_{bus} matrix requires considerable computation to form. Compared to Y_{bus} methods, there is greater motivation in the case of Z_{bus} algorithms to seek simple techniques for matrix modification to reflect system changes. In this section, line outage, bus deletion, and change of reference are considered.

Line Outage

In the analogous discussion with respect to Y_{bus}, a line outage is considered as a line addition, with the new line impedance equal to the negative of the impedance of the line to be outaged. The same technique is used to modify Z_{bus}: to outage line $i-j$ of impedance $\bar{z}$, add a new line to the system, a loop closure from i to j having impedance $-\bar{z}$. The Z_{bus} building algorithm is used to add this hypothetical line, which has the effect of outaging an existing line.

Bus Deletion

Bus voltages and currents are related by

$$V_{bus} = Z_{bus} I_{bus}.$$

Let bus k be unneeded for a study and let

$$I_k = 0.$$

Also, let row k of Z_{bus} be deleted. Thus V_k will not be calculated. Also, let column k be deleted; since this column contains the coefficients of I_k, no error will be introduced by this deletion since I_k is zero. The method to delete a bus is therefore simply to delete the k axis.

The implications of this simple algorithm are far reaching since it is often necessary to analyze a small part of a very large power system. Using Z_{bus} methods, the unwanted buses are deleted simply by deleting the appropriate matrix axes.

Change of Reference Bus

Consider the change of reference bus from bus a to bus b in Z_{bus}. It is convenient to consider first the augmented bus impedance matrix referenced to bus a, $Z_{bus}^{(a)}$, in which the matrix has been augmented to include an axis corresponding to bus a. Since V_a is exactly zero independent of any injected current,

$$\frac{\partial V_a}{\partial I_i} \equiv 0 \qquad i = 1, 2, \ldots, a, \ldots, n.$$

Therefore, the entire "a" axis of $Z_{bus}^{(a)}$ is zero [including the entry $(Z_{bus}^{(a)})_{aa}$]. It is desired to convert this matrix to $Z_{bus}^{(b)}$ where bus b is used as a reference bus. It is clear that in the matrix $Z_{bus}^{(b)}$, the b axis will be identically zero, but the a axis will not be identically zero.

To convert $Z_{bus}^{(a)}$ to $Z_{bus}^{(b)}$, the b column in $Z_{bus}^{(a)}$ is subtracted from all other columns including itself. This process is viewed as sweeping the b column across $Z_{bus}^{(a)}$, where the modified entry in the ij position is given by

$$(modified\ entry)_{ij} = (old\ entry)_{ij} - (old\ entry)_{ib} \ .$$

Having swept the b *column* across the matrix, the b *row* is also swept across the matrix

$$(modified\ entry)_{ij} = (old\ entry)_{ij} - (old\ entry)_{bj} \ .$$

The result is $Z_{bus}^{(b)}$.

To verify this procedure, note that $(Z_{bus}^{(b)})_{ij}$ is $\partial V_i^{(b)}/\partial I_j^{(b)}$. Let $V_i^{(b)}$, the bus voltage at i referenced to bus b, be written as

$$V_i^{(b)} = V_i^{(a)} - V_b^{(a)}.$$

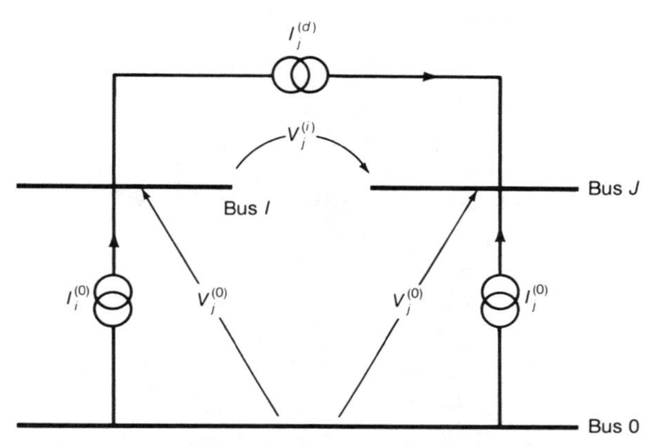

FIGURE 2.10 Voltages and currents referenced to buses 0 and *I*.

Figure 2.10 will help illustrate the voltages and currents involved. Then

$$(Z_{bus}^{(b)})_{ij} = \frac{\partial V_i^{(b)}}{\partial I_j^{(b)}} = \frac{\partial V_i^{(a)}}{\partial I_j^{(b)}} - \frac{\partial V_b^{(a)}}{\partial I_j^{(b)}} \ . \tag{2.14}$$

The injected current I_j into bus j finds return path through b. Equivalently, this current may be written as $I_j^{(a)}$ Amperes into bus j returned through bus a. The currents $I_j^{(a)}$ and $-I_b^{(a)}$ acting at the same time and both equal to $I_j^{(b)}$ Amperes is the same as $I_j^{(b)}$ injected into j returned through b. Therefore, the $\partial V_i^{(a)}$ and $\partial V_b^{(a)}$ terms in Eq. (2.14) are due to the action of two currents, $I_j^{(a)}$ and $-I_b^{(a)}$

$$(Z_{bus}^{(b)})_{ij} = \frac{\partial V_i^{(a)}}{\partial I_j^{(a)}} - \frac{\partial V_i^{(a)}}{\partial I_b^{(a)}} - \frac{\partial V_b^{(a)}}{\partial I_j^{(a)}} + \frac{\partial V_b^{(a)}}{\partial I_b^{(a)}} \ .$$

Therefore,

$$z_{ij}^{(b)} = z_{ij}^{(a)} - z_{ib}^{(a)} - z_{bj}^{(a)} + z_{bb}^{(a)} \ . \tag{2.15}$$

Equation (2.15) is equivalent to the algorithm cited above. The equivalence of (2.15) to the algorithm cited previously is verified by noting that sweeping column b across the $Z_{bus}^{(a)}$ matrix results in an intermediate matrix, Z_{bus}', whose entries are given by

$$(Z_{bus}')_{ij} = (Z_{bus}^{(a)})_{ij} - (Z_{bus}^{(a)})_{ib} \ .$$

The Z'_{bus} matrix is then modified by sweeping the b row across the matrix, resulting in $Z_{bus}^{(b)}$

$$(Z_{bus}^{(b)})_{ij} = (Z'_{bus})_{ij} - (Z'_{bus})_{bj} \ . \tag{2.16}$$

Substitute (2.15) into (2.16)

$$(Z_{bus}^{(b)})_{ij} = (Z_{bus}^{(a)})_{ij} - (Z_{bus}^{(a)})_{ib} - (Z_{bus}^{(a)})_{bj} + (Z_{bus}^{(a)})_{bb} . \tag{2.17}$$

Term for term, (2.17) is identical to (2.15).

Example 2.7

To illustrate the change in reference of a Z_{bus} matrix, consider a given bus impedance matrix referenced to bus 0 (the axis order is 1, 2, 3):

$$Z_{bus}^{(0)} = \begin{bmatrix} 1.0 & 0.5 & 0.0 \\ 0.5 & 2.0 & 0.5 \\ 0.0 & 0.5 & 1.0 \end{bmatrix} .$$

Convert the reference bus from bus 0 to bus 1.

Solution

Augment the matrix

$$Z_{bus}^{aug} = \begin{bmatrix} 0 & 0 & 0 & 0 \\ 0 & 1 & 0.5 & 0 \\ 0 & 0.5 & 2 & 0.5 \\ 0 & 0 & 0.5 & 1 \end{bmatrix} .$$

The axes of the augmented matrix are 0, 1, 2, 3. Sweep the column corresponding to bus 1. The result is

$$\text{modified matrix} = \begin{bmatrix} 0 & 0 & 0 & 0 \\ -1 & 0 & -0.5 & -1 \\ -0.5 & 0 & 1.5 & 0 \\ 0 & 0 & 0.5 & 1 \end{bmatrix} .$$

Then sweep the row corresponding to bus 1 across the matrix:

$$\text{modified matrix} = \begin{bmatrix} 1.0 & 0 & 0.5 & 1.0 \\ 0.0 & 0 & 0.0 & 0.0 \\ 0.5 & 0 & 2.0 & 1.0 \\ 1.0 & 0 & 1.0 & 2.0 \end{bmatrix} .$$

$Z_{bus}^{(1)}$ is found by deleting the bus 1 axis

$$Z_{bus}^{(1)} = \begin{bmatrix} 1.0 & 0.5 & 1.0 \\ 0.5 & 2.0 & 1.0 \\ 1.0 & 1.0 & 2.0 \end{bmatrix} ,$$

where the order of the axes is bus 0, bus 2, bus 3.

LOOP MATRICES

The Kirchhoff current law leads to the bus impedance and bus admittance matrices. These matrices are termed the ***nodal impedance and admittance matrices*** in circuit theory. The dual of this approach is based on the Kirchhoff voltage law, which leads to the loop impedance and loop admittance matrices. These matrices are not as applicable to power systems as their bus duals, and for this reason less attention is focused on the loop matrices.

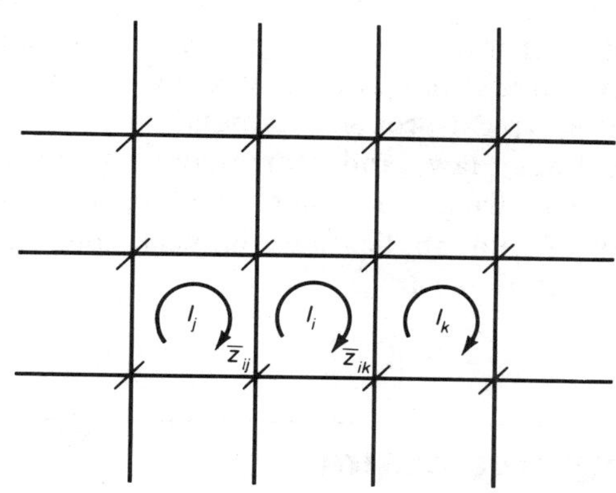

FIGURE 2.11 Loop currents used with the Kirchhoff voltage law.

Figure 2.11 shows several power system loop currents. The Kirchhoff voltage law applied to loop i is

$$E_i = \sum_q \bar{z}_{iq} I_i - \sum_{\substack{j=1 \\ \neq q}}^{n} \bar{z}_{ij} I_j \tag{2.18}$$

where I_i is the loop current for which the Kirchhoff law is applied, $\sum_q$ denotes the sum taken over all loops that touch loop i, and $\bar{z}_{ij}$ denotes the branch impedance between loops i and j. If loop i does not touch loop j, $\bar{z}$ is zero. Also, n is the total number of loops and E_i in the phasor sum of voltage sources in loop i. In vector form, (2.18) is

$$V_{loop} = Z_{loop} I_{loop}$$

and inspection of (2.18) gives the rules for building the Z_{loop} matrix. The diagonal elements of Z_{loop} are found by summing the impedances in that loop. The off diagonal elements are the negative of the impedance between loops i and j. This algorithm, and properties of Z_{loop} in general, are duals of the Y_{bus} matrix. Interesting properties of Z_{loop} include the fact that the matrix is sparse and symmetric. Also, Z_{loop} is nonsingular provided the definition of loops is such that a loop is not included within other loops.

The inverse of Z_{loop} is Y_{loop}. This matrix is the dual of Z_{bus}. The Y_{loop} matrix is full, generally diagonal dominant, and there exists a Y_{loop} building algorithm similar to the Z_{bus} building algorithm to generate the matrix directly. As an example of the duality of loop and bus methodologies, note that the active power supplied to a passive network must be positive

$$Re\ (I_{loop}^{H}\ V_{loop}) > 0$$

and

$$Re\ (I_{loop}^{H}\ Z_{loop}\ I_{loop}) > 0.$$

Thus Z_{loop} must be a positive real matrix and must satisfy the modified Sylvester's test in the same way that Y_{bus} must satisfy this test. Also, Y_{loop} must be a positive real matrix.

Most loads and generators in electric power systems are ties from a bus to ground. These are often viewed as injection currents. The use of the Kirchhoff current law is suggested by the fact that injection currents are specified in typical problem formulations. It is inconvenient to use the Kirchhoff voltage law and corresponding loop formulation because injection currents are given rather than loop voltages. The Z_{loop} and Y_{loop} matrices have found applications in very specialized instances in which loop voltage sources occur.

2.7

EFFECTS OF MUTUAL COUPLING

Lines sited on a common right-of-way are, in general, mutually coupled. The algorithms presented thus far do not model mutual coupling and it is necessary to accommodate this phenomenon, which is particularly important in instances of heavily loaded lines on rights-of-way that contain several lines. Let line ij be mutually coupled to line $k\ell$ (see

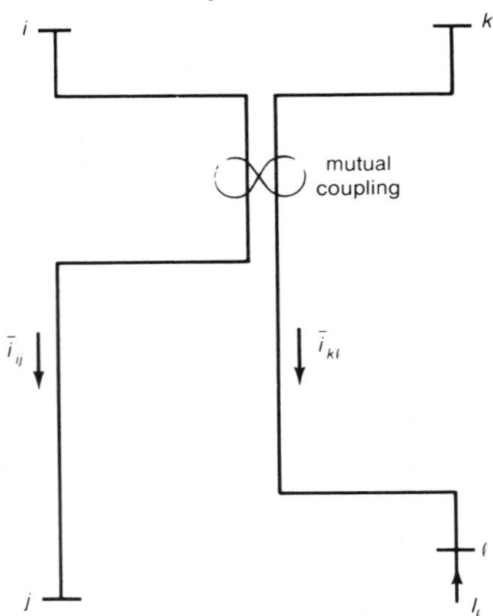

FIGURE 2.12 Mutually coupled lines.

Figure 2.12) over part or all of the length of the lines. Using the sign convention in Figure 2.12 for the line currents $\bar{i}_{ij}$ and $\bar{i}_{k\ell}$, the voltages $v_i - v_j$ and $v_k - v_\ell$ are found to have two components each, one due to the self impedance of the line and the other due to mutual coupling. In vector notation,

$$\begin{bmatrix} \bar{i}_{ij} \\ \bar{i}_{k\ell} \end{bmatrix} = \begin{bmatrix} \bar{y}_{ijij} & \bar{y}_{ijk\ell} \\ \bar{y}_{k\ell ij} & \bar{y}_{k\ell k\ell} \end{bmatrix} \begin{bmatrix} v_i - v_j \\ v_k - v_\ell \end{bmatrix} \tag{2.19}$$

where the $\bar{y}$ elements are primitive admittances that describe the self– and mutual admittances of the two lines. Equation (2.19) is rewritten

$$\bar{I} = \bar{Y}\,\bar{V} \tag{2.20}$$

where $\bar{I}$ and $\bar{V}$ are the line currents and voltages of the partial network, and $\bar{Y}$ is the primitive admittance matrix (due to reciprocity, $\bar{Y}$ is symmetric). In the absence of mutual coupling, $\bar{Y}$ is diagonal. Note that the line $k\ell$ may be coupled to several lines; in this instance $\bar{i}_{ij}$ becomes a subvector of $\bar{I}$ where the subvector is of dimension equal to the number of lines to which $k\ell$ couples. Equation (2.19) is the definition of $\bar{Y}$; a similar but not identical result would be obtained if (2.19) were written in terms of mutual impedances. The required modifications of the Z_{bus} building algorithm are presented below [4–6].

Addition of a Radial Line

Let line $k\ell$ be a radial line from existing bus k to a new bus $\ell = N+1$. The existing N by N Z_{bus} matrix must be modified to reflect the radial line addition. The bottom row of (2.19) is

$$\bar{i}_{k\ell} = \bar{y}_{k\ell ij}(v_i - v_j) + \bar{y}_{k\ell k\ell}(v_k - v_\ell). \tag{2.21}$$

If bus ℓ is at the end of a radial line (see Figure 2.12),

$$I_\ell = -\bar{i}_{k\ell}. \tag{2.22}$$

If I_ℓ is the only injection current, the voltage terms in (2.21) are readily replaced by

$$v_q = Z_{q\ell} I_\ell \qquad q = i, j, k, \ell \tag{2.23}$$

where $Z_{q\ell}$ denotes elements of the existing Z_{bus} matrix. Substitution of (2.22) and (2.23) into (2.21) yields the following when solved for $Z_{\ell m}$ and $Z_{\ell\ell}$

$$Z_{\ell m} = Z_{km} + \frac{\bar{y}_{k\ell ij}(Z_{im} - Z_{jm})}{\bar{y}_{k\ell k\ell}} \qquad m \neq \ell \tag{2.24}$$

$$Z_{\ell\ell} = Z_{k\ell} + \frac{1 + \bar{y}_{k\ell ij}(Z_{i\ell} - Z_{j\ell})}{\bar{y}_{k\ell k\ell}}. \tag{2.25}$$

Equations (2.24) and (2.25) give the off diagonal and diagonal entries of the new Z_{bus} matrix after the line addition. If bus k were the reference bus, $k=0$ and

$$Z_{km} = Z_{k\ell} = 0$$

in (2.24) and (2.25).

Equations (2.24) and (2.25) are readily found to degenerate to the formulas given earlier for the case of no mutual coupling when $\bar{y}_{k\ell ij} = 0$.

Addition of a Loop

Consider now the case in which line $k\ell$ is a loop closure. This case is identified by the fact that buses k and ℓ had already been encountered in

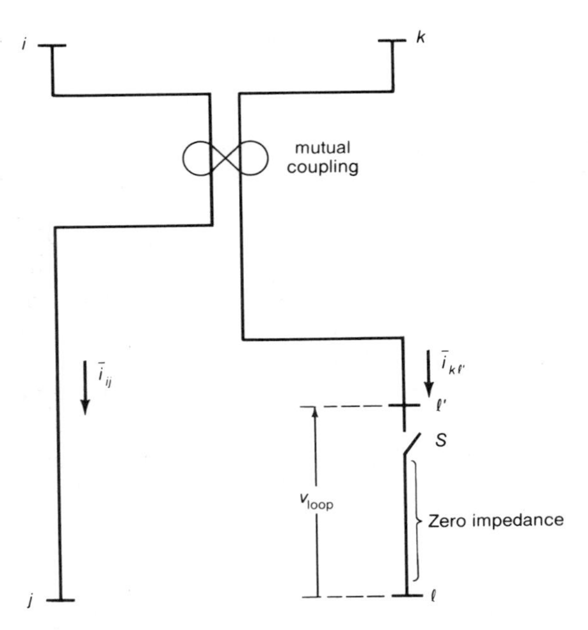

FIGURE 2.13 Loop closure for mutually coupled lines.

the bus list. This case is considered as before by considering Figure 2.13. The loop closure is completed by closing switch S. Momentarily, consider the case of S open

$$\begin{bmatrix} \bar{i}_{ij} \\ \bar{i}_{k\ell'} \end{bmatrix} = \begin{bmatrix} \bar{y}_{ijij} & \bar{y}_{ijk\ell} \\ \bar{y}_{k\ell\,ij} & \bar{y}_{k\ell\,k\ell} \end{bmatrix} \begin{bmatrix} v_i - v_j \\ v_k - v_{\ell'} \end{bmatrix} \tag{2.26}$$

$$v_{loop} = v_{\ell'} - v_{\ell}. \tag{2.27}$$

Now define Z_{bus}^{loop} implicity

$$\begin{bmatrix} V_{bus} \\ \hline v_{loop} \end{bmatrix} = Z_{bus}^{loop} \begin{bmatrix} I_{bus} \\ \hline \bar{i}_{k\ell'} \end{bmatrix}. \tag{2.28}$$

Evidently

$$Z_{bus}^{loop} = \begin{bmatrix} Z_{bus}^{old} & col_{loop}(Z_{bus}^{loop}) \\ row_{loop}(Z_{bus}^{loop}) & (Z^{loop})_{loop\ loop} \end{bmatrix} \tag{2.29}$$

since Z_{bus}^{old} simply related V_{bus} to I_{bus}. In (2.29) the loop column in the off diagonal position must be formed using a modified form of (2.24) such that the impedance corresponding to voltage $v_{\ell'} - v_{\ell}$ is found. Since S is open, the impedances in the loop axis are found from (2.24) and (2.25)

$$Z_{m\ loop}^{loop} = Z_{km} - Z_{\ell m} + \frac{\bar{y}_{k\ell\,ij}(Z_{im} - Z_{jm})}{\bar{y}_{k\ell\,k\ell}} \qquad m \neq loop \tag{2.30}$$

$$Z_{loop\ loop}^{loop} = Z_{k\ loop} - Z_{\ell\ loop} + \frac{1 + \bar{y}_{k\ell\,ij}(Z_{i\ loop} - Z_{j\ loop})}{\bar{y}_{k\ell\,k\ell}}. \tag{2.31}$$

These expressions are also valid when S is closed since the mutual terms are included. When S is closed, v_{loop} is zero, and $\bar{i}_{k\ell'}$ in (2.28) must be solved and eliminated. This is the Kron reduction of (2.28) about the loop axis,

$$(Z_{bus}^{new})_{rc} = (Z_{bus}^{old})_{rc} - \frac{(Z_{bus}^{loop})_{r\ loop}(Z_{bus}^{loop})_{loop\ c}}{(Z_{bus}^{loop})_{loop\ loop}} . \tag{2.32}$$

Note that (2.30) and (2.31) degenerate to the familiar case of no mutual coupling when $\bar{y}_{k\ell\, ij} = 0$. Also note that either bus k or bus ℓ may be the reference bus. If bus k is the reference bus ($k = 0$),

$$Z_{km} = Z_{k\ loop} = 0$$

in (2.30) and (2.31). Also note that bus i or bus j could be the reference bus.

The procedure for inclusion of mutual coupling in a new line, $k\ell$, which couples to an existing line ij is

Radial 1. Write the primitive line admittance matrix $\bar{Y}$. The diagonal entries are the self admittances and the off diagonal entries are the mutual coupling terms [Eq. (2.19)]. $\bar{Y}$ is symmetric.

2. Add a new axis to Z_{bus}. The off diagonal position is found using (2.24), where $m = 1, 2, ...,$ and $m \neq \ell$. The diagonal position of this new axis is found using (2.25).

Loop 1. Write the primitive line admittance matrix $\bar{Y}$ as in step 1 of the radial procedure.

2. Add a new axis to Z_{bus} – this is called the loop axis. Equations (2.30) and (2.31) describe entries in this axis.

3. Eliminate the loop axis using Kron reduction of each entry in the first N axes of Z_{bus}^{old} [see Eq. (2.32)]. Then discard the loop axis. The result is an N by N matrix which is Z_{bus} after the line addition.

Outage of Mutually Coupled Lines [7]

The outage of mutually coupled lines is accomplished in a way similar to the case without mutual coupling. For the case of line ij mutually coupled to line $k\ell$, consider the outage of line $k\ell$. For the case that $k\ell$ is a radial line from system bus k to terminal bus ℓ, the preoutage $\bar{Y}$ matrix is

$$\bar{Y} = \begin{bmatrix} \bar{y}_{ijij} & \bar{y}_{ijkl} \\ \bar{y}_{ijk\ell} & \bar{y}_{k\ell\, k\ell} \end{bmatrix} .$$

This primitive line admittance matrix relates the line currents, $\bar{I}$, and line voltages, $v_i - v_j$, $v_k - v_\ell$, by

$$\bar{I} = \bar{Y}\begin{bmatrix} v_i - v_j \\ v_k - v_\ell \end{bmatrix} .$$

When line $k\ell$ is outaged, it is obvious that $\bar{Y}$ becomes

$$\bar{Y} = \begin{bmatrix} \bar{y}_{ijij} & 0 \\ 0 & 0 \end{bmatrix} \tag{2.33}$$

because $\bar{i}_{ij}$ no longer depends on $v_k - v_\ell$ (thus the ij–$k\ell$ position of $\bar{Y}$ must vanish) and the self admittance, $\bar{Y}_{k\ell\, k\ell}$, of an open circuit is zero.

Note that the postoutage $\overline{Y}$ matrix may be viewed as the result of the addition of a fictitious line from k to ℓ with self admittance $-\overline{y}_{k\ell k\ell}$ and mutual admittance with line ij of $+\overline{y}_{ijk\ell}$. The postoutage $\overline{Y}$ matrix is found by the following operations on the preoutage matrix:

1. Modify the diagonal entry $k\ell\, k\ell$ by adding $-\overline{y}_{k\ell k\ell}$
2. Modify the two off diagonal entries $ijk\ell$ and $k\ell\, ij$ by subtracting $+\overline{y}_{ijk\ell}$.

The result is (2.33).

For the case of outage of mutually coupled lines in the Z_{bus} matrix, a new line is added, which has the negative of the impedance of the line to be outaged. The mutual admittance of the fictitious added line, however, is equal to the mutual admittance of the line to be outaged (without reversal of sign). The added fictitious line in combination with the original line to outaged will be an open circuit. The addition of the new, fictitious line results in the use of (2.30) and (2.31). This process is illustrated in Example 2.8.

Example 2.7

Consider the following perunitized line data:

Line			Mutual with $(k\ell)$	
From (i)	To (j)	$\overline{z}$	line $k\ell$	$\overline{y}_{ijk\ell}$
1	2	$j0.01$	line 1–4	$-j10$
2	3	$j0.01$	line 4–5	$-j5$
3	5	$j0.02$	–	–
1	4	$j0.01$	line 1–2	$-j10$
4	5	$j0.01$	line 2–3	$-j5$

The data indicated have been taken using the notation in $\overline{Y}$ [Eq. (2.19)]. Find Z_{bus} referenced to bus 1.

Solution

Let bus 1 be the reference bus. Then the first entry in the line list is a tie to reference and Z_{bus} is a 1 by 1 "matrix,"

$$Z_{bus} = 2[j0.01] .$$

The "2" at the left of the matrix indicates that the only axis indicated corresponds to bus 2. The mutual coupling of line 1–2 with 1–4 does not affect this entry. Lines 2–3 and 3–5 are the next two entries in the line list. Their mutual coupling with lines as yet not processed cause these two entries to be calculated in the conventional way. The result is

$$Z_{bus} = \begin{matrix} 2 \\ 3 \\ 5 \end{matrix} \begin{bmatrix} j0.01 & j0.01 & j0.01 \\ & j0.02 & j0.02 \\ \text{symmetric} & & j0.04 \end{bmatrix} .$$

The next line in the list, line 1–4, is a radial line to a new bus (bus 4). Equation (2.24) is used with

$$i = 1 \quad j = 2 \quad k = 1 \quad \ell = 4 \quad m = 2$$

in order to obtain the $Z_{bus\ 4\ bus\ 2}$ entry. The formula is

$$Z_{42} = Z_{12} + \frac{\bar{y}_{1412}(Z_{12} - Z_{22})}{\bar{y}_{1414}} .$$

The primitive $\bar{y}$ is

$$\bar{y} = \begin{array}{c} \\ 1\text{-}2 \\ 1\text{-}4 \end{array} \begin{array}{c} \begin{array}{cc} 1-2 & 1-4 \end{array} \\ \begin{bmatrix} -j100 & -j10 \\ -j10 & -j100 \end{bmatrix} \end{array} .$$

Calculating Z_{42},

$$Z_{42} = -j0.001.$$

The Z_{43} entry is calculated similarly,

$$i = 1 \quad j = 2 \quad k = 1 \quad \ell = 4 \quad m = 3$$

$$Z_{43} = Z_{13} + \frac{\bar{y}_{1412}(Z_{13} - Z_{23})}{\bar{y}_{1414}}$$

$$= -j0.001 .$$

The Z_{44} entry is calculated in a similar way with the result j0.0101. At this point,

$$Z_{bus} = \begin{array}{c} 2 \\ 3 \\ 5 \\ 4 \end{array} \begin{bmatrix} j0.01 & j0.01 & j0.01 & -j0.0010 \\ & j0.02 & j0.02 & -j0.0010 \\ & & j0.02 & -j0.0010 \\ \text{symmetric} & & & j0.0101 \end{bmatrix} . \tag{2.34}$$

The last line to be processed, line 4–5, is a loop closure with mutual coupling. The $\bar{y}$ matrix is

$$\bar{y} = \begin{array}{c} \\ 2-3 \\ 4-5 \end{array} \begin{array}{c} \begin{array}{cc} 2-3 & 4-5 \end{array} \\ \begin{bmatrix} -j100 & -j5 \\ -j5 & -j100 \end{bmatrix} \end{array} .$$

Equation (2.30) is used to find most entries of the loop axis; for example, the loop–2 entry is found

$$i = 2 \quad j = 3 \quad k = 4 \quad \ell = 5 \quad m = 2$$

$$Z_{loop\ 2}^{loop} = Z_{42} - Z_{52} + \frac{\bar{y}_{4523}}{\bar{y}_{4545}} (Z_{22} - Z_{32})$$

$$= -j0.0101.$$

The remaining three nondiagonal entries in the loop axis are found and the result is

$$Z_{bus}^{loop} = \begin{array}{c} \\ 2 \\ 3 \\ 5 \\ 4 \\ \text{loop} \end{array} \begin{array}{c} \begin{array}{ccccc} 2 & 3 & 5 & 4 & \text{loop} \end{array} \\ \begin{bmatrix} j0.01 & j0.01 & j0.01 & -j0.001 & -j0.0101 \\ & j0.02 & j0.02 & -j0.001 & -j0.0215 \\ & & j0.02 & -j0.001 & -j0.0215 \\ & & & j0.0101 & j0.0111 \\ \text{symmetric} & & & & \end{bmatrix} \end{array} .$$

The loop–loop entry is calculated using (2.31)

$$i = 2 \quad j = 3 \quad k = 4 \quad \ell = 5$$

$$Z^{loop}_{loop\ loop} = Z_{4loop} - Z_{5loop} + \frac{1 + \bar{y}_{4523}(Z_{2loop} - Z_{3loop})}{\bar{y}_{4545}}$$

$$= j0.04317.$$

At this point, the Z^{loop}_{bus} matrix is Kron reduced about the loop–loop position. The result (using the axis ordering 2, 3, 5, 4) is

$$Z_{bus} = j\begin{bmatrix} 0.007637 & 0.004970 & 0.004970 & 0.001597 \\ & 0.009292 & 0.009292 & 0.004528 \\ & & 0.009292 & 0.004528 \\ & & & 0.007246 \end{bmatrix}.$$

Example 2.8

In this example, the Z_{bus} matrix Example 2.7 is modified by outage of line 4–5.

Solution

The outage of a mutually coupled line is accomplished by adding an new line to the line list, line 4–5, which has self impedance –j0.01 per unit ohms and mutual admittance with line 2–3 of $-j5$ per unit mhos. Note that the sign of $\bar{y}_{ijk\ell}$ is not reversed compared with the line to be outaged.

The new line is a loop closure, and (2.30) and (2.31) are used to obtain the loop axis

$$col_{loop}(Z^{loop}) = j\begin{bmatrix} -0.00324 \\ -0.00498 \\ -0.00477 \\ 0.00255 \\ -0.00277 \end{bmatrix}.$$

Kron reduction about the loop axis, (2.32), is used to obtain the final result

$$Z_{bus} = j\begin{bmatrix} 0.01 & 0.01 & 0.01 & -0.0010 \\ & 0.02 & 0.02 & -0.0010 \\ & & 0.02 & -0.0010 \\ & & & 0.0101 \end{bmatrix}. \tag{2.35}$$

Although the negative signs in the last axis may appear unusual, this matrix is readily shown to be positive real by examination of successively larger determinants along the principal diagonal (Sylvester's test). Equation (2.35) must agree with (2.34).

2.8 SUMMARY

To put the several algorithms and properties in proper perspective, this condensed summary is provided:

	Y_{bus}	Z_{bus}
Basic properties	Complex Symmetric Sparse $I_{bus} = Y_{bus} V_{bus}$ Positive real	Complex Symmetric Full $V_{bus} = Z_{bus} I_{bus}$ Diagonal generally dominant Positive real
Matrix is singular when	Path to reference is open circuited	A bus is shorted to reference
Matrix does not exist when	A bus is shorted to reference	Path to reference open circuited
Methods of formation	Y_{bus} building algorithm	Z_{bus} building algorithm Inversion of Y_{bus} Direct formation using $\partial V_i / \partial I_j$
To delete bus j	Kron reduce using jj element as pivot	Delete j axis
To delete a line	Use Y_{bus} building algorithm with $-\overline{y}$ added	Use Z_{bus} building algorithm with $-\overline{z}$ added
To change reference to bus b	Form Y_{bus}^{aug} and delete b axis	Sweep b axis across Z_{bus}^{aug} and delete b axis
Indefinite matrix	Always singular and equal to $(Z_{bus}^{aug})^+$	Always singular and equal to $(Y_{bus}^{aug})^+$

Bibliography

[1] G. T. Heydt and J. B. Franklin, "Power System Reliability Calculations by a Rapid Method," *1978 Allerton Circuits and Systems Conference,* Monticello, Ill., October 1978.

[2] H. Brown, C. Person, L. Kirchmager and G. Stagg, "Digital Calculation of Three Phase Short Circuits by Matrix Method," *Trans. AIEE,* v. 79, pt. III, 1960, pp. 1277–1281.

[3] A. El-Abiad, "Digital Calculation of Line to Ground Faults by Matrix Method," Trans. AIEE, v. 79, pt. III, 1960, pp. 323–332.

[4] G. Stagg and A. El-Abiad, *Computer Methods in Power System Analysis,* McGraw–Hill, New York, 1968.

[5] H. Brown, *Solution of Large Networks by Matrix Methods,* Wiley, New York, 1975.

[6] H. Brown and J. Storry, "Improved Method of Incorporating Mutual Couplings in Single Phase Short Circuit Calculations," *Proc., PICA Conference, 1969, pp. 335–342.*

[7] D. Reitan and K. Kruempel, "Modification of the Bus Impedance Matrix for System Changes Involving Mutual Couplings," *Proc. IEEE,* August 1969, p. 1432.

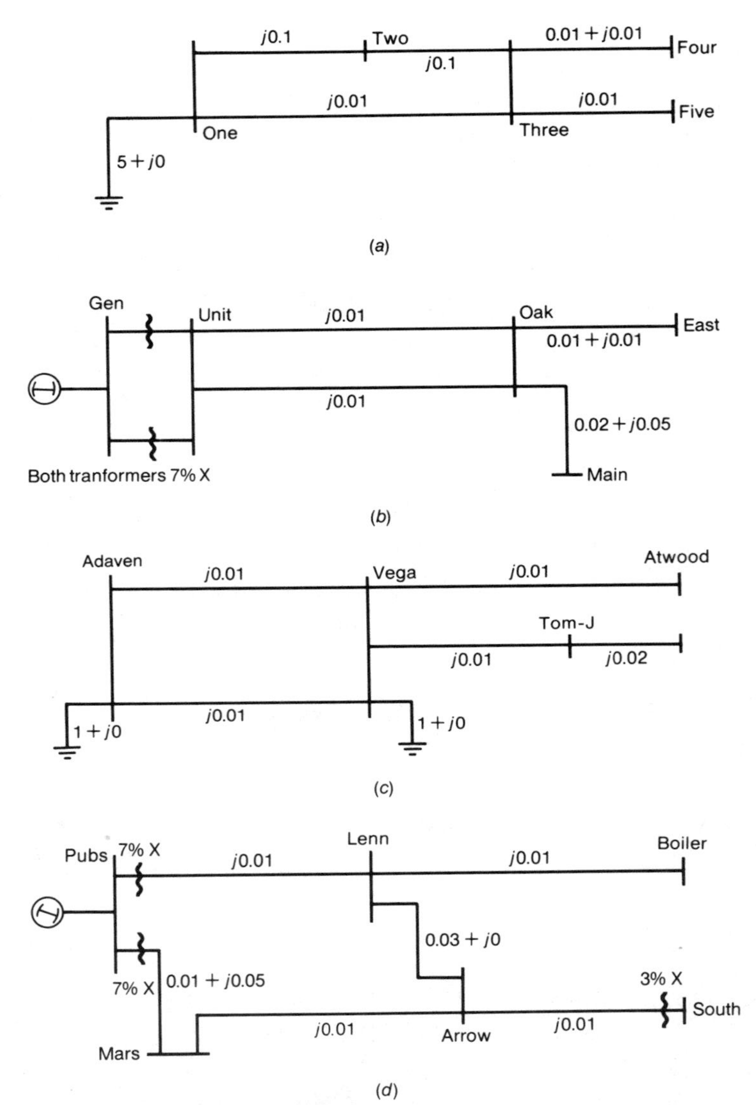

FIGURE P2.1 Four small power systems used to illustrate Y_{bus} construction.

Exercises

Exercises that do not involve computer programming

2.1 Find Y_{bus} referenced to ground for the systems depicted in Figure P2.1a through d. Note that in these figures, line impedances are shown in per unit.

2.2 For each of the systems in Figure P2.1, determine which systems posess nonsingular Y_{bus} matrices.

2.3 For the system depicted in Figure P2.1d, form Y_{bus} for the full system, then use the bus deletion algorithm to delete the LENN bus. Note that the resulting Y_{bus} is the same as that obtained if the wye connection formed by the lines PUBS-LENN, BOILER-LENN, and ARROW-LENN is replaced by a delta equivalent.

2.4 For each of the following Y_{bus} matrices, the order of the axes is bus 1, bus 2, and so on.

a. Eliminate bus 2 ($I_2 = 0$).

b. Eliminate both buses 1 and 2 ($I_1 = I_2 = 0$)

$$Y_{bus} = \begin{bmatrix} -j0.05 & j0.01 & j0.04 & 0 \\ j0.01 & 1-j0.01 & 0 & 0 \\ j0.04 & 0 & -j0.03 & j0.01 \\ 0 & 0 & j0.01 & 1-j0.01 \end{bmatrix}$$

$$Y_{bus} = \begin{bmatrix} -j0.10 & 0 & j0.01 & 0 & 0 \\ 0 & -j0.10 & j0.10 & 0 & 0 \\ j0.01 & j0.10 & 10-j.15 & j0.05 & 0 \\ 0 & 0 & j0.05 & 1-j0.15 & j0.1 \\ 0 & 0 & 0 & j0.1 & 1-j0.1 \end{bmatrix}$$

2.5 The two Y_{bus} matrices given in Exercise 2.4 are referenced to the ground bus. Convert the reference to bus 1 in each case.

2.6 In Chapter 3, a method for compressing the computer storage of Y_{bus} is discussed. In this method, only nonzero entries in the upper right triangle plus diagonal of Y_{bus} will be stored. The other (zero valued) entries will not be stored. For a 1000 bus system in which the average number of lines entering each bus is four, estimate the number of nonzero entries in the upper right triangle plus diagonal of Y_{bus}.

Exercises that involve computer programming

2.7 This exercise concerns the development of a standard subroutine for the calculation of the bus admittance matrix. The program should be of a general nature so that an arbitrary system can be processed. To test the program, a 69 kV system will be used (see Figure P2.7).

a. The required subroutine will be called as follows:

CALL FORMY(YBUS, N, LINEA, LINEB, ZLINE, NDIMY)

where YBUS is the bus admittance matrix. This is dimensioned NDIMY in the calling program and is declared complex.

N is returned to the main program as the number of system buses.

LINEA and LINEB are the names of the line terminal buses. They are formed in the calling program and consist of alphameric data read in A10 format. They are dimensioned 100 each.

ZLINE is a complex array dimensioned 100 and consists of line impedances in per unit. This array is formed in the main program.

You are to write FORMY and demonstrate its use with the Lake Superior REMC 69 kV Northern Division.

b. Your report should consist of:

- Brief introductory comments on your solution.
- A flowchart or other program description.
- A description of any special features.
- A description of the output. (How should the output be interpreted?)
- This company plans to use your program on their Southern Division, which consists of 800 buses and 943 lines. What changes will have to be made?

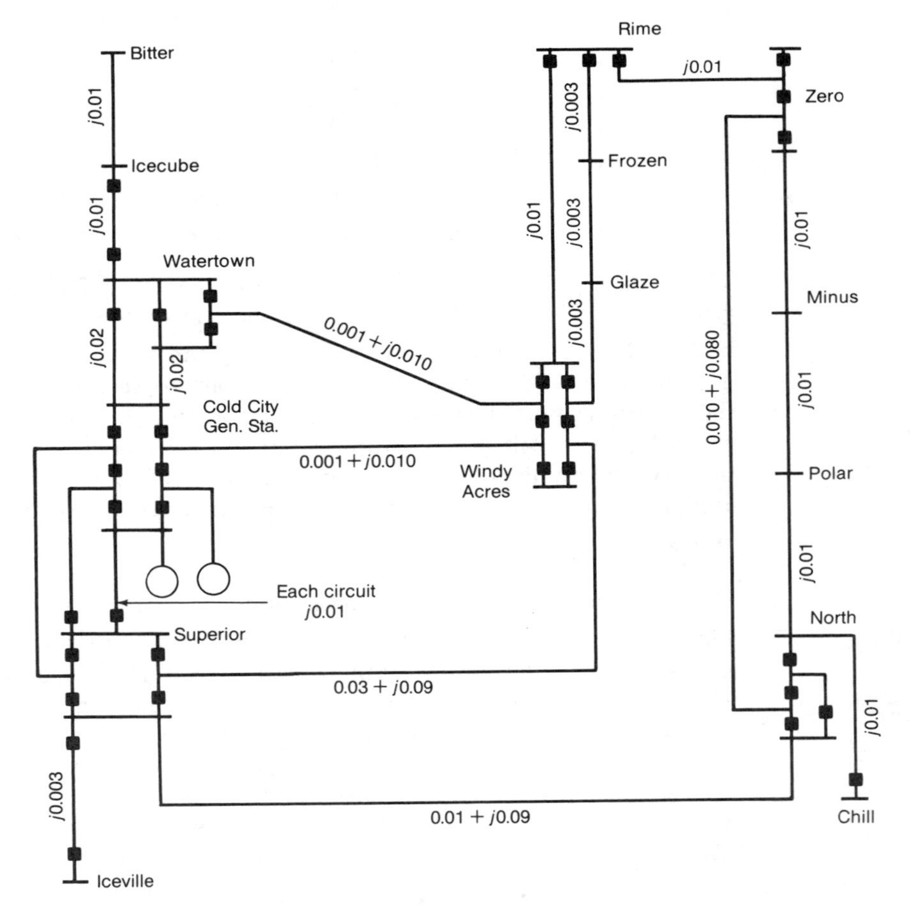

FIGURE P2.7 69-kV Nothern Division. All impedances in per unit on 50-MVA base.

2.8 This problem concerns the application of the theory of incidence matrices to practical fault studies. Consider a company policy in which, as a general guideline, if a fault occurs at bus X, only buses that are within five buses of X need to be considered. For example, with reference to Figure P2.8a, consider a fault at the Central bus. The North bus must be considered since it is one bus away from Central. Similarly, Adda, Kiam, Oakville, and N. Oakville must be considered (N. Oakville is five buses away from Central). However, Gimli S need not be considered.

a. Your task is to write a computer program that will accept line data in 2A10 format and will, for each system bus, print out a list of buses that must be considered in a fault study. Also, your program should be tested using the system shown in Figure P2.8b.

b. Your report should consist of:

- Introductory remarks.
- A description of the theory used.
- A program flowchart.
- Your program and solution.

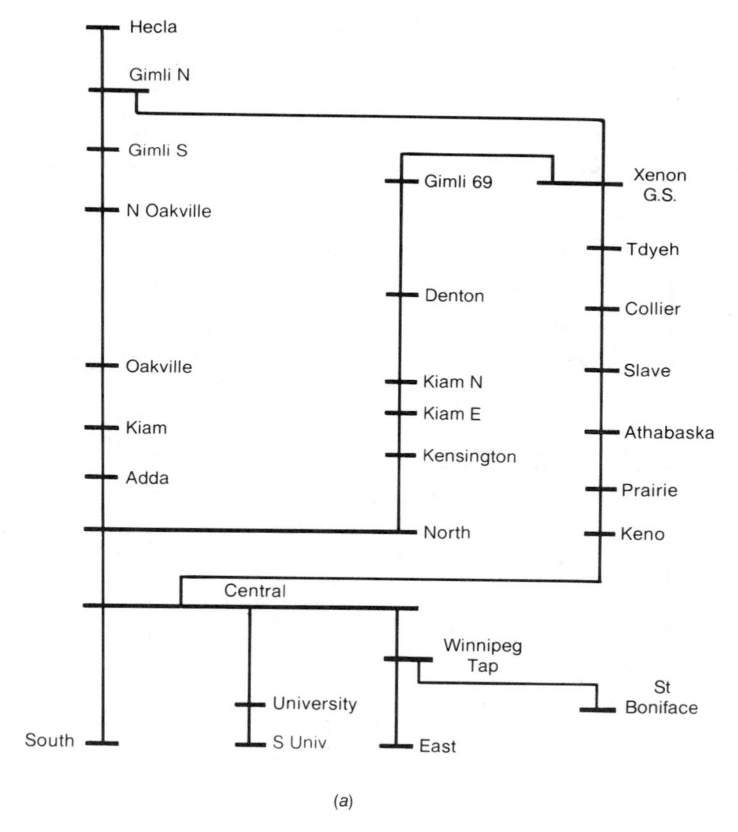

FIGURE P2.8A Twenty-seven bus system.

2.9 The Tampico Power Co. is requesting that you write a program to form the bus impedance matrix for its 69/138 kV system. Required features are:

- Line data are to be read in 2A10, 2F7.4, 2I3 formats with the line impedance in ohms (the 2F7.4 fields) and the base voltages in kilovolts (the I3 fields). The first base voltage corresponds to the first bus name given.

- Z_{bus} in per unit, 100 MVA base is to be formed.
- Z_{bus} is to be formed by inversion of Y_{bus} using the Shipley–Coleman method. Tie the swing bus to ground through $j0.01$ ohms per unit to make Y_{bus} non-singular.

Figure P2.8b shows the system. All transformers shown are shown with impedance already in per unit on a 100 MVA base and consistent with the voltage ratings. Transformers are to be input together with line data except that in column 41 a "T" is punched. Thus a transformer from OAK to MAIN of 7% reactance is shown as

OAK_ _ _ _ _ _ _ _ _MAIN_ _ _ _ _ _ _ _+0.0000 + 0.0700T

Print out Z_{bus} and comment on:

a. How your program could be expanded to find a 100 by 100 Z matrix?

b. The execution time to process a 100 bus system.

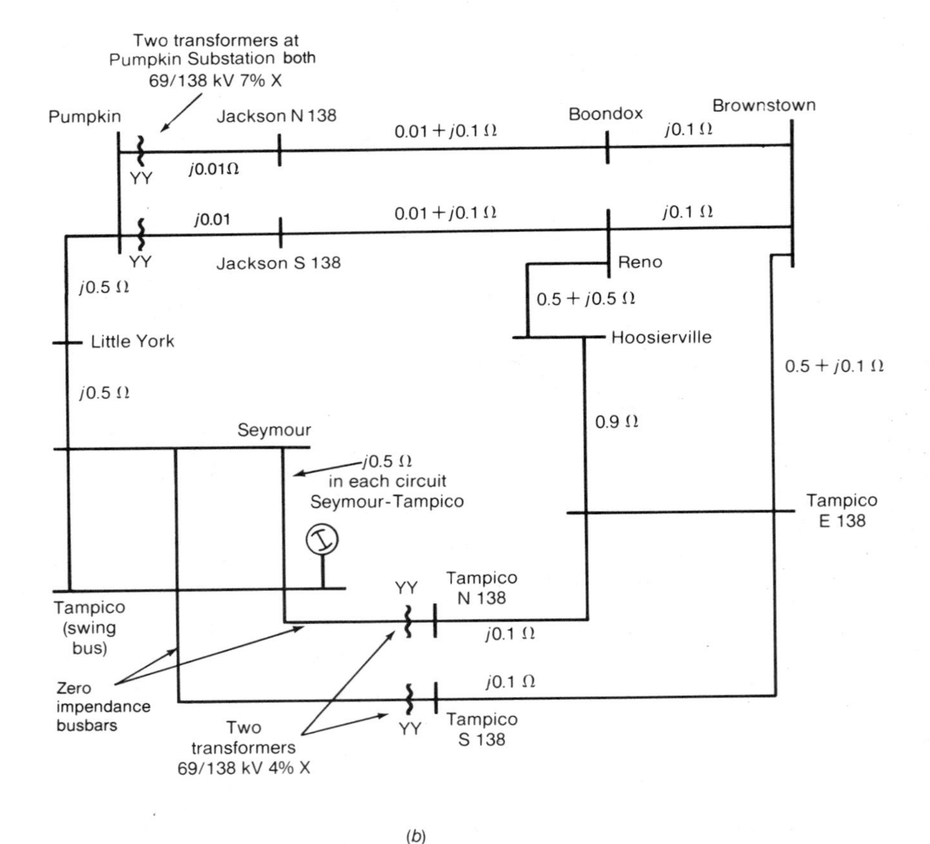

(b)

FIGURE P2.8B Tampico Power Company.

CHAPTER 3

Programming Considerations

3.1 INTRODUCTION

Digital computers generally have memory limitations imposed by the hardware used. Also, practical scheduling of use of computers usually imposes time limitations for a given application. In this chapter, attention we turn our attention to time and memory limitations and methods to alleviate difficulties imposed by these limitations. In many cases, it is possible to trade memory requirements for speed, and vice versa. Such a tradeoff may be crucial to the solution of a problem. For example, a digital computer with small memory capacity may not be suitable for the solution of a given power engineering problem due to memory constraints. If the conventional algorithm is reformulated, often at the expense of program speed, the memory requirements might be reduced sufficiently to permit a solution. In some programming techniques, both time and memory requirements are reduced.

Unfortunately, many time/memory improving programming techniques are either dependent on the computer used (i.e., computer system specific) or are programming language dependent or power system specific. Such dependence makes detailed discussion impossible in a general context. Certain algorithms used to compress stored data and to solve linear algebraic equations are well known to power engineers and are readily discussed: these are presented in this chapter. The reader interested in further detail is directed to the following references: *sparsity programming* [2–4]; *triangular factorization* [5]; *data management* [6].

As a brief example of a computer specific programming technique, the use of "overlaying" is described. Overlaying is a method by which certain portions of a computer program are held in high speed memory during all phases of the execution, and certain portions are brought into high speed memory only as needed. The latter portions are nominally not in high speed ("core") memory, but rather, are stored on slower media, such as magnetic disk or tape. This technique is both computer and language specific. Many operating systems have Fortran or Basic language compilers which permit overlaying. Figure 3.1 shows a typical configuration: the program segment or overlay labeled "0.0" is resident in core at all times. It contains calls to the other overlays. Usually, this programming segment is labeled simply

OVERLAY 0.0

The secondary overlays are brought into core only as needed. For example (see Figure 3.1), overlay 1.0 is called from overlay 0.0 and it is loaded only after the call. When overlay 1.0 returns to 0.0, overlay 1.0 is dropped

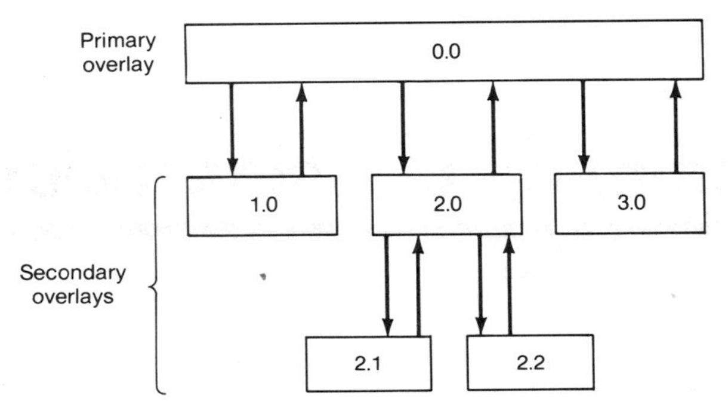

FIGURE 3.1 Overlays.

and cannot be called again. Similarly, overlay 2.0 is loaded when called from 0.0. When a return is executed, 2.0 is dropped. In some systems, several levels of secondary overlays are permitted; these are exemplified by overlays 2.0 and 2.1. When 2.1 is called from 2.0, it is loaded. A "return" causes 2.1 to be dropped. Overlaying differs from the use of subroutines in that overlays are brought into high speed memory only as needed, whereas subroutines are always in core.

The use of overlaying has diminished somewhat with the advent of large memory capacity computers and the more widespread use of algorithms that allow sparsity programming. The latter inherently have low memory requirements and overlays are not needed.

3.2 SPARSITY PROGRAMMING

Sparsity programming is a digital programming technique whereby sparse matrices are stored in compact form. The usual basis of sparsity programming techniques is the storage only of nonzero entries. Since the zero entries are not stored, there must be a mechanism to determine the row and column location of each nonzero entry stored. In some applications of sparsity programming, the techniques goes beyond the storage/access programming: in processing array data (e.g., inversion of a square matrix, the programming may be modified to process only the nonzero data. By this methodology, the sparsity programmed algorithm proceeds more rapidly than the conventional algorithm, and storage requirements are reduced. These techniques are algorithm and data specific and generally work best when very sparse data occur (such as in Y_{bus} methods). Sparsity programming is insensitive to the computer and programming language used. Most production programs intended for use with large scale systems are sparsity programmed.

There are several techniques for sparsity programming, depending on matrix symmetry, percent sparsity, and occurrence of blocks of zeros. Also, the order in which the matrix is stored is important (i.e., if matrix entries are generated in arbitrary order, it is necessary to use an algorithm that allows for insertion of entries between existing data). Two broad classes of sparsity programming techniques are the ***entry–row–column storage*** method and the ***chained data structure*** method. These are discussed below. Reference [7] presents a discussion of the impact of the network configuration on storage requirements, and [8] is a well known documentation of Y_{bus} sparsity storage methods.

Entry–Row–Column Storage Method

Perhaps the simplest sparsity storage method is the storage of nonzero elements in one array, and in a parallel array, storage of the corresponding row and column. Thus the matrix

$$A = \begin{bmatrix} 0 & 1 & 0 \\ 3 & 0 & 0 \\ 0 & 0 & 2 \end{bmatrix}$$

is stored as follows

STO	IR	IC
1	1	2
3	2	1
2	3	3

The array STO contains the nonzero entries to be stored, and IR and IC contain the corresponding rows and columns of the entries in STO. For an n by n sparse matrix with n_e nonzero entries, the total required storage is $3n_e$ entries.

There are several possibilities for the ordering of the entries in the three parallel arrays STO, IR, and IC. If all nonzero entries of A in column 1 are stored first, column 2 next, and so on, (i.e., storage in column fashion), insertion of a new entry will require pushing down entries below the insertion. A similar requirement occurs for storage in row fashion. This requirement results in long storage time per entry if A is stored in random order. If the algorithm for generating A may be designed to generate the matrix by columns, columnar–fashion storage is time efficient, since no insertions will occur.

Alternative to row– and columnar–fashion storage is arbitrary ordering of the list. If arbitrary ordering is used, the entire IR/IC lists must be scanned whenever an access is done. The result is long access time unless the access order is the same as the storage order.

Chained Data Structure Method

The row–column storage method above has unpleasant qualities upon insertion of an entry. The chained data structure avoids this difficulty through the use of a data chain which can be broken, an insertion made, and rejoined. There are many configurations of chained data structures. A fast row-by-row method is illustrated here. Consider the matrix A to be stored

$$A = \begin{bmatrix} 1 & 0 & 0 & 1 \\ 4 & 3 & 0 & 0 \\ 0 & 0 & 0 & 2 \\ 0 & 0 & 1 & 0 \end{bmatrix}.$$

Row 1 is sparsity stored first. The two nonzero entries of row 1 are stored in STO as before with a parallel array IC indicating the column number,

STO	IC
1	1
1	4

Now consider a third parallel array, NX, which tells how far down the list one will find the next entry

STO	IC	NX
1	1	1
1	4	0

The entry NX = 0 means that there are no further nonzero entries in this row. Now if it is desired to insert a nonzero entry in column 2 of row 1 (e.g., A_{12} = 200), add to the tables as follows

STO	IC	NX
1	1	2
1	4	0
200	2	-1

The first entry in this table is interpreted as an entry of 1 in column 1. The next entry is two lines down (since NX = 2). Moving two lines down to the third line, an entry of 200 is found and it is in column 2 (since IC = 2). The next entry will be found -1 line down, (i.e., one line upward). This is a 1 in column 4. Since NX = 0, this is the last entry.

Return now to the storage of the original A matrix. This is done as follows

STO	IC	NX
1	1	1
1	4	0
4	1	1
3	2	0
2	4	0
1	3	0

NFIRST
1
3
5
6

It is necessary to employ a vector NFIRST which tells where row 1 starts in the list. Row 1 starts in line 1 of STO/IC/NX. Row 2 starts in line 3 of STO/IC/NX. Rows 3 and 4 start in lines 5 and 6. Thus NFIRST contains 1, 3, 5, 6.

To insert an entry using this sparsity storage algorithm, a new line is added in the parallel arrays STO/IC/NX. This contains the new entry. Also, the chain must be modified (i.e., entry in NX modified) to show that the new entry is added to the table. Thus insertion of an entry does not involve reordering elements or "pushing down" entries to accomplish the insertion. Only the chain established by the NX pointer is modified. One advantage of this method is the ease (i.e., speed) of insertion of a new element. In some applications, however, this is an unneeded advantage since elements are formed in order and insertions are not required.
Usually, chained data structures are favored since it is convenient to use the NX array data to tell the calling program where the next nonzero element is located. This is used to skip operations on zero elements. In algorithms where such skipping is possible, *not only are memory requirements reduced, but processing speed is increased.*

Applications of sparsity programming methods are primarily in Y_{bus} methods. As we shall see later, in certain types of power flow studies, an array known as the Jacobian matrix is used. This matrix is as sparse as the Y_{bus} matrix and leads itself to sparsity programming. Generally, for large systems, the Y_{bus} matrix is over 99% sparse.

3.3

UPPER RIGHT TRIANGLE STORAGE

For reasons of symmetry, the bus impedance and admittance matrices are not stored in entirety. Only the upper right triangle (including the diagonal) is stored. To take advantage of this memory saving, a two–dimensional array must be stored in a linear array; Figure 3.2 shows

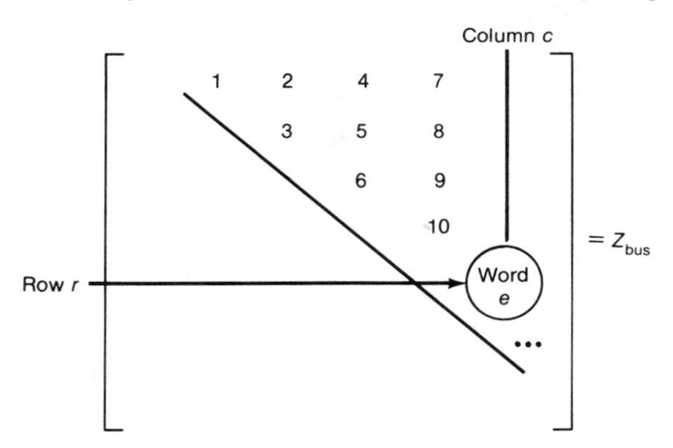

FIGURE 3.2 Upper right triangle storage.

such a method. Entry z_{11} is stored as the first word in the singly subscripted array, followed by z_{12}, z_{22}, z_{13}, z_{23} z_{33}, z_{14}, and so on. If the access or storage of a lower left triangle position is required, the row/column subscripts are interchanged to access or store in the transposed, upper right triangle position. It is necessary to determine a rule by which the location of entry z_{rc} $(r \leq c)$ is found. Let z_{rc} be stored in entry e of the singly subscripted array as illustrated in Figure 3.2. Then $z_{r+1,c}$ must be at position $e+1$,

$$e = f(r, c)$$

$$e+1 = f(r+1, c).$$

Apparently, the desired rule or function, f, is linear in the row entry, r. By similar reasoning, f is noted to be quadratic in c,

$$e = f(r, c) = Ac^2 + Bc + D + Grc + r.$$

To find A, B, D, and G, substitute

$$1 = f(1,1)$$

$$2 = f(1, 2)$$

$$3 = f(2, 2)$$

$$4 = f(1, 3).$$

The result is $A = 1/2$, $B = -1/2$, $D = G = 0$. Thus

$$e = f(r, c) = 1/2c^2 - 1/2c + r.$$

This equation gives a means of accessing z_{rc} in a singly subscripted array.

If both r and c equal N,

$$e = 1/2\ N^2 - 1/2\ N + N = 1/2\ N^2 + 1/2\ N,$$

and it is noted that $1/2(N^2 + N)$ words are required to store the entire upper right triangle (plus diagonal) of a symmetric N by N matrix.

3.4

TRIANGULAR FACTORIZATION

In power engineering, as well as many other branches of engineering, it is necessary to solve a simultaneous set of n algebraic linear equations. The general form is the solution of

$$Ax = b \qquad (3.1)$$

given n by n matrix A and n-vector b. Such is the case in the solution of

$$I_{bus} = Y_{bus} V_{bus}$$

for V_{bus} given the injection currents and admittance matrix. Using the notation of (3.1), let A be factored into two matrices,

$$A = LU, \qquad (3.2)$$

where L is lower left triangular (i.e., L_{rc} is zero for $c > r$) and let U be upper right triangular (i.e., U_{rc} is zero for $r > c$). Then

$$LUx = b. \qquad (3.3)$$

Let Ux be a new variable, w, which is an n-vector. Then

$$w = Ux \qquad (3.4)$$

$$Lw = b. \qquad (3.5)$$

Equation (3.5) is readily solved for w since the form of the equations is rather special

$$\begin{bmatrix} \ell_{11} & 0 & 0 & 0 \\ \ell_{21} & \ell_{22} & 0 & 0 \\ \ell_{31} & \ell_{32} & \ell_{33} & 0 \\ & \cdots & & \end{bmatrix} w = b \ . \tag{3.6}$$

The top row of (3.6) is solved readily for w_1 since the only nonzero entry on the left hand side is $\ell_{11}w_1$. In row 2, find

$$\ell_{21}w_1 + \ell_{22}w_2 = b_2 \ .$$

Since w_1 is known from the first equation, the only unknown is w_2. The process of forward substitution continues until all of w is found.

Having found vector w, (3.4) is used to find x. The form of (3.4) is

$$w = \begin{bmatrix} u_{11} & u_{12} & u_{13} & \cdots \\ 0 & u_{22} & u_{23} & \\ 0 & 0 & u_{33} & \\ \cdots & & & \end{bmatrix} x, \tag{3.7}$$

and the bottom row is readily solved for x_n,

$$w_n = u_{nn} \, x_n \ .$$

This value of x_n is back substituted into the $(n-1)$st row of (3.7) to find x_{n-1}

$$w_{n-1} = u_{n-1\,n-1}x_{n-1} + u_{n-1\,n}x_n \ .$$

The process of back substitution is used to find all other entries of x.

By this process of triangular factorization of A and forward and backward substitution, the solution of $Ax = b$ is found without inverting A. At first glance, it may appear remarkable that the inversion may be avoided, but there is some information that lies uncalculated in the use of triangular factorization compared to a solution using inversion. If $Ax = b$ is to be solved for many different values of b, inversion is readily used since a given vector b premultiplied by A^{-1} gives the required result. In such an applications, A^{-1} is calculated only once. If triangular factorization is used, the forward/backward substitution process would have to be repeated for each different b vector. In power engineering applications, however, it is usually necessary to solve $Ax = b$ only once. The inverse of A is not really desired – only x is sought. The triangular factorization method is favored for such applications.

The success of the triangular factorization method depends in addition on the speed of the factorization of A into L and U. Also, the sparsity of L and U for sparse A is of considerable interest since if the triangular factors were sparse, the memory advantages of sparsity programming may be added to the speed advantages of triangular factorization. These questions are considered after a detailed description of the calculation of L and U.

Formation of the Table of Factors

Let L have a unit diagonal

$$L = \begin{bmatrix} 1 & 0 & 0 & \cdots \\ \ell_{21} & 1 & 0 & \\ \ell_{31} & \ell_{32} & 1 & \\ \cdots & & & \end{bmatrix},$$

and let U be stored "superimposed" onto L (omitting the writing of the unity entries in L)

$$\begin{bmatrix} u_{11} & u_{12} & u_{13} & \cdots \\ \ell_{21} & u_{22} & u_{23} & \\ \ell_{31} & \ell_{32} & u_{33} & \\ & \cdots & & \end{bmatrix}.$$

This superimposed array is called the ***table of factors***. The upper right triangle of the table is U; the lower left triangle is L with the unit diagonal deleted. It is possible to factorize arbitrary nonsingular matrices, J, into L and U without the need for new storage areas; in other words, J is destroyed and replaced by the table of factors. The table of factors is approximately as sparse as J and therefore the table of factors does not require additional storage. Techniques that result in the destruction of the original data and the replacement of those data by the solution are known as ***in situ*** methods. Matrix J is factored into L and U in situ and J is destroyed and replaced by the table of factors.

The formulas required for the calculation of the table are readily found, noting that

$$LU = J .$$

Write L and U, and expand row 1 of J

$$\begin{bmatrix} 1 & 0 & 0 & 0 & \cdots \\ \ell_{21} & 1 & 0 & 0 & \\ \ell_{31} & \ell_{32} & 1 & 0 & \\ \cdots & & & & \end{bmatrix} \begin{bmatrix} u_{11} & u_{12} & u_{13} & \cdots \\ 0 & u_{22} & u_{23} & \\ 0 & 0 & u_{33} & \\ \cdots & & & \end{bmatrix} = J$$

$$u_{11} = J_{11}; \qquad u_{12} = J_{12} \quad \cdots \quad u_{1c} = J_{1c}.$$

Since the first row of the table of factors is $row_1(U)$, the first row of the table is simply $row_1(J)$. In row 2 of J,

$$\ell_{21} u_{11} = J_{21}$$

$$\ell_{21} u_{12} + u_{22} = J_{22}$$

$$\ell_{21} u_{13} + u_{23} = J_{23},$$

and so on. These equations yield ℓ_{21}, u_{22}, u_{23}, u_{24}, $\ldots$,

$$\ell_{21} = \frac{J_{21}}{u_{11}}$$

$$u_{22} = J_{22} - \ell_{21}u_{12}$$

$$u_{23} = J_{23} - \ell_{21}u_{13}$$

$$\vdots$$

Thus in the expansion of row 2 of J, one ℓ and $n - 1$ u's are formed. The generalization for the expansion of row r of J is straightforward

$$\ell_{rc} = \frac{(J_{rc} - \sum_{q=1}^{c-1} \ell_{rq}u_{qc})}{u_{cc}} \qquad c = 1, 2, \ldots, r-1 \tag{3.8}$$

$$u_{rc} = J_{rc} - \sum_{q=1}^{r-1} \ell_{rq}u_{qc} \qquad c = r, r+1, \ldots, n. \tag{3.9}$$

The rules for the formation of the table of factors in situ are as follows:

1. Row 1 of J is not modified.
2. Row 2 of J is modified to yield one ℓ and $n - 1$ u's. These entries are found using (3.8) and (3.9). For these equations, when the summation upper index is zero, there are no terms in the sum (i.e., ignore the summation term). When the upper and lower indices of summation are the same, there is a single term in the sum.
3. Rule (2) is repeated for each row and J is replaced by the table of factors. In row r, there will be $r - 1$ ℓ-terms calculated and $n - r + 1$ u–terms. Equations (3.8) and (3.9) are used.

Optimal Ordering

Equations (3.8) and (3.9) describe how the table of factors is calculated. The L entries (denoted ℓ) and the U entries (denoted u) replace the original matrix (denoted J). Thus the triangular factorization equations might well be written as

$$J_{rc} = \begin{cases} \dfrac{(J_{rc} - \sum_{q=1}^{c-1} J_{rq}J_{qc})}{J_{cc}} & c \leq r - 1 \\[2ex] J_{rc} - \sum_{q=1}^{r-1} J_{rq}J_{qc} & c \geq r \,. \end{cases} \tag{3.10}$$

As r progresses from 1 to n, the table of factors is formed a row at a time; in the process of forming the table, elements of the original matrix as well as elements of the table which were calculated earlier are used. Consider now the case of sparse matrices. Equation (3.10) reveals that the table of factors will also be sparse in row 1 since the original matrix is sparse in this row. In row 2, the table entries to the right of $c = 2$ (i.e., for $c > r$) are mostly zero since the J_{qc} terms are mostly zero. In (3.10), for $c \geq r$, most, if not all of the elements in the sum are zero. This is the case since the sum is carried to $q = r - 1$. Thus row 2 of the table of factors will contain mostly zeros. A similar phenomenon occurs in later rows, although as r becomes large, the lower left triangle formula (3.8) contains more terms in the sum for values of c near $r - 1$. Also, the upper right triangle formula (3.9) contains many terms in the sum.

A procedure by which the "fill-ins" in the table are minimized is termed *optimal ordering.* A fill-in results when (3.10) results in a nonzero value for J_{rc}. When the fill-ins are minimized, the table of factors is very sparse. Optimal ordering refers to renumbering the matrix axes so that fill-ins are minimized.

To illustrate the objective of optimal ordering, consider a matrix with a 1-axis which is nearly full (i.e., few zeros). Further, let the remainder of the matrix be very sparse. If renumbering of the axes is not used, (3.9) will result in a filled row for $r = 1$ in the table of factors. In row 2 of the table, the $n - 1$ upper right triangle entries calculated using (3.9) have a rather high probability of being nonzero because the sum,

$$-\sum_{q=1}^{r-1} \ell_{rq} u_{qc},$$

will have many nonzero values of u_{qc}. The fill-ins so generated in row 2 will propagate into row 3, and so on. On the other hand, if the full 1-axis is reordered (i.e., renumbered) to the last axis, this fill-in propagation will not occur. Thus a rather elementary optimal ordering procedure is to push the relatively full rows down in the matrix, thereby resulting in late processing. There are other optimal ordering algorithms [5,9], although one must be aware of a trade-off between the time spent in calculating the best ordering and that gained by the reordering. Also, it is possible that inappropriate (i.e., too small values of u_{cc} may occur in (3.8); the resulting numerical instability has been noted in certain ordering and triangular factorization methods.

Advantages Attainable Using Optimal Ordering

As described above, the objective of optimal ordering is essentially to minimize the percent fill of later rows of the table of factors. When the percent fill of the original matrix is high, optimal ordering will be rather ineffective. Typical admittance matrices of very large systems will be very sparse and optimal ordering will be highly effective: for example, if a large N-bus system has an average of four lines at each bus, the typical row of Y_{bus} consists of nonzero entries on the diagonal and two other positions in the URT. Fill-ins are likely in two positions in the next row of the table of the factors. A row with twice the number of nonzero entries corresponding generates twice as many fill-ins. The greater number of fill-ins translate into additional processing time in forward/backward substitution due to added terms in the sums in (3.8) and (3.9). Thus, the effectiveness of optimal ordering is not only dependent on overall matrix sparsity, but also on the existence of very sparse rows and less sparse rows in the original matrix. When the sparsity of the matrix is rather uniform, optimal ordering is not as effective as in the case of nonuniform sparsity. Since typical power systems contain several substations with numerous line terminations, optimal ordering is quite essential; without reordering, the percent fill-ins in the table of factors could be so high as to cause a substantial increase in both memory and execution time.

Forward and Backward Substitution

The solution of a simultaneous set of n algebraic linear equations by triangular factorization involves two main steps: formation of the table of factors and forward and backward substitution. Having found the table of factors as indicated above, (3.5) and (3.4) are used to complete the solution,

$$Lw = b \tag{3.5}$$

$$Ux = w. \tag{3.4}$$

Vector b is destroyed in this process and converted in situ to vector w. Since ℓ_{11} is unity, w_1 is simply b_1 [see Eq. (3.6)]. Subsequent rows of w are found using

$$w_r = b_r - \sum_{q=1}^{r-1} \ell_{rq} w_q . \tag{3.11}$$

Equation (3.11) is the expansion of row r of (3.5). Equation (3.11) is used for $r = 1, 2, ..., n$; as this equation is used, w_1 replaces b_1, w_2 replaces b_2, and so on.

Having found w, a similar process based on (3.4) is used to find x. The nth (last) row of (3.4) is

$$u_{nn} x_n = w_n$$

and x_n is readily found. Again in situ technique is used and x_n replaces w_n. In row $n-1$,

$$u_{n-1\,n-1} x_{n-1} + u_{n-1\,n} x_n = w_{n-1}$$

and

$$x_{n-1} = \frac{w_{n-1} - u_{n-1\,n} x_n}{u_{n-1\,n-1}} .$$

Element x_{n-1} replaces w_{n-1}. The process continues for rows $n-2$, $n-3$, ..., 1. In row r,

$$x_r = \frac{w_r - \sum_{q=r+1}^{n} u_{rq} x_q}{u_{rr}} . \tag{3.12}$$

This completes the solution.

Debugging Triangular Factorization Programs

The most commonly encountered errors in computer implementations of triangular factorization – forward/backward substitution are:

1. Failure to recognize that the sums in (3.8) and (3.9) are zero when the upper index of summation is zero.

2. Failure to initialize these sums to zero.

3. Failure to recognize that the sum in (3.12) is zero when the upper index is zero. Also, failure to initialize this sum to zero.

4. Failure to recognize that the sum in (3.11) is zero for the case

$$r + 1 > n$$

and also the initialization of this sum to zero.

Example 3.1

Consider a nonsymmetric matrix J,

$$J = \begin{bmatrix} 1 & 0 & 0 & 1 \\ 0 & 1 & 0 & 5 \\ 2 & 0 & 1 & 0 \\ 0 & 0 & 0 & 4 \end{bmatrix} .$$

Solve the set of equations

$$Jx = \begin{bmatrix} 1 \\ 3 \\ 0 \\ 1 \end{bmatrix}$$

using triangular factorization.

Solution

First J must be replaced by the table of factors. Row 1 of J is unchanged,

$$\begin{bmatrix} 1 & 0 & 0 & 1 \\ x & x & x & x \\ x & x & x & x \\ x & x & x & x \end{bmatrix} .$$

In row 2, one ℓ is found and three u's are found. The ℓ term is found using (3.8). There are no terms in the summation and ℓ_{21} is simply J_{21}. The remaining three u terms are found using (3.9). The result is

$$\begin{bmatrix} 1 & 0 & 0 & 1 \\ 0 & 1 & 0 & 5 \\ x & x & x & x \\ x & x & x & x \end{bmatrix} .$$

In a computer implementation, at this point the first two rows of J are destroyed and replaced by the first two rows of the table of factors. The lower two rows of J are intact.

Proceeding to row 3, two ℓ-terms and two u-terms are found. The calculation of u_{34} will serve to illustrate the calculation within this row,

$$\begin{aligned} u_{34} &= J_{34} - \ell_{31}u_{14} - \ell_{32}u_{24} \\ &= 0 - (2)(1) - (0)(5) \\ &= -2 . \end{aligned}$$

The results of calculation of this row are

$$\begin{bmatrix} 1 & 0 & 0 & 1 \\ 0 & 1 & 0 & 5 \\ 2 & 0 & 1 & -2 \\ x & x & x & x \end{bmatrix} .$$

Finally, in row 4, three ℓ-terms and one u-term are calculated. For example, the ℓ_{43} term is

$$\begin{aligned} \ell_{43} &= \frac{J_{43} - \ell_{41}u_{13} - \ell_{42}u_{23}}{u_{33}} \\ &= 0 . \end{aligned}$$

The completed table of factors is

$$\begin{bmatrix} 1 & 0 & 0 & 1 \\ 0 & 1 & 0 & 5 \\ 2 & 0 & 1 & -2 \\ 0 & 0 & 0 & 4 \end{bmatrix}.$$

At this point, it is instructive to note that this table is, in fact, L and U superimposed,

$$L = \begin{bmatrix} 1 & 0 & 0 & 0 \\ 0 & 1 & 0 & 0 \\ 2 & 0 & 1 & 0 \\ 0 & 0 & 0 & 1 \end{bmatrix} \qquad U = \begin{bmatrix} 1 & 0 & 0 & 1 \\ 0 & 1 & 0 & 5 \\ 0 & 0 & 1 & -2 \\ 0 & 0 & 0 & 4 \end{bmatrix}.$$

The second half of this example is the solution of $Jx = b$ using forward and backward substitution. The forward substitution process uses (3.11), in which vector w replaces vector b. Note that the table of factors does not contain the unity diagonal entries ℓ_{rr}; these entries are not needed in (3.11). In row 1,

$$w_1 = b_1 = 1.$$

Therefore, the vector b is

$$\begin{bmatrix} 1 \\ x \\ x \\ x \end{bmatrix}$$

(the lower three entries are still b_2, b_3, b_4). In row 2,

$$w_2 = b_2 - \ell_{21} w_1 = 3 - (0)(1)$$

$$= 3.$$

Similarly, rows 3 and 4 are calculated with the following result

$$w = \begin{bmatrix} 1 \\ 3 \\ -2 \\ 1 \end{bmatrix}.$$

Finally, (3.11) is used to find x_4, x_3, x_2, and x_1. In row 4,

$$x_4 = \frac{w_4}{u_{44}}$$

$$= 3/4.$$

The $b/w/x$ vector is

$$\begin{bmatrix} x \\ x \\ x \\ 0.25 \end{bmatrix}.$$

In row 3,

$$x_3 = \frac{w_3 - u_{34}x_4}{u_{33}}$$

$$= \frac{-2 - (-2)(0.25)}{1}$$

$$= -1.5.$$

After finding rows 2 and 1,

$$x = \begin{bmatrix} 0.75 \\ 1.75 \\ -1.5 \\ 0.25 \end{bmatrix}.$$

Example 3.2

For a given power system, the bus admittance matrix referenced to ground is

$$Y_{bus} = \begin{bmatrix} 1 - j200 & j100 & 0 & j100 \\ j100 & -j200 & j100 & 0 \\ 0 & j100 & -j200 & j100 \\ j100 & 0 & j100 & -j200 \end{bmatrix}.$$

Find V_{bus} for

$$I_{bus} = \begin{bmatrix} 0.9 + j0.0 \\ 0.1 + j0.1 \\ 0 \\ 0 - j0.1 \end{bmatrix}.$$

Solution

It is necessary to solve

$$I_{bus} = Y_{bus} V_{bus}$$

for V_{bus}. First, Y_{bus} must be replaced by its table of factors. Row 1 of the table is row 1 of Y_{bus}. In row 2, ℓ_{21}, u_{22}, u_{23}, and u_{24} are found; the calculation of u_{22} illustrates the process:

$$u_{22} = (Y_{bus})_{22} - \ell_{21}u_{12}$$

$$= -j200 - \frac{(j100)(j100)}{1} - j200 = 0.24999 - j150.001.$$

The first two rows of the table of factors (TOF) are

$$\begin{bmatrix} 1 - j200 & j100 & 0 & j100 \\ 0.5 - j0.0025 & 0.24999 - j150.001 & j100 & 0.25 + j50 \\ \text{x} & \text{x} & \text{x} & \text{x} \\ \text{x} & \text{x} & \text{x} & \text{x} \end{bmatrix}.$$

The remainder of the table is calculated without difficulty (note that the unit diagonal of L is never needed). The final result is

$$
\text{TOF} = \begin{bmatrix} 1-j200 & j100 & 0 & j100 \\ 0.5-j0.0025 & 0.24999-j150.001 & j100 & 0.25+j50 \\ 0 & 0.666666-j0.00111 & 0.111-j133.33 & 0.222+j133.33 \\ 0.5-j0.0025 & 0.33333-j0.002222 & 0.99999-j0.002499 & 0.9999-j0.0075 \end{bmatrix}.
$$

This completes the first half of the solution. The calculation of w is straightforward; the calculation of w_2 will illustrate the procedure

$$
\begin{aligned}
w_2 &= 0.1 + j0.1 - \ell_{21} w_1 \\
&= 0.35 + j0.10225.
\end{aligned}
$$

The result is

$$
w = \begin{bmatrix} 0.9 \\ -0.35 + j0.1225 \\ -0.2332 + j0.06856 \\ 1.0005 - j0.00725 \end{bmatrix}.
$$

Finally, x is calculated and replaces vector w. The result is (x is V_{bus})

$$
\begin{bmatrix} 1.000 + j0.0000 \\ 0.9995 + j0.0008 \\ 1.0000 + j0.0005 \\ 1.0005 + j0.0003 \end{bmatrix}.
$$

The solution is readily checked using

$$
I_{bus} = Y_{bus} V_{bus}.
$$

Although this procedure may, at first, appear formidable, the required software is straightforward since each step in the procedure is similar to the other steps. The advantages of the method are speed, low memory requirements, and applicability of sparsity programming. The disadvantage is that the inverse matrix is never calculated and repeated solutions require repeating the forward/backward substitution process (although the table of factors needs not be recalculated). Most production software in power engineering uses the triangular factorization procedure to solve equations of the form $Ax = b$.

3.5 SHIPLEY-COLEMAN MATRIX INVERSION

If it is desired to solve $Ax = b$ by matrix inversion, the use of an efficient matrix inversion method is required in view of the typically large dimension of matrix A in power engineering applications. Most efficient matrix inversion methods are based on Gaussian elimination; the Shipley–Coleman method [1] is presented here not only for its efficiency, but also because of its connection to power engineering and ease in programming.

The algorithm to invert A of dimension N by N in situ (i.e., A is destroyed and replaced by A^{-1}) is:

1. Let the pivot axis, p, be axis 1.

2. Kron reduce all elements outside the pivot axis

$$A_{ij}^{new} = A_{ij}^{old} - \frac{A_{ip}^{old} A_{pj}^{old}}{A_{pp}^{old}} \quad i \neq p; \; j \neq p.$$

3. Replace the pivot position by its negative inverse

$$A_{pp}^{new} = \frac{-1}{A_{pp}^{old}}.$$

4. Reduce elements in the pivot axis outside the p, p position according to

$$A_{ip}^{new} = (A_{ip}^{old})(A_{pp}^{new}) \quad i \neq p$$

$$A_{pj}^{new} = (A_{pj}^{old})(A_{pp}^{new}) \quad j \neq p.$$

5. Repeat steps 2 through 4 for p = 2, 3, ..., N. The result is $-A^{-1}$. Therefore, reversal of the sign of the matrix will yield A^{-1}.

The following is a Fortran implementation of the method for the case of arbitrary symmetry of A (A is stored in full in this illustration–considerable savings of memory and time are possible using URT storage):

```
      SUBROUTINE INVERT (A,N)
      DIMENSION A(N,N)
      DO 1 IP = 1, N
      DO 2 IR = 1, N
      IF (IR.EQ.IP)GO TO 2
      DO 3 IC = 1,N
      IF(IC.EQ.IP)GO TO 3
      A(IR,IC) = A(IR,IC) - (A(IR,IP)*A(IP,IC)/A(IP,IP))
    3 CONTINUE
    2 CONTINUE
      A(IP,IP) = -1./A(IP,IP)
      DO 4 I = 1,N
      IF (I.EQ.IP) GO TO 4
      A(I,IP) = A(I,IP)*A(IP,IP)
      A(IP,I) = A(IP,I)*A(IP,IP)
    4 CONTINUE
    1 CONTINUE
      DO 5 IR = 1,N
      DO 5 IC = 1,N
    5 A(IR,IC) = -A(IR,IC)
      RETURN
      END
```

The Shipley–Coleman method is the Gaussian elimination of successive variables. It is possible to avoid problems associated with zeros on the diagonal of A by altering the sequence in which pivot axes are chosen. The method requires computer central processing time which is of order $O(N^3)$ [the notation $O(N^m)$ denotes that the complexity and processing time increases approximately as the mth power of N for large N]. It is interesting to try to compare this time requirement with that of the triangular factorization method, although considerable caution in this comparison is required since the Shipley–Coleman method produces an inverse matrix and triangular factorization and forward–backward substitution does not. The triangular factorization of an N by N matrix requires $O(N^2)$ operations in each row [by examination of (3.8) - (3.9) which give the required T.O.F. entries for each of the N entries per row]. Since there are N rows, the calculation of the entire TOF requires $O(N^3)$ operations. Having found the TOF, the forward/backward substitution process requires $O(N^2)$ operations. Thus, for a small (i.e., much less than N) number of solutions of $Ax = b$, it is better to use triangular factorization *once* to obtain the TOF, then use the forward/backward substitution process to obtain x in each problem. For a large number of solutions of $Ax = b$, it may be better to use inversion. From a memory requirement point of view, for sparse A, the triangular factorization method is always favored to solve $Ax = b$. This is the case since the TOF is sparse when A is sparse.

Example 3.3

Invert A using the Shipley–Coleman method

$$A = \begin{bmatrix} 4.0 & 1.0 & 0.5 \\ 1.0 & 3.0 & 0.0 \\ 0.5 & 0.0 & 3.0 \end{bmatrix}.$$

Solution

Let $p = 1$ and reduce elements A_{22}, A_{23}, A_{32}, A_{33},

$$\begin{bmatrix} \cdot & \cdot & \cdot \\ \cdot & 2.750 & -0.125 \\ \cdot & -0.125 & 2.9375 \end{bmatrix}.$$

Replace A_{11} by $-1/A_{11}$ and multiply the resulting -0.25 value across all elements in axis 1

$$\begin{bmatrix} -0.250 & -0.250 & -0.125 \\ -0.250 & 2.750 & -0.125 \\ -0.125 & -0.125 & 2.9375 \end{bmatrix}.$$

The process is repeated for $p = 2$. The result of the Kron reduction is

$$\begin{bmatrix} -0.27272 & \cdot & -0.13636 \\ \cdot & \cdot & \cdot \\ -0.13636 & \cdot & 2.93183 \end{bmatrix};$$

completing the calculation of the 2–axis, we obtain

$$\begin{bmatrix} -0.27272 & 0.09091 & -0.13636 \\ 0.09091 & -0.36364 & 0.04545 \\ -0.13636 & 0.04545 & 2.93182 \end{bmatrix}.$$

The result of the $p = 3$ iteration is

$$\begin{bmatrix} -0.27906 & 0.09302 & 0.04651 \\ 0.09302 & -0.36434 & -0.01550 \\ 0.04651 & -0.01550 & -0.34109 \end{bmatrix}.$$

This is $-A^{-1}$.

Bibliography

[1] R. B. Shipley and D. Coleman, "A New Direct Method of Matrix Inversion," *Trans. AIEE,* v. CE-78, November 1959, pp. 568–572.

[2] W. Tinney, "Compensation Methods for Network Solutions by Optimally Ordered Triangular Factorization," *IEEE Trans. Power Apparatus and Systems,* v. PAS-91, no. 1, January–February 1972, pp. 123–127.

[3] E. Ogbuobiri, "Dynamic Storage and Retrieval in Sparsity Programming," *IEEE Trans. Power Apparatus and Systems,* v. PAS-89, no. 1, January 1970, pp. 150–155.

[4] B. Randell and C. Kuehner, "Dynamic Storage Allocation Systems," *Communications of the ACM,* v. 11, May 1968, pp. 297–306.

[5] W. Tinney and J. Walker, "Direct Solution of Sparse Network Equations by Optimally Ordered Triangular Factorization," *Proc. IEEE,* v. 55, No. 11, November 1967.

[6] G. T. Heydt, "Data Management and Allocation in Z_{bus} Methods," *IEEE Trans. Power Apparatus and Systems,* v. PAS-91, November–December 1972.

[7] E. Ogbuobiri, W. Tinney and J. Walker, "Sparsity Directed Decomposition for Gaussian Elimination on Matrices," *IEEE Trans. Power Apparatus and Systems,* v. PAS-89, no. 1, January 1970, pp. 141–150.

[8] N. Sato and W. Tinney, "Techniques for Exploiting the Sparsity of the Network Admittance Matrix," *IEEE Trans. Power Apparatus and Systems,* v. PAS-82, December 1963, pp. 944–950.

[9] I. Duff, "A Survey of Sparse Matrix Research," *Proc. IEEE,* v. 65, no. 4, April 1979.

Exercises

Exercises that do not require programming

3.1 For a 100 by 100 symmetric matrix stored in URT form, find the word number (in a linear array) for the entry at row 2, column 56. Also find the storage requirements for the entire matrix.

3.2 Consider a three bus power system with bus impedance matrix referenced to ground:

$$Z_{bus} = \begin{matrix} 1 \\ 2 \\ 3 \end{matrix} \begin{bmatrix} 0.10 & 0.01 & 0.01 \\ 0.01 & 0.05 & 0.04 \\ 0.01 & 0.04 & 0.05 \end{bmatrix}$$

(all in per unit ohms).

a. For the system given, line 21 is outaged and $(Z_{bus})_{33}$ changes from 0.05 to 0.10 per unit. Find the line impedance of the line that was outaged.

b. For the system given, change the reference bus from ground to bus 1.

c. For the system given, I_2 is identically zero. Find the bus impedance matrix reference to ground with bus 2 eliminated.

3.3 For the real matrix A,

$$A = \begin{bmatrix} 1 & 5 & 1 \\ 5 & 2 & 0 \\ 1 & 0 & 1 \end{bmatrix}$$

a. Find the table of factors.
b. Solve

$$\begin{bmatrix} 1 \\ 1 \\ 0 \end{bmatrix} = Ax$$

for the vector x.

3.4 Comment on the following:

a. In the text, the triangular factors of a matrix A are L and U. Matrix L has unity diagonal. Is it possible to select unity diagonal for U (instead of L?). Under what conditions, if any, could both L and U have unity diagonal
b. Under what conditions is $L^t = U$?
c. Let A be rectangular with dimensions m by n

$$m < n \ .$$

Also, let the factors of A be L and U, L being m by n, and U being m by n. The set of equations

$$Ax = b$$

will be solved by triangular factorization using $n - m$ selected values of x. Let these selected values of x be the last m rows of x. (Note: x is an n vector and b is an m vector). Write an algorithm to calculate x and illustrate using

$$A = \begin{bmatrix} 1 & 1 & 0 \\ 4 & 7 & 3 \end{bmatrix} \qquad b = \begin{bmatrix} 0 \\ 1 \end{bmatrix}$$

and $x_3 = -1$.

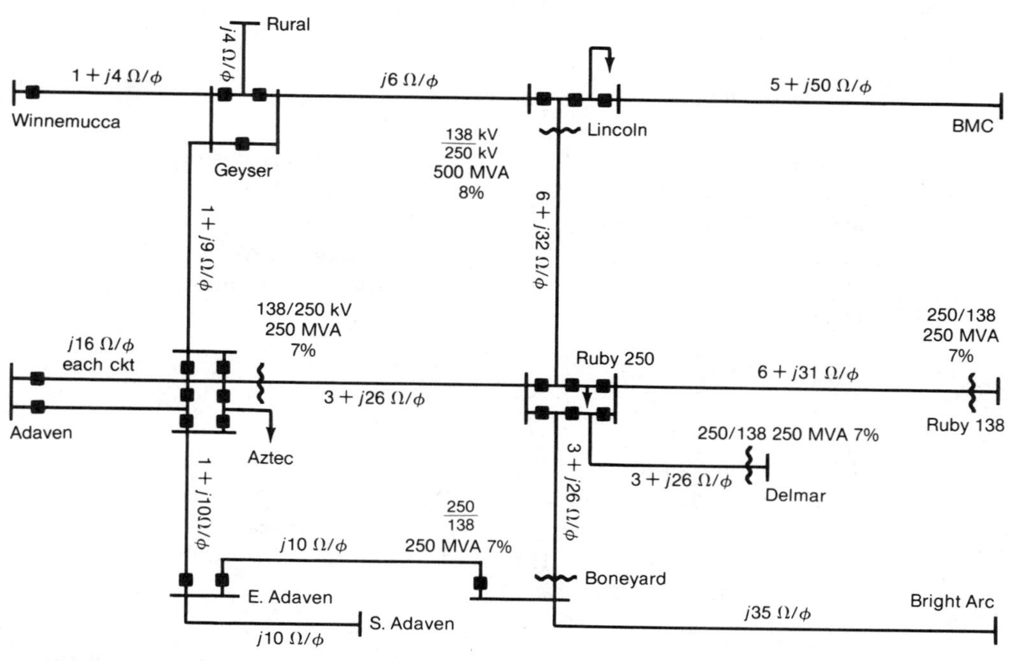

FIGURE P3.6 138 kV system. Actual line Z shown.

3.5 Using the Shipley–Coleman matrix inversion method, find A^{-1} for

$$A = \begin{bmatrix} a & b \\ c & d \end{bmatrix}.$$

Verify that $A^{-1}A = I$.

Exercises that require programming

3.6 The Central Nevada Edison Co. has contacted you to form the bus admittance matrix of their 138 kV system. The Y_{bus} matrix should be formed from a line and transformer list and should be stored in URT form. A 100 MVA base should be used for the per unit system.

Figure P3.6 shows the CNE 138 kV system. The Ruby 250 bus is a 250 kV bus located at the center of an imbedded 250 kV system. All work is to be done in per unit, thereby avoiding problems of turns ratios of transformers and making possible the addition of per unit line impedances. Thus the Boneyard–Ruby 250 kV line and transformer converted to per unit proceeds as follows. For the 250 kV system, $|Z_{base}|$ is $|V|^2/S$

$$\frac{|V^2|}{S} = \frac{(250k/\sqrt{3})^2}{100M/3}$$

$$= 625\ \Omega \textit{ per phase.}$$

Therefore, the line impedance in per unit is

$$Z_{pu} = \frac{Z_{act}}{|Z_{base}|} = 3 + \frac{j26}{625}$$

$$= 0.0048 + j0.0416 \textit{ per unit .}$$

The transformer in this line is 7% on a 250 MVA base (percent reactances are always given consistent with the unit's own power rating). Since $|Z_{base}|$ varies like $1/S$, and Z_{pu} varies like $1/|Z_{base}|$, one concludes that Z_{pu} is proportional to the S base. Therefore, to convert 7% to a 100 MVA base, multiply by 100/250

$$Z_{pu} = (j0.07)(100/250)$$

$$= j0.028 \textit{ per unit .}$$

The total per unit impedance of the line plus transformer is $0.0048 + j0.0416 + j0.028$ or $0.0048 + j0.0696$ per unit.

Your computer program should read one card giving the number of system lines (punched in I3 format). Subsequent cards give line data as follows:

Column	Format	Data
1 – 10	A10	Starting bus name
11 - 20	A10	Ending bus name
21 - 23	I3	Starting bus voltage in kV line to line
24 - 26	I3	Ending bus voltage in kV line to line
27 - 30	I4	Transformer MVA rating (three phase); if no transformer, punch zeros. The transformer is assumed to be located at the starting bus.
31 - 32	I2	Transformer percent reactance on own base, in percent
33 - 46	2F7.2	Line impedance, in ohms per phase

Thus, for Boneyard to Ruby 250, a possible coding of the line data is

BONEYARD_ _RUBY_ _250_38250025007_ _ _3.00_ _26.00

Your program should read the line data, form Y_{bus} in URT form, and print the URT.

In addition to the introduction, statement of the task at hand, and description of the solution method, your report should contain an estimate of time and memory requirements to process a 500 line 420 bus system.

3.7 Verify the program given in Section 3.5 by inverting A

$$A = \begin{bmatrix} 1.3 & 1.7 & 4.0 \\ 1.7 & 2.6 & 2.6 \\ 4.0 & 2.6 & 9.0 \end{bmatrix}.$$

Also calculate $A^{-1}A$.

3.8 To work this problem, you must have a way to obtain the central processing time required to execute a program or section of a program. For the matrix

$$(A)_{ij} = \begin{cases} 1 & i = j \\ 2 & i = 1;\; j = N \\ 0 & \text{otherwise} \end{cases}$$

a. Determine the execution time required to calculate A^{-1} for $N = 15, 20, 25, 30$.

b. Deduce a formula for execution time as a function of N.

CHAPTER 4

Power Flow Studies

4.1

THE SIMPLIFIED POWER FLOW PROBLEM

Perhaps the electric power flow problem is the most studied and documented problem in power engineering. In essence, this problem is the calculation of line loading given the generation and demand levels. Ward and Hale [1] are often credited with the first formulation of the power flow problem, although other early investigators [2,3] were drawn to this field by the increased interconnection of electric power systems. The transmission network is nearly linear and one might superficially expect the power flow problem to be a linear problem; however, because power is a *product* of voltage and current, the problem formulation is nonlinear even for a linear transmission network. Additional nonlinearities arise from the specification and use of complex voltages and currents. Also, there are transmission component nonlinearities which may be considered (such as tap changing transformers, in which the tap is adjusted to hold a given bus voltage magnitude fixed). The usual power flow problem formulation does not consider time variation of loads, generation, or network configuration. The sinusoidal steady state is assumed and as a result, equations are algebraic in form rather than differential. Sometimes, very slow load or generation variation is considered, but this variation is considered to be slow enough to justify sinusoidal steady state assumptions.

The power flow study of an electric power system is also known as a "load flow" study. Prior to 1940, there were a limited number of interconnected power systems and the servicing of the load was elementary in that systems were primarily radial circuits. In more recent years, the numerous advantages of interconnection were recognized and modern power systems are characterized by a high degree of interconnection, many "loop circuits," many load/generation buses, and high levels of power exchange between neighboring companies. The latter point relates closely to interconnection advantages. With no addition to generation capacity, it is possible to increase the possible generation available through interconnection. For example, in the case of an outage of a generating unit, power may be purchased from a neighbor. The cost savings realized from lower installed capacity usually far outweighs the cost of the transmission circuits required to access neighboring companies.

Fortunately, the advent of the interconnected power system was accompanied approximately by the development of the digital computer. The analysis of large interconnected power systems generally involves the simultaneous solution of many nonlinear algebraic equations. This solution is best done by computer, and several of the solution techniques are the subject of this chapter.

Like most engineering problems, the power flow problem has a set of given data and a set of quantities which must be calculated (i.e., knowns and unknowns). Hopefully, the number of equations available will match the number of unknowns. Counting the "knowns" and equations may not be so evident since we are dealing with complex quantities and complex equations. For example, consider the nonlinear problem,

$$(1 + j1)x + (2 + j2)y + z = j5$$

$$xy = 1 + j1$$

$$z = j10.$$

At first glance, there appear to be three equations in three unknowns. But x, y, and z are complex and could be considered as two unknowns each (i.e., real and imaginary parts). Also, each equation must balance in both real and imaginary parts. Further, one may ask whether

$$z = j10$$

constitutes an equation – or does this expression imply that z is not an unknown at all. These are semantic questions with several reasonable answers. In this chapter, each complex unknown will be counted as two real unknowns, and each complex equation will be considered as two real equations, no matter how simple the form of the equation (i.e., "$z = j10$" is two equations, z is two unknowns). The example set of equations shown is six equations in six unknowns.

Using the cited terminology, consider an electric power system with N buses (not counting the ground bus). These N buses are comprised of a single swing bus, N_{pv} buses at which active power P and voltage magnitude $|V|$ are specified (these are PV buses), and N_{pq} buses at which active and reactive power (P, Q) are specified (these are PQ buses). The single swing bus is a generation bus which is usually centrally located in the system; the voltage magnitude and phase angle are specified at this bus. The PV buses are mostly generation buses at which the injected active power is specified and held fixed by turbine settings. A voltage regulator holds $|V|$ fixed at PV buses by automatically varying the generator field excitation. This variation causes the generated reactive power to vary in such a way as to bring the terminal voltage magnitude to the specified value. A more extensive discussion of the role of reactive power at a PV bus is given in Section 4.4. The N_{pq} PQ buses are primarily load buses. Since injection notation is used, an authentic load bus will exhibit specified injected P such that

$$P < 0 .$$

Evidently,

$$N = N_{pq} + N_{pv} + 1.$$

Proceed now to examine the known and unknown variables. At PV buses,

$$P_i = \textit{specified quantity} \tag{4.1}$$

$$|v_i| = \textit{specified quantity} \qquad i = N_{pv} \text{ buses.} \tag{4.2}$$

At PQ buses,

$$P_i + jQ_i = \textit{specified quantity} \qquad i = N_{pq} \text{ buses.} \tag{4.3}$$

At the swing bus,

$$|v_i| \underline{/\delta_i} = \textit{specified quantity} \qquad i = \text{swing bus.} \tag{4.4}$$

In (4.4), δ_i denotes the phase angle of the voltage at bus i. Equations (4.1) through (4.4) are rewritten in terms of the bus voltages and injection currents

$i = N_{pv}$ buses

$$Re\ (v_i I_i^*) = \textit{specified quantity} \tag{4.1a}$$

$$|v_i| = \textit{specified quantity.} \tag{4.2a}$$

$i = N_{pq}$ buses

$$v_i I_i^* = \textit{specified quantity.} \tag{4.3a}$$

i = swing bus

$$v_i = \textit{specified quantity.} \tag{4.4a}$$

The equations are counted as follows:

Equation (4.1a and4.2a)	$2N_{pv}$ equations
Equation (4.3a)	$2N_{pq}$ equations
Equation (4.4a)	2 equations
Total	$2N_{pv} + 2N_{pq} + 2$ equations

The number of unknowns are

V	$2N$ unknowns.
I	$2N$ unknowns.

Thus there appear to be $2(N_{pv} + N_{pq} + 1)$ equations (or $2N$ equations) in $4N$ unknowns. The remaining $2N$ equations are

$$V_{bus} = Z_{bus} I_{bus} \; . \tag{4.5}$$

In this simplified formulation, the load flow problem is the simultaneous solution of (4.1a) through (4.4a) and (4.5), which is a set of $4N$ equations in $4N$ unknowns. The equations are nonlinear but algebraic.

Equations (4.1a) through (4.4a) and (4.5) are not the only possible formulation of the power flow problem. It is possible to eliminate the $2N$ injection current unknowns and reduce the problem to the solution of the bus voltage magnitudes and angles ($2N$ unknowns). The advantages of this alternative formulation are twofold: the reduction of knowns translates into a faster solution, and the resulting formulation is readily solved by a rapid solution technique based on reducing the solution error to zero iteratively. Before proceeding to describe solution methods, the cited alternative formulation is presented as well as a few remarks on the existence and uniqueness of a solution.

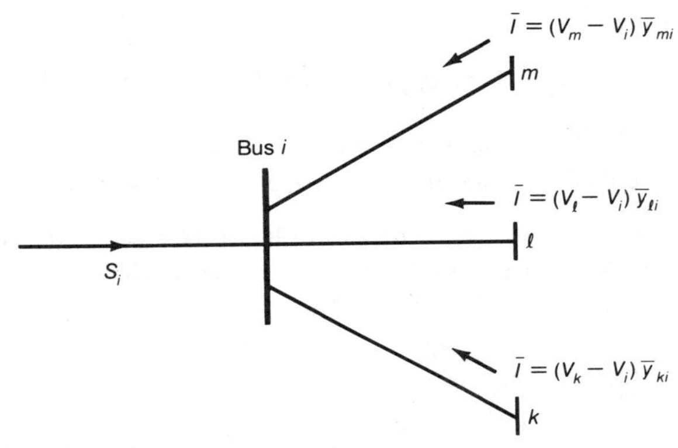

FIGURE 4.1 Mismatch at bus *i*.

Mismatch Formulation of the Power Flow Problem

Consider Figure 4.1, which depicts bus i of a large power network. Bus i is connected to several neighboring buses illustrated pictorially as buses k, ℓ, and m. The total current arriving from remote buses via lines is

$$(v_k - v_i)\bar{y}_{ki} + (v_\ell - v_i)\bar{y}_{\ell i} + (v_m - v_i)\bar{y}_{mi} \; .$$

The bar above the line admittances denotes that these are primitive admittances, not Y_{bus} matrix entries. Let $\bar{S}_i$ denote the total complex power arriving into bus i via lines,

$$\bar{S}_i = v_i[(v_k - v_i)\bar{y}_{ki} + (v_\ell - v_i)\bar{y}_{\ell i} + (v_m - v_i)\bar{y}_{mi}]^* .$$

Manipulation gives

$$\bar{S} = v_i[v_i(-\bar{y}_{ki} - \bar{y}_{\ell i} - \bar{y}_{mi}) + v_k(\bar{y}_{ki}) + v_\ell(\bar{y}_{\ell i}) + v_m(\bar{y}_{mi})]^*$$

$$= v_i[-v_i(Y_{bus})_{ii} - v_k(Y_{bus})_{ik} - v_\ell(Y_{bus})_{i\ell} - v_m(Y_{bus})_{im}]^* .$$

Since k, ℓ, and m are the only buses to which bus i is connected,

$$\bar{S}_i = [(-row_i\, Y_{bus})V_{bus}]^* v_i \; .$$

Define the mismatch complex power at bus i as M_i,

$$M_i = \overline{S}_i + S_i \ . \tag{4.6}$$

The $\overline{S}_i$ term is the complex power arriving via lines to bus i and the S_i term is the specified injected power at that bus. This quantity is frequently referred to simply as "mismatch." Since the bus itself generates no power, the mismatch should be zero; that is, the power arriving via lines plus that arriving via the injection must sum to zero.

The definition of bus power mismatch given results in the following alternative formulation of the power flow problem

$$M_i = [(-row_i \, Y_{bus}) V_{bus}]^* v_i + S_i \quad i = \text{all } PQ \text{ buses} \tag{4.7}$$

$$v_{swing} = specified. \tag{4.8}$$

For a system with only a swing bus and PQ buses, (4.7) gives $2(N - 1)$ equations and (4.8) gives two equations. Hence (4.7) and (4.8) reveal $2N$ equations. Only V_{bus} is unknown ($2N$ unknowns), and the problem is well posed. There is a simple extension to include PV buses – a subject to be presented later.

The simultaneous solution of (4.7) and (4.8) for V_{bus} requires the solution of $2N$ nonlinear algebraic equations. An iterative solution method will be presented in this chapter.

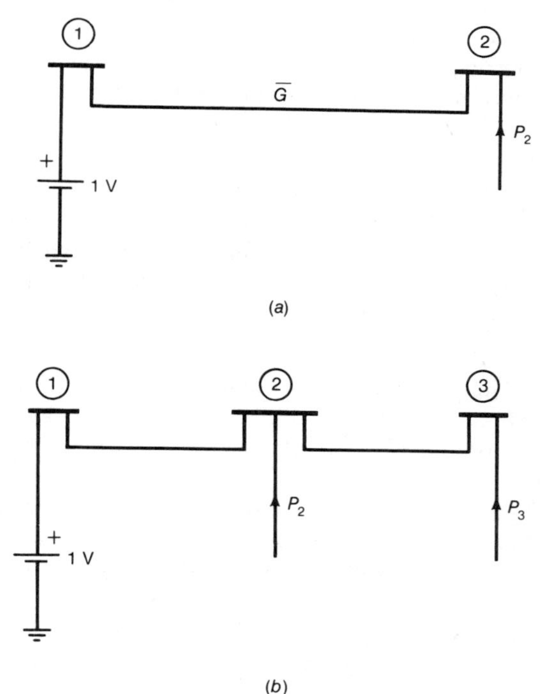

FIGURE 4.1A Two-bus dc power system; 4.1B three-bus dc power system.

Existence and Uniqueness of Solution

As described earlier, the power flow problem requires the simultaneous solution of many nonlinear algebraic equations. The existence and uniqueness of the solution is a matter of concern. Even if the equations were linear, it is possible that no solution exists. If there were only two unknowns, the problem might be visualized as plotting each equation as a locus on a two-dimensional plane. The intersections represent solutions. It is possible that the loci will not intersect. For example, a very simple case depicted in Figure 4.2a gives

$$1(1 - v_2)v_2 + P_2 = 0.$$

Expanding this dc equation yields

$$v_2^2 - v_2 - P_2 = 0.$$

The solutions for v_2 are

$$v_2 = 1/2(1 \pm \sqrt{1+4P_2}) \ . \qquad (4.9)$$

Equation (4.9) gives the following cases:

Two solutions	$P_2 > 1/4$
One solution	$P_2 = -1/4$
No real solution	$P_2 < -1/4.$

For the dc circuit shown (which may be viewed as a Thevenin equivalent circuit of a more general circuit), it is a simple matter to show that a finite, maximum power is delivered to the load when the load resistance equals the Thevenin resistance. This property, known as the maximum power transfer property (or theorem) is readily extended to the ac case. Depending on the injection power, there may be two, one, or no solutions. The "no solution" case is not a surprise since, at least for simple cases, the maximum power transfer theorem effectively prescribes the maximum possible delivered active power. It is not possible to deliver more power than the *maximum* power. If a problem is stated in which the delivered power is greater than the maximum, *no* solution exists.

In Figure 4.2b, a three bus dc system is depicted. It is readily shown that the load flow equations at buses 2 and 3 when solved yield as many as four solutions (the equations are a pair of coupled quadratic expressions). Depending on the values of P_2 and P_3, there may be no solution.

In general, the power flow problem has many possible solutions or no solution. The case at hand depends on the loading condition. The existence of many possible solutions may give concern since it is desirable to discern which of these solutions is expected in the field. Fortunately, it is usually the case that the solution V_{bus} consists nearly of all $1\angle 0^\circ$ entries. Usually, only one solution will be near the unity bus voltage profile – the other solutions usually contain one or more bus voltages near zero. A bus voltage solution near $1\angle 0^\circ$ is referred to as an open circuit solution and a solution near zero is termed a short circuit solution. The short circuit solutions, while valid mathematically, are rejected on the grounds of violation of usual operating practice.

4.2

GAUSS AND GAUSS-SEIDEL METHODS

For the remainder of this chapter, attention turns to the solution of the power flow equations, which are simultaneous, nonlinear, algebraic equations. The two formulations given above ([(1) $V = ZI$ plus $S=VI^*$;

(2) the mismatch formulation] lead to numerous other forms. For example, in the $V=ZI$ formulation, Y_{bus} may be used in place of Z_{bus}. Also, it is possible to solve for some unknowns in terms of others and, using substitution, reduce the number of equations and unknowns. For these reasons, there is no single, universally accepted form of the power flow equations. The most commonly used form is the mismatch formulation, which is amenable to the Newton–Raphson method of solution; this is the subject of the next section. Two older, simpler–to–program solution methods are presented here although these have execution time disadvantages compared with the Newton–Raphson method.

The Gauss and Gauss–Seidel methods are very similar and they rely on the sequential solution of the simultaneous equations updating some of the unknowns at each equation. A simple example from outside power engineering will illustrate the sequential procedure. Consider the simultaneous, nonlinear algebraic equations

$$y - 3x + 1.9100 = 0.0$$

$$x^2 + y - 1.8425 = 0.0.$$

Let the first be solved for x and the second for y:

$$x = \frac{y}{3} + 0.6367$$

$$y = 1.8425 - x^2 .$$

Starting with an initial guess at $x_0 = 1.000$, $y_0 = 1.000$, x is updated with the first equation, then y is updated with the second

$$x_1 = \frac{y_0}{3} + 0.6367 = 0.9700$$

$$y_1 = 1.8425 - x_0^2 = 0.8425.$$

The subscript indicates the iterate. The process is repeated using

$$x_{n+1} = \frac{y_n}{3} + 0.6367$$

$$y_{n+1} = 1.8425 - x_n^2 .$$

Representative results are as follows:

i	x_i	y_i
2	0.9175	0.9016
3	0.9372	1.0006
4	0.9702	0.9642
...	...	...
9	0.9532	0.9245
10	0.9448	0.9338
...	...	...
20	0.9505	0.9406
21	0.9502	0.9390
...	...	...

Note that the iterative formulas for x_{n+1}, y_{n+1} are nonlinear difference equations. The convergence of these formulas to the solution ($x = 0.9500$, $y = 0.9400$ in this case) depends on the nonlinearity and coefficients. Also, the convergence is dependent on the starting value. To illustrate this point, consider

$$x_0 = 100 \qquad y_0 = 100$$

in the example. Then

$$\begin{aligned} x_1 &= 33.9700 & y_1 &= -9998.1575 \\ x_2 &= -3332.0825 & y_2 &= -1152.184 \\ x_3 &= -383.4028 & y_3 &= -1.1103 \times 10^7, \end{aligned}$$

and the solution diverges.

This sequential solution method is known as the *Gauss* method. To solve the power flow equations by the Gauss method, arrange the M load flow equations such that the first is solved for unknown x_1, the second for $x_2, \cdots,$ to x_M. Using an "educated guess" as a starting point, sequentially update x_1 through x_M. Repeat the process as necessary (stopping criteria will be discussed below).

The *Gauss–Seidel algorithm* is similar to the method just described. In the Gauss method, equations 1 through M are formulas for updating the unknowns using only data from the previous iteration. In the Gauss–Seidel method, equation 1 is used to update the first unknown using data from the previous iteration, but equation 2 employs both previous iterates *and* the update from equation 1. Equation 3 uses the updates from equations 1 and 2 as well as previously values calculated. This sequential updating continues using the most recently calculated values. The previous example will illustrate the difference between the Gauss and Gauss–Seidel methods:

Gauss Method

$$x_{n+1} = \frac{y_n}{3} + 0.6367$$

$$y_{n+1} = 1.8425 - x_n^2$$

Gauss–Seidel Method

$$x_{n+1} = \frac{y_n}{3} + 0.6367$$

$$y_{n+1} = 1.8425 - x_{n+1}^2 .$$

Note that in the Gauss–Seidel method, the right hand side of the update formulas will, in general, contain terms that have already been updated in the present iteration. The practice of updating all variables as soon as possible heuristically results in an improvement in solution speed.

Example 4.1

Solve the following power flow equations (for a dc system) by the

Gauss and Gauss–Seidel methods

$$V_{bus} = \begin{bmatrix} 1.0 & 1.0 & 1.0 \\ 1.0 & 1.1 & 1.1 \\ 1.0 & 1.1 & 1.2 \end{bmatrix} I_{bus}$$

$$V_2 I_2 = -0.10 \qquad V_3 I_3 = -0.05 \qquad V_1 = 1.$$

Solution–Gauss Method

To illustrate the Gauss solution method, rewrite the six equations as the following five equations (eliminating V_1)

$$1 = I_1 + I_2 + I_3$$

$$V_2 = I_1 + 1.1I_2 + 1.1I_3$$

$$V_3 = I_1 + 1.1I_2 + 1.2I_3$$

$$V_2 I_2 = -0.10$$

$$V_3 I_3 = -0.05.$$

Use the first equation to update I_1, the second and third to update V_2 and V_3, and the remaining two to update I_2 and I_3

$$I_1 = 1 - I_2 - I_3$$

$$V_2 = I_1 + 1.1I_2 + 1.1I_3$$

$$V_3 = I_1 + 1.1I_2 + 1.2I_3$$

$$I_2 = \frac{-0.10}{V_2}$$

$$I_3 = \frac{-0.05}{V_3}.$$

The left hand sides of these equations are the updated values, the right hand sides are the not-yet-updated values. Let the initial guess at a solution be

$$I_1 = 1.0 \quad I_2 = 0 \quad I_3 = 0 \quad V_2 = 1 \quad V_3 = 1.$$

Then update I_1, V_2, V_3, I_2, I_3 in that order

$$\left.\begin{aligned} I_1 &= 1 - 0 - 0 = 1.0 \\ V_2 &= 1 + 0 + 0 = 1.0 \\ V_3 &= 1 + 0 + 0 = 1.0 \\ I_2 &= -0.10/1.0 = -0.10 \\ I_3 &= -0.05/1.0 = -0.05 \end{aligned}\right\} \text{first iteration.}$$

The updated values are now used in the right hand side of the update formulas in the second iteration,

$$\left.\begin{aligned} I_1 &= 1 - (-.10) - (-.05) = 1.15 \\ V_2 &= 1 + 1.1(-.10) + 1.1(-.05) = 0.835 \\ V_3 &= 1 + 1.1(-.10) + 1.2(-.05) = 0.830 \\ I_2 &= -0.10/1.0 = -0.10 \\ I_3 &= -0.05/1.0 = -0.05 \end{aligned}\right\} \text{second iteration.}$$

The results of the next few iterations are as follows:

Iteration	I_1	V_2	V_3	I_2	I_3
3	1.150	0.985	0.980	-0.120	-0.060
4	1.180	0.952	0.946	-0.102	-0.051
5	1.153	1.012	1.007	-0.105	-0.053
6	1.158	0.979	0.974	-0.099	-0.050
7	1.148	0.995	0.990	-0.102	-0.051
8	1.154	0.980	0.974	-0.101	-0.051
9	1.151	0.987	0.982	-0.102	-0.051
10	1.153	0.982	0.977	-0.101	-0.051

Although there is still considerable variation in V_{bus} from iteration to iteration, there are indications of stabilization to a final solution. Note that I_{bus} stabilizes fairly quickly.

Solution – Gauss–Seidel Method

In the Gauss–Seidel method, the update formulas for I_1, V_2, V_3, I_2, and I_3 use updated values as they become available. To make the best use of the updated values, currents and voltages are usually alternately updated. Let $I_{1,n}$ be I_1 at iteration n, $V_{2,n}$ be V_2 at iteration n, and so on. Then the interleaved update formulas are

$$I_{1,n+1} = 1 - I_{2,n} - I_{3,n}$$

$$V_{2,n+1} = I_{1,n+1} + 1.1 I_{2,n} + 1.1 I_{3,n}$$

$$I_{2,n+1} = \frac{-0.10}{V_{2,n+1}}$$

$$V_{3,n+1} = I_{1,n+1} + 1.1 I_{2,n+1} + 1.2 I_{3,n}$$

$$I_{3,n+1} = \frac{-0.05}{V_{3,n+1}}.$$

The following initial values are used

$$I_{1,0} = 1 \quad V_{2,0} = 1 \quad I_{2,0} = 0 \quad V_{3,0} = 1 \quad I_{3,0} = 0.$$

Results are as follows:

i	$I_{1,i}$	$V_{2,i}$	$I_{2,i}$	$V_{3,i}$	$I_{3,i}$
1	1.000	1.000	-0.100	0.890	-0.056
2	1.156	0.984	-0.102	0.977	-0.051
3	1.153	0.985	-0.102	0.980	-0.051
4	1.153	0.985	-0.102	0.980	-0.051
5	1.153	0.985	-0.102	0.980	-0.051

It is clear that the solution stability is improved over that of the Gauss method. The performance of the Gauss and Gauss–Seidel methods is system dependent and firm comparisons are not possible, but some generalizations may be made based on typical power flow studies. This section concludes with these generalizations as well as a few practical considerations.

Stopping Criteria for the Gauss and Gauss–Seidel Methods; Mismatch Stopping Criterion

It is desirable to stop the iterative procedure in the Gauss and Gauss–Seidel methods when the iterates are at or near the solution. This philosophy leads to several candidate stopping criteria. The simplest stopping criterion is allowing the process to proceed with a fixed number of iterations; this method, however, should be dismissed because the objective of arrival close to the solution is met to an unknown degree. Alternatively, it is possible to keep track of the degree of correction of the bus voltages and currents. By this technique, $| V_{bus}^{new} - V_{bus}^{old} |$, $\max_i (| v_i^{new} - v_i^{old} |)$, or similar quantities using bus currents may be used to assess the size of the correction at an iteration. If the correction at an iteration is sufficiently small, one concludes that the iterations should be stopped. Note that this appears to be attractive in the Gauss–Seidel solution in Example 4.1. The use of a stopping criterion based on correction size at an iteration has the advantage of simplicity in programming but the disadvantage of unknown proximity to a solution. The latter is illustrated by noting that reactive power flow in a transmission line is a sensitive function of the difference in bus voltage magnitude across that line. Also, the active power flow in a line is a sensitive function of the difference in bus voltage phase angle across the line. Thus rather small errors in $| V |$ or δ can have totally unacceptable consequences.

The most widely used stopping criterion in all types of power flow studies is based on the reduction of bus power mismatch to near zero. The mismatch complex power at a bus is the total complex power entering that bus from lines and loads. This has been evaluated earlier as M_i, the mismatch at bus i

$$M_i = [(-row_i \, Y_{bus}) V_{bus}]^* v_i + S_i . \tag{4.10}$$

The term S_i is the specified complex power injected at bus i. Since S_i is not specified at the swing bus, M_i is not calculated for the swing bus. If all M_i were zero, the load flow solution would be exact; usually, M_i is reduced to tolerable levels at all buses. Two common stopping criteria based on mismatch are as follows:

1. Reduce M_i so that

   $$\max_i | M_i | < \epsilon$$

   where ϵ is the load flow tolerance (typically, 0.01 per unit). The left hand side of this expression is called the infinite norm of vector M.

2. Reduce M_i so that

   $$\max \left(\max_i | Re \, (M_i) | , \max_k | Im \, (M_k) | \right) < \epsilon,$$

   where ϵ is the load flow tolerance.

The notation $|\cdot|$ denotes the magnitude of a complex number in criterion 1 and absolute value in criterion 2.

The advantage of the use of a mismatch stopping criterion relates to the relationship between mismatch and proximity to a solution. If the worst mismatch is 0.01 per unit, one may assume that the power flow study is accurate approximately to 1 MW for a 100 MVA base. The term "1 MW load flow" refers to a tolerance (ϵ) of 0.01 on a 100 MVA base.

Number of Iterations Required for a Gauss or Gauss–Seidel Solution

The required number of iterations for a Gauss or Gauss–Seidel power flow study depends on the system, loading conditions, and the mismatch tolerance. Generally speaking, the number of iterations is insensitive to the number of system buses. For typical systems, reduction of worst mismatch to a magnitude of 0.01 per unit is accomplished in 80 iterations (Gauss) or 40 iterations (Gauss–Seidel).

Acceleration Factors

In both the Gauss and Gauss–Seidel methods, the unknown voltages and currents are updated at each iteration:

$$\left[\frac{V_{bus,n}}{I_{bus,n}}\right] \xrightarrow{update} \left[\frac{V_{bus,n+1}}{I_{bus,n+1}}\right].$$

It is possible to apply an acceleration factor to the amount of this update. To accomplish this, let Δ_{n+1} denote the correction applied at iteration $n+1$,

$$\Delta_{n+1} = \left[\frac{V_{bus,n+1}}{I_{bus,n+1}}\right] - \left[\frac{V_{bus,n}}{I_{bus,n}}\right].$$

Then the usual update is

$$\left[\frac{V_{bus,n+1}}{I_{bus,n+1}}\right] = \left[\frac{V_{bus,n}}{I_{bus,n}}\right] + \Delta_{n+1}\,.$$

If it is desired to accelerate (or decelerate) this step, apply a factor α to Δ_{n+1},

$$\left[\frac{V_{bus,n+1}}{I_{bus,n+1}}\right]_{accelerated} = \left[\frac{V_{bus,n}}{I_{bus,n}}\right] + \alpha\Delta_{n+1}.$$

If $\alpha > 1$, the correction magnitudes are increased. If $\alpha < 1$, a deceleration factor is applied.

It is also possible to apply different acceleration factors to V_{bus} and I_{bus}. Typical values of α lie in the range $0.7 \leq \alpha \leq 1.5$.

Initialization

The convergence of the Gauss and Gauss–Seidel methods are sensitive to the starting values. Typically, bus voltages are initialized to $1\angle 0^\circ$. If other information is available (such as a previously converged solution), this may be used to initialize V_{bus}. If V_{bus} entries are all near $1\angle 0^\circ$, then I_{bus} entries will be near $S^*_{specified}$ at PQ buses. If a Z_{bus} formulation is used, a convenient initialization for the swing bus current is $1/(Z_{bus})_{11}$ (bus 1 is the swing bus).

Y_{bus} Formulation

There are potential advantages in reduction of memory requirements to be obtained by reformulating the power flow problem in terms of Y_{bus} rather than Z_{bus}. If a Y_{bus} algorithm is used, sparsity programming may be employed. For large systems, this is not a luxury – it is a necessity since Z_{bus} requires considerable storage space. There are several alternatives for reformulating the stated power flow problem in terms of Y_{bus}; one method is given here.

The solution of the $S_i = V_i I_i^*$ equations at PQ buses for current gives

$$I_i = \left(\frac{S_i}{V_i}\right)^* .$$

Substitution into the $I = YV$ equations gives

$$V_i = \frac{1}{Y_{ii}}\left(I_i - \sum_{k=1 \neq i}^{N} V_k Y_{ik}\right)$$

$$= \frac{1}{Y_{ii}}\left(\frac{S_i^*}{V_i^*} - \sum_{k=1 \neq i}^{N} V_k Y_{ik}\right) . \qquad (4.11)$$

In (4.11), V_i is updated using an old value of V_i on the right hand side. Equation (4.11) is written for all PQ buses and the swing bus voltage is used as given in the summation term. In the absence of PV buses, there will be $2(N - 1)$ real equations in $2(N - 1)$ real unknowns.

The advantage of this Y_{bus} formulation is one of modest memory requirements. The principal disadvantage is slow convergence. Also, in some systems, the convergence is insufficiently strong to allow a solution by this method.

Calculation of Line Power Flows

Most power flow study methods entail the calculation of V_{bus}. The solution bus voltages are important in the identification of high and low bus voltage magnitude conditions. Also, in some cases, the phase angle of the bus voltages may be examined to identify potential stability problems (which may occur for large phase angles). The line power flows are also very important. These line flows are calculated for each line in the line list (e.g., the line from bus k to bus ℓ) as

$$\bar{S}_{\ell k} = \bar{y}_{k\ell}^* \, (v_\ell - v_k)^* v_k \qquad \text{(metered at } k\text{)}$$

$$\bar{S}_{k\ell} = \bar{y}_{k\ell}^* \, (v_k - v_\ell)^* v_\ell \qquad \text{(metered at } \ell\text{)} .$$

The output of a power flow study always includes line loading because a principal use of a power flow study is to assess the impact of specific operating conditions on line loads; line loads that exceed line ratings may be "flagged" in the output to attract the attention of the user. Often, the line loading in line ℓk is output two times, once as $\bar{S}_{\ell k}$ metered at k and once as $\bar{S}_{k\ell}$ metered at ℓ. These two complex powers differ by the complex power loss in the line.

Typical Power Flow Study Solutions

The solution bus voltage vector is typically within 15% of rated value (1.00 per unit). At buses where capacitive reactive loads are present (e.g., shunt capacitors), bus voltage magnitudes are often higher than at other PQ buses. At buses where large inductive reactive loads are present, bus voltages will be of lower magnitude. The phase angle of the solution bus voltage tends to be more leading at generation buses and more lagging at load buses. It is not unusual to find all phase angles within $\pm 10^\circ$ of the swing bus.

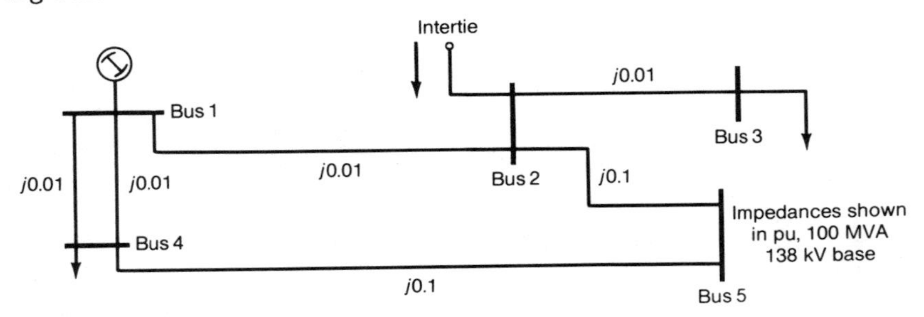

FIGURE 4.3 System for Example 4.2.

Example 4.2

For the system in Figure 4.3, let the per unit bases be 100 MVA and 138 kV. Bus 1 is the swing bus and is at rated voltage, bus 2 is an intertie with a neighboring utility from which 41 MW, 19.5 Mvar is obtained and the remaining buses are load buses as follows (*loads* are specified in per unit):

$$S_3 = 1.4\angle 44^\circ \quad S_4 = 2.4\angle 48^\circ \quad S_5 = 0.9\angle 24^\circ \ .$$

The diagram shows line impedances in per unit. During a Gauss load flow study, the following bus voltage profile is obtained (in per unit)

$$v_1 = 1.00 + j0.00$$

$$v_2 = 0.99 - j0.01$$

$$v_3 = 0.98 - j0.02$$

$$v_4 = 0.99 - j0.01$$

$$v_5 = 0.97 - j0.05.$$

Calculate the line power flows and the bus power mismatches. Also determine whether 1 MVA accuracy is obtained at this point.

Solution

The per unit injected complex voltamperes are

$$S_2 = 0.410 + j0.195 = 0.4540\angle 25.436$$

$$S_3 = -1.007 - j0.973 = 1.4000\angle 224.000$$

$$S_4 = -1.606 - j1.784 = 2.4004\angle 228.006$$

$$S_5 = -0.822 - j0.366 = 0.9000\angle 204.000 \ .$$

The line flows are calculated as

$$\bar{S}_{\ell k} = \bar{y}_{k\ell}^{*}\,(v_\ell - v_k)^{*}\,v_k \quad \text{(metered at } k\text{)}.$$

Thus

$$\bar{S}_{21} = (-j100)^{*}(0.99 - j0.01 - 1 + j0)^{*}(1 + j0)$$

metered at bus 1. This line flow is

$$\bar{S}_{21} = -1.0 - j1.0.$$

Similarly, the flow at the other end of this line is

$$\bar{S}_{12} = (-j100)^{*}(-0.99 + j0.01 + 1.00 - j0.00)^{*}(0.99 - j0.01)$$

$$= +1.0 + j0.98.$$

Note that $\bar{S}_{12} \neq -\bar{S}_{21}$ since there is a line loss. This line is purely reactive and therefore $\bar{S}_{12}$ and $\overline{-S_{21}}$ differ only in the imaginary part (only reactive voltamperes are "lost" in the line). The other line flows are calculated in a similar way and shown in Figure 4.4.

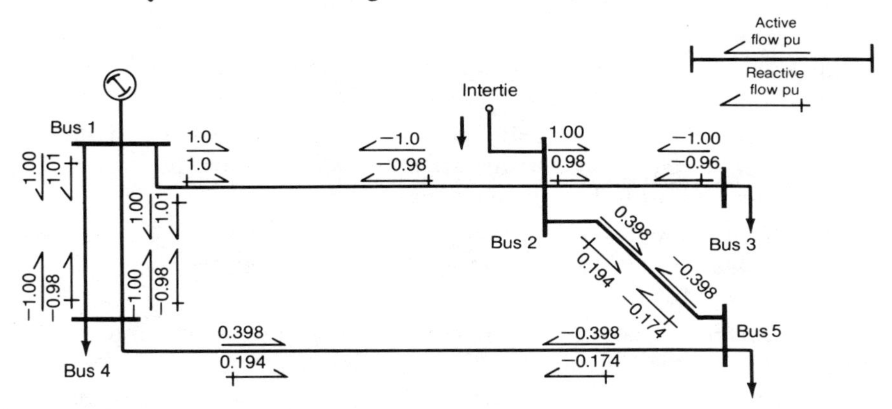

FIGURE 4.4 Calculated line flows for Example 4.2.

The bus power mismatches could be calculated using (4.10). Since line flows have already been calculated, it is probably more convenient simply to sum the complex power entering each bus. For example, at bus 2, the complex power entering the bus is M_2

$$M_2 = -(-1.00 - j0.98) - (1.00 + j0.98) - (0.398 + j0.194) + S_2.$$

where S_2 is the specific bus injection at bus 2 (0.410 + j0.195). The sum is

$$M_2 = 0.012 + j.001 \quad (|M_2| = 0.0120)\ .$$

Similarly, the mismatches at the other buses are

$$M_3 = -0.007 - j0.013 \quad (|M_3| = 0.0148)$$

$$M_4 = -0.004 - j0.018 \quad (|M_4| = 0.0184)$$

$$M_5 = -0.026 - j0.018 \quad (|M_5| = 0.0316).$$

The worst (largest) mismatch magnitude is at bus 5, where a 3.16 MVA mismatch occurs. The desired 1 MVA accuracy has *not* yet been obtained.

A Few Observations on Gauss and Gauss–Seidel Solutions, Debugging Programs

It is very difficult to draw truly general conclusions about power flow solutions since the solution line flows and bus voltages depend on the loading schedule. Different loading schedules produce different (and sometimes surprising) effects. Nonetheless, a few remarks will provide some practical insight and expectations relating to power flow solutions. In Example 4.2, note that the worst mismatch occurs at bus 5. It is common to find the worst mismatch at heavily loaded buses particularly where there is a large reactive power demand/source. Also, high mismatches tend to occur at the end of long feeders and far from the swing bus. Further, low or high bus voltage conditions will cause the initial bus voltage profile to be far from the solution and high bus power mismatch is expected. In Example 4.2, bus 5 has the lowest voltage magnitude (97.1%) and it is electrically far from the swing bus. Apparently, these factors result in the high mismatch at bus 5. What can be said about bus 4, which is very near rated voltage and is electrically closely tied to the swing bus? One may conjecture that the heavy load at bus 4 causes the second largest mismatch power there.

Because power flow solutions often follow an expected format, debugging digital computer implementations is not too difficult. It is possible to identify errors and divergent cases in power flow solutions by observing

1. Line flows at opposite ends of the same line which do not obey $\bar{S}_{ab} \simeq -\bar{S}_{ba}$ (look for miscalculation of line flows and errors in accessing and using line list data and subscripts).
2. Bus voltages out of range (i.e., far from $|v| = 1.0$; possible error in using the swing bus voltage or faulty perunitization of load and/or line data).
3. Mismatch powers which are diverging (becoming larger and larger). Check for miscalculation of mismatch (particularly the *sign* of the mismatch). Also check for consistent use of injection notation for load power and current.
4. Bus voltage phase angles which are opposite to the expectation that load buses have lagging phase and generation buses have leading phase (this observation frequently implies that the load power data were not prepared using injection notation).
5. Unexpectedly high or low line power flows (this may be due to faulty perunitization of line or transformer data).
6. Apparent convergence to a solution (i.e., V_{bus} stabilizes at a reasonable value), yet the power flow study does not stop. This is due to an error in the stopping criterion – check the mismatch calculation; be sure that *no* mismatch calculation is done at the swing bus).

Convergence Characteristics of the Gauss and Gauss–Seidel Methods

The convergence characteristics of the Gauss and Gauss–Seidel methods may be examined empirically by examining typical power flow studies. Alternatively, it is possible to approach questions of convergence from a theoretical point of view by examining the properties of the difference equations, which are the discretized, finite mathematical representation of the numerical solution. Each of these approaches has its impasses, but in combination, some insight is obtained of the convergence properties.

Examination of typical power flow studies indicates that the number of iterations required to obtain a solution is insensitive to the number of buses but is sensitive to the system impedances and configuration. When the system is highly interconnected with low impedance lines, the Z_{bus} matrix has similar entries in each position. Thus when v_k is calculated at iteration ℓ,

$$v_k^{(\ell)} = \sum_{i=1}^{N} (Z_{bus})_{ki} (I_{bus}^{(\ell-1)})_i \ ,$$

small errors in I_{bus} entries will potentially result in large errors in $v_k^{(\ell)}$. Thus slow convergence is expected in tightly interconnected networks.

Radial networks also result in slowed convergence. This case is examined by considering a string of $c+1$ buses, b, $b+1$, ..., $b+c$ in a radial configuration (b to $b+1$ to $b+2$ ...). If $(I_{bus}^{(\ell)})_{b+c}$ contains an error, the bus voltages at all buses in the string will experience an error component. But the update in $(I_{bus}^{(\ell)})_{b+c}$ will come late in the iteration. Reordering the buses with the end of the radial feeder first in the list will not obviate the difficulty since the injection current at bus b also affects the bus voltages at the end of the feeder (e.g., at bus $b+c$). If the solution were formulated using Y_{bus} rather than Z_{bus}, the convergence problems of radials would be reduced, but the characteristically large entries of Y_{bus} would result in difficulties of a different sort. Errors in V_{bus} entries are potentially "magnified" in I_{bus}.

The theoretical approach to the convergence properties of the Gauss and Gauss–Seidel methods is hampered by the nonlinearity of the power flow equations. Before proceeding to the nonlinear case, momentarily consider the much simpler, linearized formulation using real numbers only (i.e., a dc power flow)

$$V_{bus} = R_{bus} I_{bus}$$

$$(I_{bus})_i = \frac{P_i}{(V_{bus})_i} \qquad i = 2,3,\ldots,N$$

$$(V_{bus})_1 = 1 \ .$$

Substituting the bus currents into bus voltage equation yields

$$\begin{bmatrix} 1 \\ v_2 \\ v_3 \\ \vdots \\ v_N \end{bmatrix} = R_{bus} \begin{bmatrix} i_1 \\ \dfrac{P_2}{v_2} \\ \dfrac{P_3}{v_3} \\ \vdots \\ \dfrac{P_N}{v_N} \end{bmatrix} .$$

This vector equation is rewritten by bringing the vector on the left to the right hand side; the result is the form

$$0 = G(B), \qquad (4.12)$$

where G is a vector valued function of the vector argument $B = (i_1, v_2, v_3, \ldots, v_N)^t$. The function G is linearized by examining the first two terms of the Taylor series around the value of B at iteration ℓ (i.e., $B^{(\ell)}$):

$$G(B) = AB + C, \tag{4.13}$$

where A is an N by N matrix of partial derivation of G with respect to B evaluated at $B^{(\ell)}$

$$(A)_{ij} = \begin{cases} R_{i1} & j = 1 \\ \dfrac{-P_j R_{ij}}{v_j^2} & j \neq i \\ -1 - \dfrac{P_j R_{jj}}{v_j^2} & i = j \neq 1 . \end{cases} \tag{4.14}$$

Equation (4.13) is to be considered as solved by the Gauss method when G is near zero in each row. There are several criteria which guarantee convergence of the Gauss method; two are examined briefly here.

Diagonal dominant A. The Gauss method will converge if

$$|(A)_{kk}| \geq \sum_{\substack{i=1 \\ \neq k}}^{N} |(A)_{ki}| \quad all\ k . \tag{4.15}$$

Equation (4.15) is a sufficient but not necessary condition. Applying (4.14) to (4.15) gives us

$$R_{ll} \geq \sum_{i=2}^{N} \frac{|P_i| R_{1i}}{v_i^2}$$

$$\left|1 + \frac{P_k R_{kk}}{v_k^2}\right| \geq \sum_{\substack{i=1 \\ \neq k}} \frac{|P_i| R_{ki}}{v_i^2} . \tag{4.16}$$

In practice, it is difficult to check (4.16) since the values of solution voltage are not known. However, for near unity bus voltage profile, these conditions more or less amount to dominant diagonal of R_{bus}. Remember that this is a sufficient requirement (not a necessary requirement) of the linearized problem.

Positive definite A. A second test for convergence of a Gauss formulation of the linearized problem [of the form (4.13)] is that matrix A must be positive definite. This is a sufficient but not a necessary condition. Also, matrix A must be Hermitian to apply this test. For the case of nearly uniform voltage profile and bus loading, A is symmetric and Sylvester's test may be applied

$$(A)_{11} > 0 \qquad det \begin{bmatrix} A_{11} & A_{12} \\ A_{21} & A_{22} \end{bmatrix} > 0 \qquad det \begin{bmatrix} A_{11} & A_{12} & A_{13} \\ A_{21} & A_{22} & A_{23} \\ A_{31} & A_{32} & A_{33} \end{bmatrix} > 0$$

and so on. If A is not symmetric, it is possible to rewrite (4.13) in terms of a sum of a symmetric coefficient matrix and a skew symmetric coefficient matrix. Similar results are obtained. This approach to examining convergence is presented for theoretical interest only. Sylvester's test is difficult to apply and conclusions are invalid for the nonlinear problem. Note, however, that Sylvester's test holds when R_{bus} is strongly diagonal dominant, thus yielding agreement with the former approach.

Examination of convergence characteristics of the Gauss and Gauss–Seidel methods in the nonlinear formulation is much more difficult and no general results are known. References [4], [5] contain additional details on the convergence characteristics of these methods. If the power flow equations are expanded into a three term Taylor series [in Eq. (4.13), a Hessian term is included which depends on quadratic functions of B]; higher order terms cause equivalent diagonal entries of the linearized problem [matrix A in Eq. (4.13)] to become smaller. This suggests that cases of marginal convergence may become nonconvergent. The reader is cautioned to note that the off-diagonal entries of A also change, and convergence is related not to $(A)_{ii}$ alone, but to the magnitude of $(A)_{ii}$ in relation to the other entries in row i. Results are highly system and load specific, and no general results have been published.

The questions of power flow study convergence will be revisited in connection with the Newton–Raphson method of solution. For that method, somewhat more conclusive statements can be made but a number of general questions relating to the convergence of power flow studies remain open.

Methods to Accelerate Convergence of the Gauss and Gauss–Seidel Methods

A brief discussion of acceleration factors has already been given in connection with the Gauss and Gauss–Seidel power flow studies. Other methods have been suggested to improve convergence speed of these methods, although in recent years attention has turned away from these power flow study methods. This is the case since the Newton–Raphson method has largely replaced the Gauss and Gauss–Seidel methods in the industry. Nonetheless, methods to accelerate convergence of these algorithms hold contemporary interest from both a theoretical and a practical point of view since there are potential applications in modern power flow solution methods. Beyond acceleration factors, methods to accelerate convergence include the modelling and location of the swing bus, the sequence in which updates are made, and the techniques used to model loads. A load modelling alternative will serve to illustrate these approaches.

A technique known as ***Brown's Z_{bus} load flow study algorithm*** [11] entails the modelling of the load as a ground tie ($\bar{z}_{0i}$) in parallel with a current source (I_i). This parallel combination replaces the simple current injection used in formulations described to this point. Brown's Z_{bus} power flow study is a departure from the load modelling philosophy presented thus far since it includes loads, in some sense, in the network model. The conventional formulation does not include load models in the transmission model ($V_{bus} = Z_{bus} I_{bus}$ or $I_{bus} = Y_{bus} V_{bus}$). Let the load at bus i be a parallel configuration of z_{0i} and I_i (injected). Thus the total injection current will be

$$-\frac{V_i}{\bar{z}_{0i}} + I_i \ ,$$

and the specified complex power, S_i, at bus i must be

$$S_i = V_i(-\frac{V_i}{\bar{z}_{0i}} + I_i)^* \ .$$

Thus,

$$I_i = \left(\frac{S_i}{V_i}\right)^* + \frac{V_i}{\bar{z}_{0i}}.$$

Let $\bar{z}_{0i}$ be *included* in Z_{bus} (or Y_{bus} if the admittance formulation is used). Then the power flow equations to be solved are

$$V_{bus} = Z_{bus} I_{bus}$$

$$I_i = I_i = \left(\frac{S_i}{V_i}\right)^* + \frac{V_i}{\bar{z}_{0i}}$$

$$V_1 = specified\ swing\ bus\ voltage$$

where i denoted load buses (compare this formulation with the conventional load flow equations

$$V_{bus} = Z_{bus} I_{bus}$$

$$I_i = \left(\frac{S_i}{V_i}\right)^*$$

$$V_1 = specified\ swing\ bus\ voltage.$$

The essential difference in these formulations lies in the inclusion of the $\bar{z}_{0i}$ ground tie. This tie causes Z_{bus} to be more diagonally dominant and convergence is improved. Note that in Brown's Z_{bus} formulation, the I_{bus} at the solution is a vector of "correction currents" in the sense that they, in conjunction with the ground tie $\bar{z}_{0i}$, model the load at bus i. The ground tie impedance is usually chosen numerically as $-1/S_i^*$. Thus, if the solution $|V_i|$ is unity, the solution "correction current" I_i will be zero.

4.3

NEWTON-RAPHSON METHOD

Although the Gauss–Seidel method offers some advantages in power flow problem solution (principally ease in programming and low memory requirements when the Y_{bus} formulation is sparsity programmed), the principal disadvantage, slow convergence, is motivation to examine alternative methods. The Newton–Raphson method (also known simply as Newton's method or Newton's point form) is such an alternative which relies on a gradient-like approach. Van Ness [6] is credited with one of the first applications of this method to the solution of the power flow problem. In this technique, derivatives of the mismatch form of the power flow equation are used to force the bus mismatches to zero. It is instructive to examine the general approach before proceeding to the power engineering application. Consider the general problem of finding a vector X such that

$$F(X) = 0, \tag{4.17}$$

where F is a vector valued function of a vector argument of same dimension. Let X_s be such a solution vector. Expand (4.17) in a Taylor series about $X = X_s$,

$$F(X) = F(X_s) + J(X - X_s) + \text{higher order terms.} \qquad (4.18)$$

Note that F vanishes at X_s and (4.18) may be solved for X_s if the higher order terms are ignored

$$X_s \simeq X - J^{-1}F(X) \, . \qquad (4.19)$$

Due to the error in neglecting the higher order terms in the Taylor series, (4.19) is only approximate. Equation (4.19) is used to iteratively determine X_s (the superscript denotes iteration number)

$$X^{(n+1)} = X^{(n)} - J^{-1}F(X^{(n)}) \, . \qquad (4.20)$$

In (4.18) through (4.20), J is a square matrix of same dimension as X and F, and this matrix contains the partial derivatives

$$(J)_{ij} = \left. \frac{\partial F_i}{\partial X_j} \right|_{X^{(n)}} \, .$$

This matrix is the Jacobian matrix of the system of equations in (4.17). Note that the scalar expansions of F_i, $i = 1, 2, \ldots, N$, may be used to obtain (4.18) if these expansions are written in vector form.

Example 4.3

Solve the simultaneous set of real nonlinear equations:

$$f_1 = 0 = 6x_2 \cos x_1 - 4\sqrt{2}x_1$$

$$f_2 = 0 = 12x_1x_2 - 3\pi x_2 \, .$$

Use a starting "guess" $X = (1 \ 1)^t$.

Solution

The Jacobian matrix is

$$J = \begin{bmatrix} \frac{\partial f_1}{\partial x_1} & \frac{\partial f_1}{\partial x_2} \\ \frac{\partial f_2}{\partial x_1} & \frac{\partial f_2}{\partial x_2} \end{bmatrix} = \begin{bmatrix} -6x_2 \sin x_1 - 4\sqrt{2} & 6 \cos x_1 \\ 12x_2 & 12x_1 - 3\pi \end{bmatrix}$$

and the residuals, $F(X^{(1)})$, are

$$F\left(\begin{bmatrix} 1 \\ 1 \end{bmatrix}\right) = \begin{bmatrix} -2.415 \\ 2.575 \end{bmatrix} .$$

The next iterate, $X^{(2)}$, is

$$X^{(2)} = X^{(1)} - J^{-1}F(X^{(1)})$$

$$= \begin{bmatrix} 1 \\ 1 \end{bmatrix} - \begin{bmatrix} -10.706 & 3.242 \\ 12.000 & 2.575 \end{bmatrix}^{-1} \begin{bmatrix} -2.415 \\ 2.575 \end{bmatrix}$$

$$= \begin{bmatrix} 0.781 \\ 1.021 \end{bmatrix} .$$

The next iterate is found similarly. Note that $F(X^{(2)})$ gives an accurate measure of the validity of the solution. As $\max_i(|F_i|)$ goes to zero, a solution is obtained. To obtain $X^{(3)}$,

$$X^{(3)} = X^{(2)} - J^{-1}\Big|_{X^{(2)}} F(X^{(2)})$$

$$= \begin{bmatrix} 0.781 \\ 1.021 \end{bmatrix} - \begin{bmatrix} -9.970 & 4.262 \\ 12.255 & -0.055 \end{bmatrix}^{-1} \begin{bmatrix} -0.065 \\ -0.056 \end{bmatrix}$$

$$= \begin{bmatrix} 0.785515 \\ 1.047342 \end{bmatrix} .$$

Calculation of $F(X^{(3)})$ shows further reduction in the residual

$$F(X^{(2)}) = \begin{bmatrix} -0.000564 \\ 0.001464 \end{bmatrix} .$$

Note that the exact solution in this problem is $X = (\pi/4 \ \pi/3)^t$ and the error in $X^{(3)}$ is about $(0.000116 \ \ 0.000144)^t$.

General Solution Characteristics of the Newton–Raphson Method

Let the scalar equation

$$f(x) = 0$$

have the solution x_s $[f(x_s) = 0]$. Then the Newton–Raphson method is

$$x^{(\ell+1)} = x^{(\ell)} - \left[\frac{-df}{dx} \Big|_{x^{(\ell)}} \right]^{-1} f(x^{(\ell)})$$

where the superscripts denote the iteration number. The Taylor series expansion for $x_s - x^{(\ell+1)}$ is

$$x_s - x^{(\ell+1)} = x_s - x^{(\ell)} - \frac{f(x_s) - f(x^{(\ell)})}{df/dx|_{x^{(\ell)}}} + \text{higher order terms}$$

$$= x_s - x^{(\ell)} - \frac{(x_s - x^{(\ell)})df/dx|_z}{df/dx|_{x^{(\ell)}}} ,$$

where z is between x_s and x^ℓ (this is obtained from the mean value theorem of the calculus). Therefore, the error at iteration $\ell+1$ is related to that at iteration ℓ as follows

$$\epsilon^{(\ell+1)} = |x_s - x^{(\ell+1)}|$$

$$\epsilon^{(\ell)} = |x_s - x^{(\ell)}|$$

$$\epsilon^{(\ell+1)} \leq \epsilon^{(\ell)} \frac{|1 - df/dx|_z}{|(1 - (dv/dx)_{x^{(\ell)}}|} .$$

If the mean value theorem is applied to second derivatives and the same analysis is applied,

$$\epsilon^{(\ell+1)} = -[\epsilon^{(\ell)}]^2 \frac{(d^2f/dx^2)_z}{df/dx|_{x^{(\ell)}}} . \tag{4.21}$$

One concludes from (4.21) that if the first and second derivatives are nearly constant from iteration to iteration, the error in the iterate will decrease as the square of the error in the previous step. This type of convergence is called *quadratic convergence,* and the Newton–Raphson algorithm exhibits quadratic convergence near the solution. If $\epsilon^{(\ell)}$ is plotted versus ℓ with $\epsilon^{(\ell)}$ on a logarithmic scale, quadratic convergence implies a straight line characteristic. For this reason, an alternative term for quadratic convergence is *logarithmic convergence.*

In the vector case, the result is similar and obtained noting that the vector equation

$$F(X) = 0$$

has solution X_s. The Newton–Raphson iterative formula

$$X^{(\ell+1)} = X^{(\ell)} - [J^{(\ell)}]^{-1} F(X^{(\ell)}),$$

where $J^{(\ell)}$ is the Jacobian at iteration ℓ. The Taylor series expansion for $X_s - X^{(\ell+1)}$ is

$$X_s - X^{(\ell+1)} = X_s - X^{(\ell)} - [J^{(\ell)}]^{-1}[F(X_s) - F(X^{(\ell)})] + \text{higher order terms.}$$

Applying the mean value theorem to each row of X results in the use of Z

$$X_s \leq Z \leq X^{(\ell)}, \tag{4.22}$$

where (4.22) inequalities hold row by row, that is,

$$row_i(X_s) \leq row_i(Z) \leq row_i(X^{(\ell)})$$

for all i. Then $J(Z)$ is the Jacobian at Z, and

$$X_s - X^{(\ell+1)} = X_s - X^{(\ell)} - [J^{(\ell)}]^{-1} J(Z)[X_s - X^{(\ell)}] \ .$$

Let the Euclidean norm of $X_s - X^{(q)}$ denote the magnitude of the error at iteration q, $\epsilon^{(q)}$; then

$$[\epsilon^{(\ell+1)}]^2 = (X_s - X^{(\ell)})^t Q^t Q (X_s - X^{(\ell)}), \tag{4.23}$$

where

$$Q = I - [J^{(\ell)}]^{-1} J(Z) \ .$$

Equation (4.23) is not as readily interpreted as in the scalar case; however, the size of $\epsilon^{(\ell+1)}$ is readily shown to be related to $(X_s - X^{(\ell)})^t (X_s - X^{(\ell)})$ through the eigenvalues of $Q^t Q$. When the eigenvalues of $Q^t Q$ are in the range $-1 < \lambda < 1$, the error at iteration $\ell + 1$ is a quadratic function of the term–by–term errors at iteration ℓ. As in the scalar case, the terms *quadratic* or *logarithmic convergence* apply.

A similar, but more conclusive result is found by considering the Newton–Raphson update formula,

$$X^{(n+1)} = X^{(n)} - J^{-1} F(X^{(n)}),$$

written as

$$\Delta X^{(n)} = \Xi F^{(n)} \tag{4.24}$$

where $\Delta(X^{(n)})$ is the update on X at iteration n, Ξ is the inverse of the Jacobian matrix, and $F^{(n)}$ is a shorthand notation for $F(X^{(n)})$. Let the eigenvalues of the Jacobian inverse (Ξ) be $\lambda_1, \lambda_2, \ldots, \lambda_N$ and let the corresponding eigenvectors be $e_1, e_2, \ldots, e_N$. Further, let the modal matrix, M, be defined as

$$M = (e_1 \ e_2 \cdots e_N)$$

and let

$$\Lambda = diag(\lambda_i) \quad i = 1, 2, \ldots, N \ .$$

Consider only the case of linearly independent eigenvectors (if the eigenvalues of Ξ repeat, it is possible to rewrite the development below in terms of generalized eigenvectors and the result is similar but slightly more complicated). Since the e_i span Euclidean N-space (R^N), both $\Delta X^{(n)}$ and $F^{(n)}$ may be expressed as linear combinations of the e_i

$$\Delta X^{(n)} = \sum_{i=1}^{N} b_i e_i \tag{4.25}$$

$$F^{(n)} = \sum_{i=1}^{N} c_i e_i \ . \tag{4.26}$$

Let B denote the N–vector of scalar elements b_i and let C denote the N–vector of scalar elements c_i,

$$\Delta X^{(n)} = MB \tag{4.27}$$

$$F^{(n)} = MC \ . \tag{4.28}$$

Then (4.24) is rewritten using (4.27) and (4.28),

$$MB = \Xi MC.$$

Hence

$$B = (M^{-1} \Xi M) C \ . \tag{4.29}$$

Equation (4.29) is possible since the e_i are linearly independent and M^{-1} exists. Since M is the modal matrix of Ξ: the term $M^{-1} \Xi M$ is a similarity transformation on Ξ which diagonalizes Ξ,

$$B = \Lambda C \ . \tag{4.30}$$

Examine (4.30) in some detail. The magnitude of vector B may be thought of as the "length" of the correction term and the magnitude of vector C as the "length" of the residuals at iteration n. The term "length" used in this context is broad, but note that for the case of orthonormal M,

$$||\Delta X^{(n)}||^2 = (\Delta X^{(n)})^t \ \Delta X^{(n)}$$

$$= B^t M^t MB$$

$$= ||B||^2.$$

Even if M is not orthonormal, $||B||$ is a measure of the norm of $\Delta X^{(n)}$. A similar argument applies to C. Hence (4.30) states that the correction applied at iteration n is related to the residual at iteration n by the eigenvalues of Ξ (i.e., the eigenvalues of the inverse Jacobian). When these eigenvalues are smaller than 1,

$$|\lambda_i| < 1 \quad i = 1, 2, \ldots, N \ ,$$

the corrections will be expected to be monotone decreasing in amplitude (if J is constant from step to step). One concludes that the largest eigenvalues of J^{-1} degrade and control the convergence at each iteration.

The numerous provisions and simplifications of the foregoing should be viewed with caution and suspicion in light of the nonlinear formulation of the power flow problem. Perhaps the most disturbing fact is that the Jacobian changes from iteration to iteration, thereby resulting in considerable difficulty in making general conclusions about the size of $\Delta X^{(n)}$. The general problem of convergence of the Newton–Raphson method will be discussed one more time in connection with an interesting "test" on the Jacobian and initialized values of X to guarantee convergence. Unfortunately, this test has been found to be very conservative and difficult to apply in practice. For these reasons, the practical views of convergence of a power flow study are generally relegated to the realm of numerical calculation. That is, if convergence is obtained from the numerical implementation, there is little motivation to pursue the question further.

Guarantees of Convergence

A few further remarks are provided here concerning the convergence of the Newton–Raphson method. As stated above, when $|\lambda_i|$ are confined to less than unity, convergence is guaranteed. It is possible to find the largest eigenvalue of Ξ. Consider the expansion of an arbitrary N vector, y, in terms of the eigenvectors of Ξ

$$y = \sum_{i=1}^{N} d_i e_i .$$

Premultiply by Ξ

$$\Xi\, y = \sum_{i=1}^{N} d_i\, \Xi\, e_i$$

and use the property that

$$\Xi\, e_i = \lambda_i e_i .$$

Thus

$$\Xi\, y = \sum_{i=1}^{N} d_i \lambda_i e_i .$$

Repeat the premultiplication process

$$\Xi^2 y = \sum_{i=1}^{N} d_i \lambda_i \Xi e_i$$

$$= \sum_{i=1}^{N} d_i \lambda_i^2 e_i$$

$$\vdots$$

$$\Xi^k y = \sum_{i=1}^{N} d_i \lambda_i^k e_i .$$

If k is large, λ_i^k will dominate on the right hand side. Thus, if $|\lambda_\ell|$ is the largest magnitude eigenvalue,

$$||\Xi^k y|| = (|d_\ell \lambda_\ell^k|)||e_\ell||$$

$$||\Xi^{k+1} y|| \simeq (|d_\ell \lambda_\ell^{k+1}|)||e_\ell||$$

and

$$|\lambda_\ell| \simeq \frac{||\Xi^{k+1}y||}{||\Xi^k y||} .$$

Thus a sufficient but not necessary condition for convergence is, for large k,

$$\frac{||\Xi^{k+1}y||}{||\Xi^k y||} < 1 .$$

A second convergence test, based on the starting value of the iterate X, is known as the *Newton–Kantorovich criterion* [10]. This criterion employs three parameters, a, b, and c, which are defined as

$$a = \max_i\left(\sum_{j=1}^{N} |(\Xi)_{ij}| \right)$$

$$b = \max_i\left(|(X^{(1)} - X^{(0)})_i| \right)$$

$$c = N \sum_{j=1}^{N} \left| \frac{\partial^2 F_i}{\partial x_j \, \partial x_i} \right| ,$$

where $X^{(\ell)}$ refers to the ℓth iterate on X, and F_i is the ith element of vector F. The Newton–Kantorovich criterion states that if

$$abc \leq 1/2,$$

then iterates on X will be such that

$$\max_i \left(|X^{(\ell)} - X^{(0)}| \right) \leq 2b$$

for all ℓ and the iterates will converge to some vector X^* such that

$$\max_i \left(|X^{(\ell)} - X^*| \right) \leq \frac{2b}{2^\ell} .$$

These results apply when Ξ is constant or, at least, the a and c parameters are calculated at iteration ℓ [i.e., $a^{(\ell)}bc^{(\ell)} \leq 1/2$]. The proof of the Newton–Kantorovich criterion is based on properties of infinite norms and the Taylor series expansion of $\partial F_i / \partial x_j$ around $X = X^{(\ell)}$ in which the sum of all terms except the first is replaced by a sum of second derivative terms evaluated at a point between $X^{(\ell)}$ and $X^{(\ell+1)}$. This is valid using the mean value theorem of the calculus. A complete proof is shown in [10].

Both the eigenvalue and Newton–Kantorovich test have primary value in research applications rather than in practical power engineering power flow studies. The reason for the lack of practicality in most power applications is that the Newton–Kantorovich product is often in the range 10^2 to 10^5 for initial actual Jacobian matrices in convergent cases. The Newton–Kantorovich test gives no information for such cases. It is reasonable to say that the Newton–Kantorovich test is far too conservative in most power engineering applications.

Contraction Constants and Convergence of the Newton-Raphson Method

An alternative view point to the study of convergence of any iterative solution of

$$F(X) = 0$$

is to consider iterates of X, for example $X^{(\ell)}$, as functions of the previous iterate,

$$X^{(\ell)} = \Phi(X^{(\ell-1)}) .$$

The solution, X^*, is attained when [8]

$$X^* = \Phi(X^*) .$$

When sequential iterates of X represent a contraction, that is, when

$$||X^{(\ell+1)} - X^{(\ell)}|| \leq c||X^{(\ell)} - X^{(\ell-1)}||$$

$$0 < c < 1$$

for all X in region R of Euclidean N–space, it can be shown that the iterative process converges in R [7]. Wu [8] has also shown that

$$||X^{(\ell)} - X^*|| \leq \frac{c}{1-c}||X^{(\ell)} - X^{(\ell-1)}||$$

when $0 < c < 1$. These expressions indicate how fast the process converges, and they apply for any iterative process in which $X^{(\ell)}$ is a function only of $X^{(\ell-1)}$ and in which the solution occurs when $X^* = \Phi(X^*)$. The application of these results lies primarily in the area of theoretical analysis of iterative processes since it is not usually practical to find c and R. The constant c is termed a contraction constant.

If the convergence tolerance is ϵ, the rate of convergence of any iterative process may be defined as R_c,

$$R_c = \frac{ln\ ||F(X^{(0)})||_\infty/\epsilon)}{n_\epsilon},$$

where $||\cdot||_\infty$ is the infinite order norm defined by

$$||F(X^{(0)})||_\infty = \max_i(|F_i(X^{(0)})|)$$

and F_i is the ith component of vector F, and n_ϵ, is the number of iterations required to obtain convergence to tolerance ϵ. In other words,

$$\epsilon \leq ||F(X^{(n_\epsilon)})||_\infty .$$

It can be shown that for a Newton–Raphson solution of a linear set of algebraic equations, the rate of convergence (with tolerance ϵ) is

$$R_c = -ln\ \rho$$

where ρ is the magnitude of the largest eigenvalue of the inverse Jacobian matrix, Ξ. The parameter ρ is termed the spectral radius of Ξ.

Again, the application of these results lie primarily in theoretical applications, particularly in assessing the effectiveness of proposed modifications of Newton's method. The logarithmic character of the convergence observed earlier in connection with a discussion of the term *quadratic convergence* is confirmed by examining the relation between R_c and ρ.

Newton–Raphson Solution of the Power Flow Equations

Equation (4.7) gives the expression for the mismatch active and reactive power at all system buses other than the swing bus. This expression was found by summing all complex power which *enters* a system bus. Note that there is an ambiguity of sign: a similar definition and expression could be used for complex power which leaves a bus. Using the sign convention of (4.7), and breaking the mismatch M_i into a real part, ΔP_i, and an imaginary part, ΔQ_i,

$$\Delta P_i = Re\ \{[(-row_i\, Y_{bus})V_{bus}]^* v_i\} + P_i \tag{4.31}$$

$$\Delta Q_i = Im\ \{[(-row_i\, Y_{bus})V_{bus}]^* v_i\} + Q_i, \tag{4.32}$$

one states the simplified power flow problem as finding V_{bus} such that ΔP_i and ΔQ_i in (4.34) and (4.32) are forced to zero. Note that P_i and Q_i are the specified active and reactive powers at buses 2, 3, ..., N. The right hand sides of (4.31) and (4.32) may be thought of as vector functions of V_{bus}

$$\Delta P_i = F_p(V_{bus}) \tag{4.33}$$

$$\Delta Q_i = F_q(V_{bus})\ . \tag{4.34}$$

Newton's method could be applied directly to (4.33) and (4.34), but experience in practical power engineering indicates that the active power flow is closely related to the bus voltage phase angles, $\delta = (\delta_2\ \delta_3\ \cdots\ \delta_N)^t$ and the reactive power flow is closely related to the bus voltage magnitudes $V = (|v_2|\ |v_3|\ \cdots\ |v_N|)^t$. For this reason, (4.33) and (4.34) are usually rewritten in terms of δ and $|V|$,

$$\Delta P_i = F_p(\delta, |V|) \tag{4.35}$$

$$\Delta Q_i = F_q(\delta, |V|)\ . \tag{4.36}$$

Let the vector $[\delta\ |V|]^t$ be the $2(N-1)$ vector of unknown bus voltage phase angles and magnitudes. This unknown vector is found using the Newton–Raphson method using

$$\begin{bmatrix} \delta \\ |V| \end{bmatrix}^{(\ell+1)} = \begin{bmatrix} \delta \\ |V| \end{bmatrix}^{(\ell)} - J^{-1} \begin{bmatrix} F_p(\delta^{(\ell)}, |V|^{(\ell)}) \\ F_q(\delta^{(\ell)}, |V|^{(\ell)}) \end{bmatrix},$$

where J is the Jacobian matrix of the simultaneous set of (4.35) and (4.36) evaluated at iteration ℓ,

$$J = \begin{bmatrix} \dfrac{\partial F_p}{\partial \delta} & \dfrac{\partial F_p}{\partial V} \\ \dfrac{\partial F_q}{\partial \delta} & \dfrac{\partial F_q}{\partial V} \end{bmatrix}. \tag{4.37}$$

In (4.37), the notation used for the matrix entries is pictorial in that $\partial F_p / \partial \delta$ denotes an $(N-1)$ by $(N-1)$ submatrix of partial derivatives of mismatch active power at buses 2, 3, ..., N with respect to $\delta_2, \delta_3, \ldots, \delta_N$

$$\frac{\partial F_p}{\partial \delta} = \begin{bmatrix} \frac{\partial \Delta P_2}{\partial \delta_2} & \frac{\partial \Delta P_2}{\partial \delta_3} & \cdots & \frac{\partial \Delta P_2}{\partial \delta_N} \\ \frac{\partial \Delta P_3}{\partial \delta_2} & \frac{\partial \Delta P_3}{\partial \delta_3} & \cdots & \frac{\partial \Delta P_3}{\partial \delta_N} \\ \cdots & & & \\ \frac{\partial \Delta P_N}{\partial \delta_2} & \frac{\partial \Delta P_N}{\partial \delta_3} & \cdots & \frac{\partial \Delta P_N}{\partial \delta_N} \end{bmatrix}.$$

The other three submatrices in (4.37) are defined similarly. Also in (4.37), the magnitude sign is dropped from $|V|$ for compactness. The formulas for $\partial F_p/\partial \delta$, $\partial F_p/\partial V$, $\partial F_q/\partial \delta$, and $\partial F_q/\partial V$ are readily evaluated by differentiating (4.31) and (4.32). For convenience, these equations are rewritten in polar form using the notation

$$(Y_{bus})_{ij} = Y_{ij} \angle \theta_{ij} \,.$$

Then (4.31) and (4.32) become

$$\Delta P_i = F_p(\delta, V) = -\sum_{j=1}^{N} |Y_{ij}| \, v_j v_i \cos(-\theta_{ij} - \delta_j + \delta_i) + P_i$$

$$\Delta Q_i = F_q(\delta, V) = -\sum_{j=1}^{N} |Y_{ij}| \, v_j v_i \sin(-\theta_{ij} - \delta_j + \delta_i) + Q_i$$

and the required partial derivatives are

$$J_1(i,\, i) = \frac{\partial P_i}{\partial \delta_i} = \sum_{\substack{j=1 \\ \neq i}}^{N} v_i v_j |Y_{ij}| \sin(\delta_i - \delta_j - \theta_{ij}) \tag{4.38}$$

$$J_1(i,\, k) = \frac{\partial P_i}{\partial \delta_k} = -v_i v_k |Y_{ik}| \sin(\delta_i - \delta_k - \theta_{ik}) \quad (k \neq i) \tag{4.39}$$

$$J_2(i,\, i) = \frac{\partial P_i}{\partial v_i} = -\sum_{\substack{j=1 \\ \neq i}}^{N} v_j |Y_{ij}| \cos(\delta_i - \delta_j - \theta_{ij}) - 2v_i |Y_{ii}| \cos(-\theta_{ii}) \tag{4.40}$$

$$J_2(i,\, k) = \frac{\partial P_i}{\partial v_k} = -v_i |Y_{ik}| \cos(\delta_i - \delta_k - \theta_{ik}) \quad (k \neq i) \tag{4.41}$$

$$J_3(i,\, i) = \frac{\partial Q_i}{\partial \delta_i} = -\sum_{\substack{j=1 \\ \neq i}}^{N} v_i v_j |Y_{ij}| \cos(\delta_i - \delta_j - \theta_{ij}) \tag{4.42}$$

$$J_3(i,\, k) = \frac{\partial Q_i}{\partial \delta_k} = v_i v_k |Y_{ik}| \cos(\delta_i - \delta_k - \theta_{ik}) \quad (k \neq i) \tag{4.43}$$

$$J_4(i,\, i) = \frac{\partial Q_i}{\partial v_i} = -\sum_{\substack{j=1 \\ \neq i}}^{N} v_j |Y_{ij}| \sin(\delta_i - \delta_j - \theta_{ij}) - 2v_i |Y_{ii}| \sin(-\theta_{ii}) \tag{4.44}$$

$$J_4(i, k) = \frac{\partial Q_i}{\partial v_k} = -v_i | Y_{ik} | \sin (\delta_i - \delta_k - \theta_{ik}) \quad (k \neq i) . \tag{4.45}$$

In (4.38) through (4.45), the notation ∂P and ∂Q are used rather than $\partial \, \Delta P$ and $\partial \, \Delta Q$. Although the latter is correct, the former is more compact and in common usage. Hopefully, notation such as $\partial P_i / \partial \delta_j$ will be interpreted as the partial of the mismatch power, ΔP_i, and not the partial of the specified power, P_i. Note that the Jacobian is partitioned as

$$J = \begin{bmatrix} J_1 & J_2 \\ J_3 & J_4 \end{bmatrix}$$

in (4.38) through (4.45), and bus 1 is assumed to be the swing bus.

The procedure for performing a Newton–Raphson power flow study for a system with only PQ buses and one swing bus is as follows:

1. Read system and load data.
2. Form Y_{bus}.
3. Initialize δ, V.
4. Calculate $F_p(\delta, V) = \Delta P_i$ and $F_q(\delta, V) = \Delta Q_i$. If these mismatches are sufficiently small, proceed to step 7. Various stopping criteria may be used – a common mismatch stopping criterion is to terminate the iterative procedure when $\max_i(\Delta P_i, \Delta Q_i)$ is smaller than 0.01 per unit (on a 100 MVA base).
5. Calculate J using (4.38) through (4.45).
6. Solve

$$\begin{bmatrix} F_p(\delta, V) \\ F_q(\delta, V) \end{bmatrix} = -J \begin{bmatrix} \Delta \delta \\ \Delta V \end{bmatrix}$$

for $[\Delta\delta \; \Delta V]^t$. This may be done by inverting J or, alternatively, by the method of triangular factorization of J and forward and backward substitution. Having found $[\Delta\delta \; \Delta V]^t$, update $[\delta \; V]$ using

$$\begin{bmatrix} \delta \\ V \end{bmatrix}^{(\ell+1)} = \begin{bmatrix} \delta \\ V \end{bmatrix}^{(\ell)} - J^{-1} \begin{bmatrix} F_p(\delta^{(\ell)}, V^{(\ell)}) \\ F_q(\delta^{(\ell)}, V^{(\ell)}) \end{bmatrix}$$

$$= \begin{bmatrix} \delta \\ V \end{bmatrix}^{(\ell)} + \begin{bmatrix} \Delta \delta \\ \Delta V \end{bmatrix} .$$

Proceed to step 4.

7. At this point, the procedure has converged to sufficiently close to a solution. The line flows must be calculated using

$$\bar{S}_{kj} = \bar{y}_{kj}^*(v_k \angle \delta_k - v_j \angle \delta_j)^* v_k \angle \delta_k$$

$$\bar{S}_{jk} = \bar{y}_{kj}^*(v_j \angle \delta_j - v_k \angle \delta_k)^* v_j \angle \delta_j .$$

Note that $\bar{S}_{kj}$ is metered at k and $\bar{S}_{jk}$ is metered at j.

8. In some power flow studies, out-of-range bus voltage magnitudes are "flagged" in the output. Similarly, if line ratings were part of the output data, values of $|\bar{S}|$ above these ratings may be flagged. Also, line losses may be calculated and printed. The standard power flow study outputs $\bar{S}$, V, and δ should always be printed.

9. Stop.

Additional Practical Considerations on the Calculation and use of the Jacobian Matrix

Examination of (4.38) through (4.45) reveals that there is generally a diagonal entry in each row of J, but the off diagonal entry will only occur corresponding to nonzero entries in Y_{bus}. Thus J is as sparse as Y_{bus}.

Sparsity programming of the Jacobian matrix has the obvious advantage of memory savings. It may be a surprise to learn that through sparsity programming, it is also possible to speed up the triangular factorization and forward/backward substitution. This is accomplished as follows. In the triangular factorization and forward/backward substitution procedures, rows of J of the table of factors are processed one at a time. In row i, one would normally think that every entry of that row would have to be processed. But the use of a "next pointer", (i.e., a register containing the location of the next nonzero entry in row i), will allow processing only nonzero entries in row i. Thus an element is processed and the next pointer indicates where the next nonzero entry is to be found. Thus only nonzero entries are processed, and the zero entries are skipped. A distinctive value of the next pointer, for example *zero,* is used to denote no further nonzero entries in a row.

The reader should also note a potential pitfall in programming (4.38) through (4.45): the swing bus does not appear as a row in submatrices J_1 nor J_2 because there is no ΔP_i mismatch at the swing bus. Additionally, the swing bus does not appear as a row in J_3 nor J_4 because there is no ΔQ_i mismatch at the swing bus. Also, there is no column in J_1 nor J_3 corresponding to $\Delta\delta_i$ at the swing bus since the swing bus voltage is the reference phasor and there is no column in J_2 or J_4 corresponding to $\Delta|V_i|$ at the swing bus since the bus voltage magnitude there is fixed. Hence J is $2(N-1)$ by $2(N-1)$. In the sums in (4.38), (4.40), (4.42), and (4.44), j *does* assume the value "1" (i.e., "swing") once. Even though the swing bus is missing in the rows and columns of J_1, J_2, J_3, and J_4, it does occur in the calculation of the Jacobian entries. Observations that further relate to this point are the following:

1. Row 1 of J corresponds to bus 2.
2. Row 2 of J corresponds to bus 3, and so on.
3. Row $N-1$ of J corresponds to bus N.
4. Row N of J corresponds to bus 2, and so on.
5. The columns of J correspond to the buses 2 through N in the same way as the row correspondence.
6. It is inadvisable to delete row 1 from the $|V_{bus}|$ vector since it occurs in the calculation of entries of J. Therefore, row 1 of $|V_{bus}|$ corresponds to bus 1, row 2 corresponds to bus 2, and so on.
7. It is similarly inadvisable to delete row 1 from the δ_{bus} vector since it occurs in the calculation of entries of J. The row/bus correspondence in δ_{bus} is the same as that in $|V_{bus}|$.
8. When the correction $[\Delta\delta \ \ \Delta|V|]^t$ is calculated, the row/bus correspondence in this $2(N-1)$ vector is the same as that in J

and not the same as that in δ_{bus}, $|V_{bus}|$. Therefore, some caution must be used in making the update on bus angle and voltage.

Example 4.4

In this example, a very simple power system is studied using the Newton–Raphson power flow study method. The objective of the example is to present in detail each step of the process. If a student has difficulty in obtaining convergence in a newly programmed Newton–Raphson power flow program, the data of this small example may be used to test each step of the program. A second objective of this small example is to illustrate the convergence characteristics of the technique

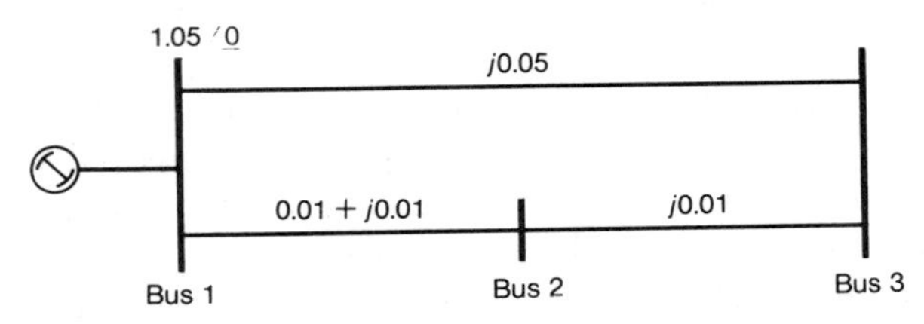

FIGURE 4.5 Simple power system used in Example 4.4.

The system to be studied is shown in Figure 4.5, in which perunitized line impedances are shown. Other given data are

$$v_1 = 1.05 + j0.00 = \text{swing bus voltage}$$

(using injected notation)

$$s_2 = -0.96 + j2.07 \quad \text{per unit}$$

$$s_3 = -3.15 - j2.85 \quad \text{per unit.}$$

In this contrived but hopefully illustrative example, the *exact* solution V_{bus} is known to be

$$V_{bus} = \begin{bmatrix} 1.05 + j0.00 \\ 1.02 - j0.03 \\ 1.00 - j0.05 \end{bmatrix}.$$

It is desired to find V_{bus} using the Newton–Raphson technique, find the line power flow, and study the convergence properties.

Solution

Let the starting V_{bus} be flat at $1.05 + j0$. Hence

$$|V_{bus}| = \begin{bmatrix} 1.05 \\ 1.05 \\ 1.05 \end{bmatrix} \qquad \delta_{bus} = \begin{bmatrix} 0 \\ 0 \\ 0 \end{bmatrix}.$$

The bus admittance matrix is easily calculated and the entries of this 3 by 3 matrix are separated into a magnitude, $|Y_{bus}|$, and angle, θ

$$|Y_{bus}| = \begin{bmatrix} 86.023 & 70.711 & 20.00 \\ & 158.114 & 100.00 \\ \text{symmetric} & & 120.00 \end{bmatrix} \quad \text{per unit}$$

$$\theta = \begin{bmatrix} -54.462 & 135.000 & 90.000 \\ & -71.565 & 90.000 \\ \text{symmetric} & & -90.000 \end{bmatrix} \quad \text{degrees}$$

$$\theta = \begin{bmatrix} -0.9505 & 2.3562 & 1.5708 \\ & -1.2490 & 1.5708 \\ \text{symmetric} & & -1.5708 \end{bmatrix} \quad \text{radians.}$$

The Jacobian matrix is $2(N-1)$ by $2(N-1)$ or 4 by 4. The entries are found using (4.38) through (4.45)

$$J = \begin{bmatrix} -165.3750 & 110.2500 & -52.5000 & 0 \\ 110.2500 & -132.3000 & 0 & 0 \\ 55.1250 & 0 & -157.5000 & +105.0000 \\ 0 & 0 & +105.0000 & -126.0000 \end{bmatrix}.$$

Note that in J,

Row 1 → *bus* 2 *Column* 1 → *bus* 2

Row 2 → *bus* 3 *Column* 2 → *bus* 3

Row 3 → *bus* 2 *Column* 3 → *bus* 2

Row 4 → *bus* 3 *Column* 4 → *bus* 3.

However, in $|Y_{bus}|$, θ, $|V_{bus}|$, and δ, row 1 corresponds to bus 1, row 2 corresponds to bus 2, and row 3 corresponds to bus 3. All calculations are done in radians, and in particular, updates of δ are in radians. The requirement to use radians internally comes from the fact that the derivatives of $\sin u$ and $\cos u$ are $\cos u$ and $-\sin u$ respectively only when u is in radians. Note that J_1, J_2, J_3, and J_4 are symmetric *only* if the bus voltage profile is flat (i.e., $v_1 = v_2 = v_3$). This usually occurs at the initial iteration only – *in general, J is not symmetric.*

The mismatches are calculated using (4.31) through (4.32):

$$\begin{aligned} \Delta P_2 = -[&|Y_{21}| v_1 v_2 \cos(-\theta_{21} - \delta_1 + \delta_2) \\ &+ |Y_{22}| v_2 v_2 \cos(-\theta_{22} - \delta_2 + \delta_2) \\ &+ |Y_{23}| v_3 v_2 \cos(-\theta_{23} - \delta_3 + \delta_2)] + P_2 \\ = -&0.9600 \end{aligned}$$

$$\begin{aligned} \Delta Q_2 = -[&|Y_{21}| v_1 v_2 \sin(-\theta_{21} - \delta_1 + \delta_2) \\ &+ |Y_{22}| v_2 v_2 \sin(-\theta_{22} - \delta_2 + \delta_2) \\ &+ |Y_{23}| v_3 v_2 \sin(-\theta_{23} - \delta_3 + \delta_2)] + Q_2 \\ = \ &2.0700; \end{aligned}$$

and ΔP_3, ΔQ_3 are found similarly

$$\Delta P_3 = -3.1500 \qquad \Delta Q_3 = -2.8500.$$

The first iteration involves the solution of

$$\begin{bmatrix} -0.9600 \\ -3.1500 \\ 2.0700 \\ -2.8500 \end{bmatrix} = J \begin{bmatrix} \Delta\delta_2 \\ \Delta\delta_3 \\ \Delta|v_2| \\ \Delta|v_3| \end{bmatrix}$$

for the last term, which is the update on $[\delta\ V]^t$. This may be accomplished by triangular factorization of J and forward/backward substitution. The result is

$$\begin{bmatrix} \Delta\delta_2 \\ \Delta\delta_3 \\ \Delta|v_2| \\ \Delta|v_3| \end{bmatrix} = \begin{bmatrix} 0.02922 \\ 0.04816 \\ 0.02737 \\ 0.04543 \end{bmatrix}.$$

These $\Delta\delta$ and $\Delta|V|$ terms are subtracted from the initial values to yield iterate 2 (the starting values being iterate 1),

$$\begin{bmatrix} \delta^{(2)} \\ |V^{(2)}| \end{bmatrix} = \begin{bmatrix} -0.02922 \\ -0.04816 \\ 1.02263 \\ 1.00457 \end{bmatrix}.$$

The mismatch formulas are applied, yielding

$$\begin{bmatrix} \Delta P \\ \hline \Delta Q \end{bmatrix}^{(2)} = \begin{bmatrix} 0.03993 \\ -0.18887 \\ 0.01284 \\ -0.16619 \end{bmatrix}.$$

The next steps are similar: evaluate the Jacobian, solve for the corrections, and apply the correction. The numerical results are

$$J = \begin{bmatrix} -154.8082 & 102.7119 & -50.1536 & -1.9366 \\ 102.7119 & -123.7835 & 1.9024 & 2.9477 \\ 53.2884 & 1.9455 & -155.4059 & 102.2445 \\ -1.9455 & 2.9611 & 100.4391 & -117.8770 \end{bmatrix}$$

$$\begin{bmatrix} \Delta\delta \\ \Delta V \end{bmatrix} = \begin{bmatrix} 0.0001818 \\ 0.0017890 \\ 0.0021786 \\ 0.0033082 \end{bmatrix}$$

$$\begin{bmatrix} \delta \\ V \end{bmatrix}^{(3)} = \begin{bmatrix} -0.029406 \\ -0.049952 \\ 1.020450 \\ 1.001263 \end{bmatrix}.$$

At this point, the mismatch powers are all less than 0.0011 per unit.

This small example gives the opportunity to examine not only the reduction in mismatch power from iteration to iteration, but also the error in $[\delta\ V]^t$. The latter is possible since in this contrived example, the exact solution vector is known. Figure 4.6 shows both the reduction in mismatch and solution error. These three characteristics each exhibit faster convergence as the iterations (ℓ) progress toward the solution. The

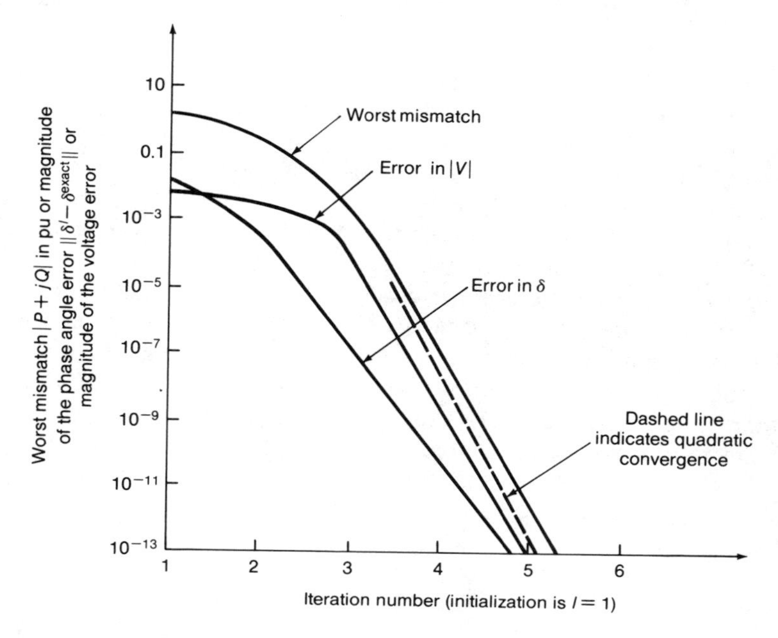

FIGURE 4.6 Worst mismatch and error in δ and $|V|$ in Example 4.4.

dashed line in the figure illustrates quadratic convergence from 10^{-6} at $\ell = 4$ to 10^{-12} at $\ell = 5$. Apparently, each characteristic exhibits approximately quadradic convergence for $\ell > 3$. Beyond the fifth iteration, the error and worst mismatch become similar to the order of magnitude of computer round–off error. (The order of the round–off error in decimal places is approximately the mantissa length in bits multiplied by $\log_{10}2$; thus for a 48 bit mantissa, one would expect about 14 decimal place accuracy.)

4.4

TAP CHANGING TRANSFORMERS AND VOLTAGE CONTROLLED BUSES

A frequent objective of power system operation is to maintain bus voltage magnitude at or near rating. There are several ways in which this voltage regulation policy is accomplished. In this section two design techniques are discussed which involve voltage controlled buses. The necessary modifications to the Newton–Raphson power flow technique are also discussed. The first of these topics is the use of tap changing transformers to control bus voltage. The technique is used when it is desired to retain a load bus or intertie bus at a specified bus voltage level. The technique is also used to control reactive power flow in the system.

The close relationship between bus voltage magnitude and reactive power flow makes the use of tap changing transformers an effective control for Q. The second type of voltage control is quite different: the voltage control of a *generation* bus is considered. This type of bus voltage control is related to synchronous generator excitation and the injection of reactive power at a generator to control $|V|$.

A conventional transformer is included in power flow studies simply by adding the transformer impedance to the transmission line in which the transformer is located. For such conventional transformers, use care in working with the correct (i.e., consistent) per unit base for both the transmission line and the transformer reactance. On the other hand, if a line terminates at a bus whose voltage is controlled by a *load tap changing* (LTC) *transformer,* also known as a *tap changing under load* (TCUL) *transformer,* a control system is used at that bus to sense the bus voltage, calculate the difference between the bus voltage and the desired set point, and initiate tap changing to raise or lower the voltage. Usually, the tap changing is discrete – typically in increments of 5/8% of the winding voltage rating. The discrete tap positions are rather difficult to analyze, and the assumption of continuous variation of tap setting is often made.

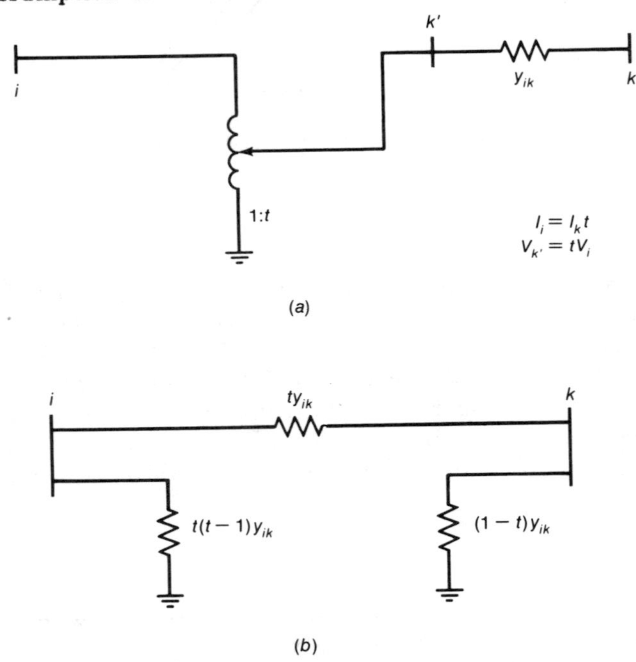

FIGURE 4.7 (a) TCUL model; (b) transformer off-nominal tap representation.

Figure 4.7a shows an equivalent circuit of a continuously variable TCUL transformer regulating bus i. The tap position is denoted t_i. Let the transformer admittance be y_{ik}. Figure 4.7b shows an equivalent circuit with the same voltampere response as that of Figure 4.7a. The case of TCUL regulation of bus i voltage is handled in the Newton–Raphson power flow study by examination of Figure 4.7b and calculation of the mismatch complex power at buses i and k

$$\Delta P_i + j\Delta Q_i = -v_i v_i^*(t_i^2 y_{ik} + Y_{ii})^* + v_i v_k^* t_i y_{ik}^* - \sum_{\substack{\ell=1 \\ \ell \neq i \\ \ell \neq k}}^{n} v_i v_\ell^* Y_{i\ell}^* \qquad (4.46)$$

$$\Delta P_k + j\Delta Q_k = -v_k v_k^*(y_{ik} + Y_{kk})^* + v_k v_i^* t_i y_{ik}^* - \sum_{\substack{\ell=1 \\ \ell \neq i \\ \ell \neq k}}^{n} v_k v_\ell^* Y_{k\ell}^* \qquad (4.47)$$

where Y_{ii} and Y_{kk} are the sums of line admittances connected to buses i and k, respectively, excluding the transformer line ik. The voltage magnitude v_i is known and does not appear in the unknown $[\delta\ V]^t$ vector, but t_i is unknown and this parameter does appear in the unknown vector. One possible ordering of unknowns is

$$\begin{bmatrix} \delta \\ \\ |V_{bus}| \\ \\ t \end{bmatrix} \begin{matrix} \} N-1 \\ \\ \} N-1-N_{tcul}, \\ \\ \} N_{tcul} \end{matrix}$$

where N is the total number of buses (including the swing bus) and N_{tcul} is the number of TCUL controlled buses. Then

$$\begin{bmatrix} \Delta P \\ \Delta Q \end{bmatrix} = -J \begin{bmatrix} \Delta\delta \\ \Delta|V_{bus}| \\ \Delta t \end{bmatrix},$$

where

$$J = \begin{bmatrix} J_1 & J_2 & J_5 \\ J_3 & J_4 & J_6 \end{bmatrix}$$

and J_1, J_2, J_3, and J_4 are the conventional Jacobian submatrices and J_5 and J_6 are the derivatives of (4.46) and (4.47)

$$J_5(i, i) = \frac{\partial P_i}{\partial t_i} = -2t_i v_i^2 |y_{ik}| \cos(-\theta_{ik}^p) + v_i v_k |y_{ik}| \cos(\delta_i - \delta_k - \theta_{ik}^p) \qquad (4.48)$$

$$J_6(i, i) = \frac{\partial Q_i}{\partial t_i} = -2t_i v_i^2 |y_{ik}| \sin(-\theta_{ik}^p) + v_i v_k |y_{ik}| \sin(\delta_i - \delta_k - \theta_{ik}^p) \qquad (4.49)$$

$$J_5(k, i) = \frac{\partial P_k}{\partial t_i} = +v_k v_i |y_{ik}| \cos(\delta_k - \delta_i - \theta_{ik}^p) \qquad (4.50)$$

$$J_6(k, i) = \frac{\partial Q_k}{\partial t_i} = +v_i v_k |y_{ik}| \sin(\delta_k - \delta_i - \theta_{ik}^p) \qquad (4.51)$$

and other entries of J_5, J_6 are zero. The entries in J_1, J_2, J_3, and J_4 must be revised to account for the tap at bus i. The admittance matrix elements must reflect the value of t_i in accordance with Figure (4.7). The

θ_{ik}^{p} of (4.48) through (4.51) is the angle of the transformer admittance y_{ik}.

Note that the update formula in the presence of a TCUL transformer,

$$\begin{bmatrix} \delta \\ |V_{bus}| \\ t \end{bmatrix}^{(\ell+1)} = \begin{bmatrix} \delta \\ |V_{bus}| \\ t \end{bmatrix}^{(\ell)} - \begin{bmatrix} J_1 & J_2 & J_5 \\ J_3 & J_4 & J_6 \end{bmatrix}^{-1} \begin{bmatrix} \Delta P \\ \Delta Q \end{bmatrix}, \tag{4.52}$$

does not contain updates of TCUL bus voltages since these voltages are fixed and do not require updating. This completes preliminary remarks about TCUL buses, but some additional complications will be considered in Section 4.5 concerning tap limits.

The case of voltage regulation of a generation bus is considered next. Most generators are operated with excitation controllers which hold the bus voltage magnitude fixed. This is accomplished essentially by increasing the field current of the generator to increase the generated reactive power and thereby increase the bus voltage (or vice-versa). The voltage regulator controller may also be designed to consider certain transient and abnormal operating conditions, but for power flow studies, these are ignored. The data specified at a voltage controlled generation bus are generated active power, P, and bus voltage magnitude, $|V|$. This type of bus is termed a PV bus. Most generation buses are PV buses, and most load buses are PQ buses.

The inclusion of PV buses in a Newton–Raphson power flow study does not require any additional Jacobian matrix formulas. The technique involves the deletion of $|V|$ update information. To formalize the procedure, let the PQ buses be tabulated first in $[\Delta\delta\ \Delta|V_{bus}|]^t$ and $[\Delta P\ \Delta Q]^t$:

$$\begin{bmatrix} \Delta P_{PQ} \\ \Delta P_{PV} \\ \Delta Q_{PQ} \end{bmatrix} \begin{matrix} \} N_{pq} \\ \} N_{pv} \\ \} N_{pq} \end{matrix} \quad \begin{bmatrix} \Delta\delta_{PQ} \\ \Delta\delta_{PV} \\ \Delta|V_{bus}|_{PQ} \end{bmatrix} \begin{matrix} \} N_{pq} \\ \} N_{pv}, \\ \} N_{pq} \end{matrix}$$

where N_{pq} is the number of PQ buses (not counting the swing bus) and N_{pv} is the number of PV buses (not counting the swing bus). Note that there is no ΔQ_{PV} formula because Q is not specified at a PV bus. Also, there is no $\Delta|V_{bus}|_{PV}$ entry since bus voltage magnitudes at PV buses do not require updating. The update formula is

$$\begin{bmatrix} \Delta\delta_{PQ} \\ \Delta\delta_{PV} \\ \Delta|V_{bus}|_{PQ} \end{bmatrix}^{(\ell+1)} = \begin{bmatrix} \Delta\delta_{PQ} \\ \Delta\delta_{PV} \\ \Delta|V_{bus}|_{PQ} \end{bmatrix}^{(\ell)} - \begin{bmatrix} J_a & J_b & J_c \\ J_d & J_e & J_f \\ J_g & J_h & J_i \end{bmatrix}^{-1} \begin{bmatrix} \Delta P_{PQ} \\ \Delta P_{PV} \\ \Delta Q_{PQ} \end{bmatrix}$$

$$J_a = \left[\frac{\partial P_i}{\partial \delta_j}\right] \quad i = PQ \text{ bus};\ j = PQ \text{ bus}$$

$$J_b = \left[\frac{\partial P_i}{\partial \delta_j}\right] \quad i = PQ \text{ bus};\ j = PV \text{ bus}$$

$$J_c = \left[\frac{\partial P_i}{\partial v_j}\right] \quad i = PQ \text{ bus};\ j = PQ \text{ bus}$$

$$J_d = \left[\frac{\partial P_i}{\partial \delta_j}\right] \quad i = PV \text{ bus}; \; j = PQ \text{ bus}$$

$$J_e = \left[\frac{\partial P_i}{\partial \delta_j}\right] \quad i = PV \text{ bus}; \; j = PV \text{ bus}$$

$$J_f = \left[\frac{\partial P_i}{\partial v_j}\right] \quad i = PV \text{ bus}; \; j = PQ \text{ bus}$$

$$J_g = \left[\frac{\partial Q_i}{\partial \delta_j}\right] \quad i = PQ \text{ bus}; \; j = PQ \text{ bus}$$

$$J_h = \left[\frac{\partial Q_i}{\partial \delta_j}\right] \quad i = PQ \text{ bus}; \; j = PV \text{ bus}$$

$$J_i = \left[\frac{\partial Q_i}{\partial v_j}\right] \quad i = PQ \text{ bus}; \; j = PQ \text{ bus}$$

and the formulas for J_1 are used for J_a, J_b, J_d, and J_e, the formulas for J_2 are used for J_c and J_f, the formulas for J_3 are used for J_g and J_h, and the formulas for J_4 are used for J_i.

4.5 PHASE SHIFTING TRANSFORMERS

Tap changing is a means of voltage control and accompanying reactive power control. Transformers may also be used to control phase angle and therefore active power flow. Such specially constructed transformers are termed phase shifting transformers or simply phase shifters. Tap changing for voltage control, and the use of phase shifters for phase control, have numerous practical limits. Beyond the obvious limits imposed by the physical device (its rating, tap setting range, phase shift range), there are economic considerations. Phase shifters have particularly unfavorable economic tradeoffs which arise mostly from high cost and the numerous economically attractive alternative methods of active power control. The principal use of phase shifters occurs at major intertie buses where the control of active power exchange is especially important.

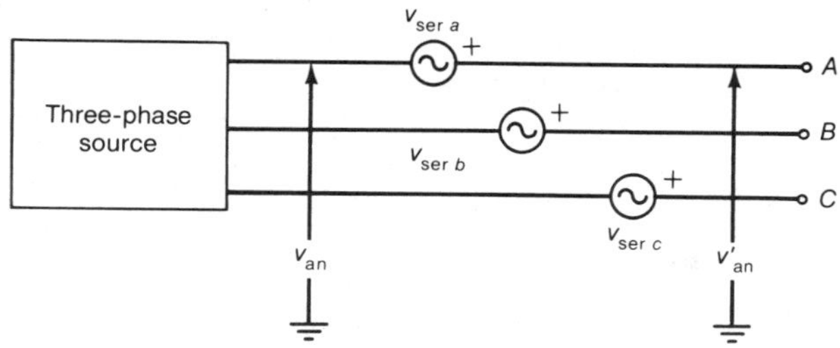

FIGURE 4.8 Pictorial of a phase shifter.

A phase shifter is a three phase device which inserts a voltage, v_{ser}, into each line (see Figure 4.8). If v_{sera} is out of phase with the line-to-neutral voltage of the source, v_{an}, the phase shifter output voltage, v'_{an}, will be controllable. Thus the phase of v'_{an} depends on v_{sera} – unfortunately, the magnitude of v'_{an} also varies with v_{sera}. It is possible to construct a controller which will cause only the phase of v'_{an} to vary – not the magnitude. The details of the dependence of $|v'_{an}|$ with the amount of phase shift depends on the phase shifter design.

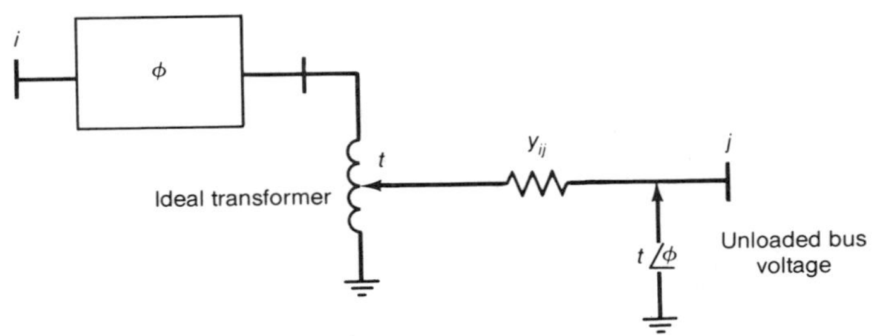

FIGURE 4.9 Generalized model of a phase shifter.

Figure 4.9 shows a generalized model of phase shifter in which both voltage magnitude and phase angle vary. The equivalent tap position is $t(\phi)$ which is only a function of the phase setting ϕ. If no voltage variation occurs, t = 1. The phase shifter admittance is y_{ij}.

Figure 4.10 shows a further equivalent. Note that in Figure 4.10,

$$|V_{i'}| = |V_i|$$
$$\delta_{i'} = \delta_i + \phi$$

and the complex voltamperes entering the phase shifter at bus i is delivered to bus i'. Let bus i' be introduced into the $[\delta \ V]^t$ vector, but

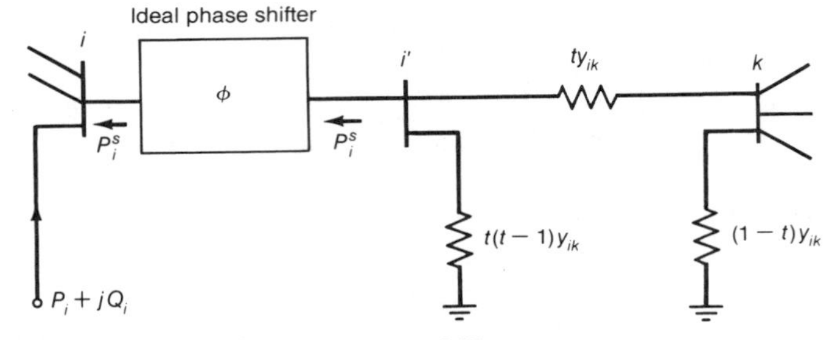

FIGURE 4.10 Equivalent circuit of a phase shifter.

only in the δ subvector. Also, rather than write $\delta_{i'}$ in this subvector, introduce the unknown phase shift, ϕ:

$$\begin{bmatrix} \delta_2 \\ \delta_3 \\ \vdots \\ \delta_i \\ \phi \\ \vdots \\ \delta_N \\ v_2 \\ v_3 \\ \vdots \\ v_i \\ v_i+1 \\ \vdots \end{bmatrix} .$$

Note that $v_{i'}$ is missing from the $|V_{bus}|$ subvector. Also note that the partial derivations $\frac{\partial P}{\partial \delta_{i'}}$ and $\frac{\partial Q}{\partial \delta_{i'}}$ are equal to $\partial P/\partial\phi$ and $\partial Q/\partial\phi$, respectively, since

$$\delta_{i'} = \delta_i + \phi .$$

The missing voltage magnitude element is "counterbalanced" by a missing $\Delta Q_{i'}$ expression. Thus J is $2(N-1) - n_\phi$ by $2(N-1) - n_\phi$, where n_ϕ is the number of phase shifters.

The procedure to handle a phase shifter bus where ϕ adjusts itself to give P_i^s (see Figure 4.10) injected into bus i is as follows:

1. Construct Y_{bus} using the conventional algorithm but do not connect buses i and i'. Also, do not connect buses i' and k.
2. Use the equivalent circuit in Figure 4.10 to connect buses i' and k. Use the initialized value of ϕ to find t. The equivalent tap setting is a function of ϕ in general (this depends on the phase shifter design).
3. Construct the Jacobian matrix. Eliminate the ΔQ expression at bus i and eliminate $|v|$ at bus i.
4. Consider P_i^s, the specified power through the phase shifter, as injected into i and as the only load at i'.
5. Perform one iteration in the power flow study and repeat as required.

4.6 ADDITIONAL MODELLING CONSIDERATIONS

The power system model discussed thus far is a very simplified version of the network which appears in the field. In the three broad parts of the system, generation, transmission, and distribution–load subsystems, there are some physical complexities which deserve second consideration. In this section these subsystems are reexamined to identify conditions which warrent more detailed modeling.

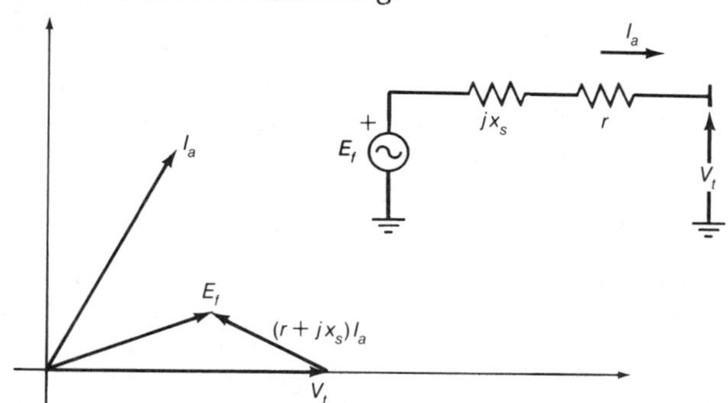

FIGURE 4.11 Phasor diagram for a synchronous generator.

Generation Subsystems

The generation bus model presented thus far is a PV model in which the turbine controls hold the generated power fixed and the machine voltage regulator holds the bus magnitude fixed. Under steady state conditions, this model may be unrealistic when the voltage regulator is unable to set the field current to produce the desired bus voltage. Figure 4.11 shows a simplified phasor diagram of the generator. To increase $|V_t|$, the field current is increased such that $|E_f|$ increases. Under constant generated power conditions, the torque angle δ will decrease slightly, but the phase of the armature current, I_a, will rotate such that the power factor angle ϕ increases. Thus more reactive power is injected into the bus. The limit on the control of $|V_t|$ occurs primarily due to the limit on the field current (and hence the limit on Q). In most cases, generators inject Q into the system since most loads have lagging power factor. Thus the salient limitation is related to the upper field current limit. In cases where the generator absorbs reactive power, the field current must be decreased. The limitations in this region are twofold: limits imposed by the maximum value of $|I_a|$ and practical stability limits associated with low field currents. In this case, the dominant limitation may depend on the circuit power factor and always depends on the generator controller designs. Additionally, the generator complex voltampere rating, $|S|$, may come into play. The generator $|S|$ rating implies possible further limitation of Q dependent on the active power setting.

A simplified approach to generator field controller limits is usually followed for power flow studies: a single upper limit of injected Q for each generator is imposed. Under the formulation of an ideal PV bus, the reactive power mismatch is not calculated since there is no specified reactive power. However, the Q generated by the machine must equal the total line flow reactive powers,

$$Q_{\text{generated at } i} = \sum_{j=1}^{N} |Y_{ij}| v_j v_i \, sin \, (-\theta_{ij} + \delta_i - \delta_j).$$

To check whether the generated Q lies below the upper limit,

$$Q_{\text{generated at } i} < Q^{u}_{\text{limit } i}$$

must hold. Similarly, the lower limit ($Q^{\ell}_{\text{limit } i} < 0$) must be checked,

$$Q^{\ell}_{\text{limit } i} < Q_{\text{generated at } i} < Q^{u}_{\text{limit } i} \,. \tag{4.53}$$

If (4.53) is violated at the upper or lower limit, the PV bus model must be modified to reflect the physical system. This usually entails conversion of the PV bus to a PQ bus where the specified Q is set at $Q^{\ell}_{\text{limit } i}$ or $Q^{u}_{\text{limit } i}$ as appropriate. The origin of Q limits of machines comes from the several operating conditions cited earlier, including limits on armature current, limits on field current and voltage, and operational stability requirements.

Before leaving the topic of generation subsystem modeling in power flow studies, the concept and treatment of the swing bus warrants further attention. The swing bus (also known as the "slack bus") is both a mathematical fiction and an operating reality. The formulation of the power flow problem precludes specification of active and reactive power at *all* buses since the $P+jQ$ lost in the transmission network is unknown. A power flow problem in which P is specified at all buses is overspecified and probably inconsistent with the transmission losses. This remark is a mathematical consideration which translates to an operating condition: one machine in the system is usually operated such that the active power generated is set to hold the area error at zero. The term "area error" refers to the generation error in a given system which causes a frequency error. If the net generation in a given system (an "area") is greater than the load plus losses, the excess injected power will result in a net injected energy. This energy must go somewhere – and it causes the machines in the system to accelerate. In other words, the integral of the error power (i.e., the error energy) becomes rotating kinetic energy. Similarly, if the net generation is too low, the system kinetic energy (and hence, frequency) will decrease. The operational turbine setting is obtained at the swing machine from the area error: a *raise power* signal is required when the system frequency falls below the standard and a *lower power* signal is required when the system frequency is too high.

In recent years, the concept of "swing machine" has been generalized to distribute area power error over several machines. Thus the swing power is made up not at a single machine, but at several generators. Usually, a production computer program still uses a single bus as a swing bus for load flow studies.

It is also possible to encounter several swing buses in one load flow study. These are usually studies involving more than one "area" (for purposes of this discussion, the term "area" refers to a coordinated power transmission network and generation in which a single swing bus is present). There are several techniques for considering multiple swing buses, perhaps the simplest being separate load flows in which intertie buses are separated. It is possible to model the swing power in a distributed fashion (i.e., at several generators) by employing several swing machines.

Transmission Subsystems

The transmission network is a deceptively simple subsystem which possesses complexities in busbar interties, transformers, and the conductors themselves. At substations, busbars are often interconnected by circuit breakers in such a way that switching flexibility is possible (i.e., single or groups of circuits may be outaged without undesirable outaging of key buses or other circuits). A bus tie circuit breaker presents somewhat of a mathematical difficulty in that opening of the breaker in the computer study requires "outaging" of a zero impedance tie. The problem is obviated by considering adjacent buses tied by a bus–tie–breaker as being tied by a very low impedance "line" (e.g., $j0.00001$ per unit impedance).

Transformers present potential modeling problems in that the magnetizing branch occasionally must be included in the power flow study. This is occasionally required in EHV studies at off peak hours when the magnetizing current may be a considerable fraction of the total transformer current. If the transformer is located at bus i, inclusion of the magnetizing reactance as a lumped ground tie at bus i is usually adequate.

Models for tap changing transformers and phase shifters should also include tap and phase shift limits. These limits may not be symmetrical about the nominal setting. If a TCUL voltage regulating transformer or phase shifter hits a limit, that limit should be retained no matter what $\Delta\phi$ and Δt are calculated.

The transmission line itself is usually modeled as a lumped series impedance and two lumped shunt capacitive susceptances. The latter occur on each line terminal and represent line charging capacitance. If the total line charging susceptance is jB, a tie of $jB/2$ mhos per unit occurs at each line terminal. The value of B depends on the line configuration and is usually obtained from a computer program in the case of cable systems. The following expression may be used for overhead lines,

$$B = \frac{2.096 \cdot 10^{-5}}{ln(D_{eq}/D_{sc})},$$

where the total lumped capacitive susceptance, B, is given in mhos/km per phase at 60 Hz and D_{eq} and D_{sc} denote the geometric mean spacing of the three phase conductors and geometric mean radius of the conductors respectively.

The series resistance of transmission circuits is usually taken as the 50°C, 60 Hz value which generally is obtained from tabular data. Approximate values of R may be obtained from the elementary expression

$$R = \frac{\rho \ell}{A}$$

where ρ is the conductor resistivity, ℓ is the length, and A is the cross-sectional area. Tabular values are generally preferred to this theoretical expression since complex geometries and effects of conductor spiralling and current skin penetration are more conveniently considered. Note that the depth of penetration of the line current is often a phenomenon which is not negligible in power system analysis. The skin depth, s, is the distance from the conductor surface to the radius at which the current density drops to $1/\epsilon$ of its surface value. The skin depth is given by

$$s = \frac{1}{\sqrt{\pi f \mu \sigma}}$$

where f is the frequency of operation (usually 60 Hz), and μ and σ are the conductor permeability and conductivity respectively. The equivalent resistance of an ac line considering the skin effect is R_{eq},

$$R_{eq} = \frac{\rho \ell}{2\pi r s}$$

where r is the conductor radius.

If the transmission circuit is very long, occasionally the long, line model is used to account for the fact that the series and shunt transmission line elements are, in fact, distributed parameters rather than lumped. At 60 Hz, this factor is significant only for the longest lines (e.g., 625 km, which is approximately the 1/8 wavelength). In some special studies, higher frequency signals are of interest, and the corresponding shorter 1/8 wavelength point is approximately $37{,}500/f$ (where f is in Hertz) kilometers. The long line equations for voltage $V(x)$ and current $I(x)$ at a point x measured from bus 1 are

$$V(x) = V_1 \cosh \gamma x - I_1 Z_c \sinh \gamma x \tag{4.54}$$

$$I(x) = -I_1 \cosh \gamma x + \frac{V_1}{Z_c} \sinh \gamma x, \tag{4.55}$$

where the subscript 1 denotes bus 1 and γ and Z_c are the propagation constant and characteristic impedance of the line. The sign convention in

(4.54) and (4.55) is such that $+I_1$ flows in the direction of $+x$. Let bus 2 be the other line terminal located at $x = \ell$, and let the voltage at bus 2 be V_2 and the current be $+I_2$,

$$V_2 = V_1 \cosh \gamma\ell - I_1 Z_c \sinh \gamma\ell$$

$$I_2 = -I_1 \cosh \gamma\ell + \frac{V_1}{Z_c} \sinh \gamma\ell .$$

Solving for $(I_1\ I_2)^t$,

$$\begin{bmatrix} I_1 \\ I_2 \end{bmatrix} = \begin{bmatrix} \dfrac{1}{Z_c \tanh \gamma\ell} & \dfrac{-1}{Z_c \sinh \gamma\ell} \\ \dfrac{-1}{Z_c \sinh \gamma\ell} & \dfrac{1}{Z_c \tanh \gamma\ell} \end{bmatrix} \begin{bmatrix} V_1 \\ V_2 \end{bmatrix} . \tag{4.56}$$

The coefficient of $[V_1\ V_2]^t$ in (4.56) is recognized as the bus admittance matrix of the two-port pi section shown in Figure 4.12. The bus 1 to ground tie is the row sum 1, the bus 2 ground tie is identical, and the off–diagonal entry in (4.56) is the negative of the admittance between buses 1 and 2.

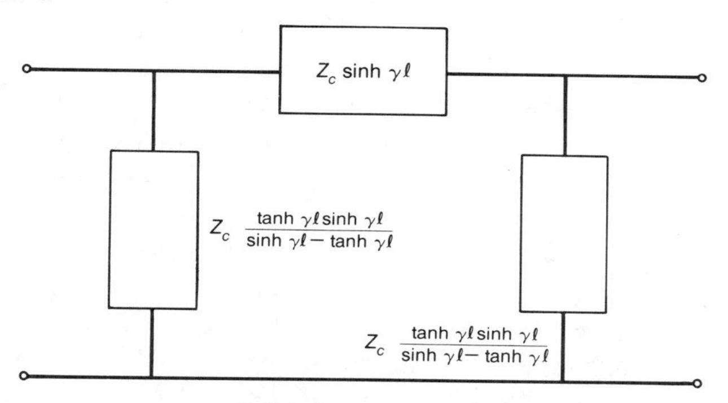

FIGURE 4.12 Impedances in a π-equivalent of a long line.

Distribution Circuit and Load Modeling

The complexities of distribution circuit and load modeling are greater than those of the generation and transmission systems because the variety of distribution equipment is very diverse and loads are not only diverse in nature, but also often very nonlinear. Most power flow studies do not include distribution circuit voltages. In fact, most large studies are for the transmission network above 69 kV. We will have more to say about distribution power flow studies with regard to applications, but the following should be noted with regard to distribution circuit models:

1. The importance of accurate charging capacitance modeling decreases as voltage level decreases and as line lengths decrease. At distribution voltages, charging capacitance is often omitted.
2. The R/X ratio of distribution branches may be quite high compared to transmission circuits. Often, the only significant reactance occurs in the distribution transformer.
3. Numerous radial circuits occur. If loads may be lumped at a substation, considerable time and memory savings may be obtained in computer solutions.

4. Several single phase distribution circuits are very commonly lumped at the substation to permit balanced three phase studies. If it is desired to calculate full three phase detail, a three phase power flow study could be performed with considerable increase in modeling complexity. The subject of three phase power flow studies will be considered again later.

5. Due to the considerable change in operating voltage level at the point of energization of a distribution circuit, the single phase circuit may often be considered as being energized at an infinite bus. A power flow study of the distribution circuit under these conditions uses either the substation bus as the swing bus, or the substation bus may be modeled as a bus separated from a fictitious swing bus by the positive sequence driving point impedance seen at the substation bus.

Load modeling may be done at several levels of sophistication. To this point, only fixed S loads have been considered, and this is the usual practice for bulk power load centers. The fixed S load (i.e., independent of bus voltage) is a reasonable representation for buses with a large percentage of rotating machines loads and voltage regulated loads. Other alternative load models include:

1. Loads whose $|S|$ demand is proportional to $|V|^2$. This type of load is a fixed impedance load and is common when the load contains significant incandescent lighting and resistive heating.

2. Loads whose $|S|$ demand is proportional to $|V|^1$. This type of load is a fixed current load and is common when the load contains significant rectifier loads and certain type of synchronous machine loads.

It has been suggested to model the actual load as a linear combination of each of the load types mentioned (constant S load in which $|S| \propto |V|^0$, constant current load in which $|S| \propto |V|^1$, and constant impedance load in which $|S| \propto |V|^2$). Such an approach would require considerable knowledge of the load composition or a knowledge of the active and reactive power variation with $|V|$ (from historical data).

A further degree of sophistication may be introduced in load modeling through the inclusion of variation of load with the hour of the day, transient response, and inclusion of nonlinear loads. The first of these is considered with regard to load forecasting and also power flow study applications. Transient response is occasionally considered in transient and small signal stability studies, but usually the additional accuracy attainable is not warranted in view of the accuracy of the other system models. Nonlinear loads are occasionally considered in special purpose studies, which include fluorescent, rectifier, and "electronic" loads. This subject is considered in Chapter 5.

4.7

PRACTICAL APPLICATIONS OF POWER FLOW STUDIES

To this point, we have presented the essentials of the power flow problem and much of its solution. Two remaining areas are discussed in this section with regard to practical applications of power flow studies: (1)

how are power flow studies used in the electric utility industry, and (2) what further practical considerations should be made to prepare and run power flow studies.

The way in which power flow studies are used is conveniently divided into two systems areas: system planning and system operation. System planning is engineering with the objective of designing an electric power system capable of providing reliable bulk electric supply. Major electric utility companies usually have a system planning department. Common tasks of the system planner include transmission planning (the design, analysis, sizing, and documentation of future transmission circuits), interchange studies, generation adequacy studies, cost-to-benefit analyses of system additions, and numerous other tasks related to the system design and future adequacy. In some cases, generation planning is also considered as part of system planning. System operation is engineering with the objective of setting, adjusting, and otherwise operating the given power system to produce reliable and economical electric energy. Included in system operation tasks are economic dispatch of the generating stations (i.e., calculating the power levels at each generating unit such that the system is operated most economically), contingency analysis (analysis of outages and other forced operating conditions), studies that ensure power pool coordination, and other tasks that relate to the day-to-day and week-to-week operation of the system. Sometimes, generation production is also considered in system operations.

The two cited areas, system planning and system operations, although not divorced from each other, do have distinct objectives and philosophies. In most phases of each of these areas, load flow studies are used to assess system performance under given conditions.

In the area of system planning, an important application of power flow studies lies in transmission planning. Models of the future system are prepared and, typically, peak load conditions are used to run a study. In this way components (conductors, transformers, reactors, shunt capacitors) may be sized. Also, transmission siting may be considered. In some cases, series capacitors, phase shifters, and TCULs are studied where control of power flow is important. A second general area of system planning in which power flow studies are indispensaible is interchange studies. Interties with neighbors are planned to meet forecast needs of the system consistent with reliability requirements. The topic of interchange capacity of a power system will be considered in some detail in Chapter 5. Load flow studies in these applications are generally used to identify line loads and bus voltages out of range, inappropriately large bus phase angles (and the potential for stability problems), component loads (principally transformers), proximity to Q-limits at generation buses, and other parameters which have the potential of creating operating difficulties. While peak load studies are frequently done, sometimes intermediate loads and off-peak (minimum) load conditions are used. Off–peak loads may result in high voltage conditions which are not identified during the peak load.

System operation tasks which use power flow studies include retaining all bus voltages and transmission component loads within range, loss calculations, area coordination, calculation of fixed tap settings, and various types of contingency checks. The latter refers to checks on the effects of line and component outages (sometimes multiple contingencies are considered). Contingency studies are usually done "off-line" (i.e., in advance of operating the system under the conditions considered). A typical line outage contingency study consists of a base case power flow study (with all lines in service) followed by the contingency cases in which key lines are outaged. Key lines are identified in a variety of ways: those lines which are heavily loaded or which have small load margin (rating minus operating load); lines with high angular difference in terminal bus voltage phases; extra high voltage (EHV) circuits; or lines which are known to the operators to be essential to system integrity. The latter comes from system operating experience and is often the most reliable means of identifying key lines. In some cases, approximate load flow studies will suffice to identify problems in contingency studies – several suitable approximate methods will be presented in the next chapter. Contingency studies are particularly important to maintaining reliable service: if a line outage causes other circuits to become heavily loaded, those circuits may trip out by the action of protective relays. These additional outages can result in further outages in an uncontrolled way. The term *cascade tripping* applies to this undesirable operating condition.

Another area of operational use of power flow studies relates to generation dispatch. Because the operating costs of an electric power system are quite high, and the profit margin is often rather small, the selection of generator operating levels (i.e., the *generation dispatch*) has considerable economic importance. The economic dispatch of a power system is a very important load flow application which is considered in some detail in Chapter 6.

The solution of very large networks in a power flow study requires some techniques which lie in the realm of "art" rather than "science." Experience with a particular network will often dictate the best initialization procedure, acceleration factor, output format options, and other considerations. A few remarks presented here will help in these areas.

Initialization of a Newton–Raphson Power Flow Study

The simplest and perhaps most common initialization procedure for a Newton–Raphson power flow study is the flat start (all phase angles equal to the swing phase angle which is usually zero, and all bus voltage magnitudes equal to the swing bus). Alternatively, a previously converged solution may be used for initialization.

The Newton–Raphson algorithm has quadratic convergence near the solution. Far from the solution, the convergence may be erratic. On the other hand, the Gauss–Seidel algorithm exhibits generally linear convergence in a smooth fashion. At points far from the solution, this technique may be superior to the Newton–Raphson algorithm. Production Newton–Raphson programs have been written in which the Gauss–Seidel method is used for one or two steps to initialize the bus angles and voltages. In some instances, such initialization will result in convergence of the Newton–Raphson algorithm to a prescribed tolerance one step sooner.

An alternative which has more fruitful applications in areas outside Newton-Raphson initialization is based on the familiar formulas for the active and reactive power flow in a line from bus i to bus j metered at j,

$$\overline{P}_{ij} = \frac{v_i v_j \sin \delta_{ij}}{\overline{x}_{ij}} \tag{4.57}$$

$$\overline{Q}_{ij} = \frac{v_j^2 - v_i v_j \cos \delta_{ij}}{\overline{x}_{ij}} . \tag{4.58}$$

Equations (4.57) and (4.58) apply when the line $i - j$ is largely reactive with reactance $\overline{x}_{ij}$. The "bars" over $\overline{P}_{ij}$, $\overline{Q}_{ij}$, and $\overline{x}_{ij}$ denote that these are line parameters. The notation δ_{ij} refers to $\delta_i - \delta_j$. In (4.57), note that

$$\overline{P}_{ij} \simeq \frac{\delta_{ij}}{\overline{x}_{ij}} = (-B_{bus})_{ij}\delta_{ij}$$

when the voltage profile is flat at unity and $\sin \delta_{ij} \simeq \delta_{ij}$ (i.e., δ_{ij} is small and in degrees). The notation B_{bus} denotes the bus susceptance matrix which is Y_{bus} for a purely reactive transmission system. The total active power injected into bus i is $\sum_j \overline{P}_{ij}$. Therefore,

$$P_i = \sum_j \overline{P}_{ij} = -B_{bus}\delta_{bus} \tag{4.59}$$

where δ_{bus} is the N–vector of bus phase angles. Similarly, (4.58) yields the approximation

$$\overline{Q}_{ij} \simeq \frac{v_i - v_j}{\overline{x}_{ij}} . \tag{4.60}$$

Using the same notation, we obtain

$$Q_i = \sum_j \overline{Q}_{ij} = -B_{bus} | V_{bus} | . \tag{4.61}$$

Equations (4.59) and (4.61) may be used to initialize and estimate δ_{bus} and $| V_{bus} |$ for given loading conditions. These expressions are valid only for small bus voltage phase angles and nearly flat bus voltage magnitudes.

Acceleration Factors

There is less motivation for the use of acceleration factors in Newton–Raphson power flow studies as compared with Gauss or Gauss–Seidel solutions due to the inherently rapid convergence rate of the Newton–Raphson solution. The more usual reason for using acceleration factors in a Newton–Raphson solution is to render a divergent case as convergent. The technique is

$$\begin{bmatrix} \delta \\ | V | \end{bmatrix}^{(\ell+1)} = \begin{bmatrix} \delta \\ | V | \end{bmatrix}^{(\ell)} - \alpha J^{-1} \begin{bmatrix} \Delta P \\ \Delta Q \end{bmatrix}^{(\ell)}$$

at iteration ℓ. The quantity α is the acceleration factor and it generally lies in the range $0.7 < \alpha < 1.4$ for power flow studies. Acceleration factors below 1.00 actually slow convergence: these are used in studies that do not converge readily. Such cases sometimes occur when high active or reactive loads are present.

The effects of acceleration factors on the eigenvalues of the inverse of the Jacobian matrix, ξ, give a qualitative insight to the mathematics of these factors. Equation (4.24) gives the correction on the $[\delta V]^t$ vector at iteration n in terms of the mismatch vector F

$$\Delta X^{(n)} = \Xi F^{(n)} .$$

The application of acceleration factor α modifies this relationship to

$$\Delta X^{(n)} = \alpha \Xi F^{(n)} .$$

Thus the effective inverse Jacobian matrix is $\alpha\Xi$ rather than Ξ. The eigenvalues of $\alpha\Xi$ are simply α times the eigenvalues of Ξ [which is readily demonstrated by noting that

$$det\ (\alpha\Xi - \gamma I) = 0,$$

where γ are eigenvalues of $\alpha\Xi$, then allow

$$\gamma = \alpha\lambda;$$

thus

$$det\ (\alpha\Xi - \alpha\lambda I) = 0$$

$$\alpha^N det\ (\Xi - \lambda I) = 0$$

and λ are eigenvalues of Ξ]. If the large eigenvalues of the unaccelerated Jacobian matrix control convergence of the unaccelerated formulation, these large eigenvalues are further increased if $\alpha > 1$ and are decreased if $\alpha < 1$. Qualitatively, when $\alpha < 1$, the nonconvergent problem might become convergent because the controlling large eigenvalues of $\alpha\Xi$ may result in sufficiently reducing these parameters. These remarks are reinforced by examining the effect of α on the contraction constant, c. Let

$$||X^{(\ell+1)} - X^{(\ell)}|| \leq c||X^{(\ell)} - X^{(\ell-1)}||$$

$$0 < c < 1$$

for all X in region R. For the case of application of acceleration factor α at update $\ell + 1$ the contraction constant effectively becomes c_{eq},

$$c_{eq} = c\alpha.$$

For fractional α, the contraction constant is smaller, and one expects enhanced convergence. Unfortunately, in cases of practical interest, the contraction constant can rarely be identified and a firm theoretical conclusion cannot be reached. The effects of acceleration factors in simplified cases on the rate of convergence, R_c, are reconsidered in an exercise at the end of the chapter.

Studied from the point of view of the Newton–Kantorovich criterion, the application of acceleration factor α causes a multiplicative factor of α to be applied to parameters a, b, and c. Thus the criterion that

$$abc \leq 1/2$$

becomes

$$\alpha^3 abc \leq 1/2 .$$

Clearly, from a qualitative point of view, for fractional α, the Newton–Kantorovich inequality is enhanced [9].

Typical Bus Voltages and Phase Angles

Elementary considerations indicate that active power flows in the direction of lagging bus voltage phase angles. The usual choice of the swing bus as the reference phasor ($\delta = 0$) results in further lagging (i.e., negative) phase angles at adjacent load buses. For phase angles through about 30°, the active power flow in a purely reactive transmission line is approximately proportional to $(\delta_i - \delta_j)/\bar{x}$. Therefore, one would expect that the phase difference in voltages across a transmission line will be nearly directly proportional to the line loading. In a simple radial circuit, this results in predictable bus phase angles, but in interconnected networks the problem in more complex. It is possible to conclude, however, that heavily loaded buses will have lagging phase angle in such a way that this phase lags behind that of adjacent buses roughly proportional to the supply line active loading. A similar remark can be made concerning generation and leading phase angles.

Bus voltage magnitude is controlled by – and controls – reactive power flow. Reactive power flows in the direction of lower voltage. Thus, if a system swing bus is operated at unity bus voltage, a simple radial circuit from this bus will experience progressively lower bus voltages as one moves away from the swing bus toward loads which are lagging power factor. In fact, the amount of bus voltage drop below the adjacent supply bus will be roughly proportional to the reactive power load in the supply line. The result is that heavy Q loads will result in depressed $|V|$; Q sources will result in higher $|V|$.

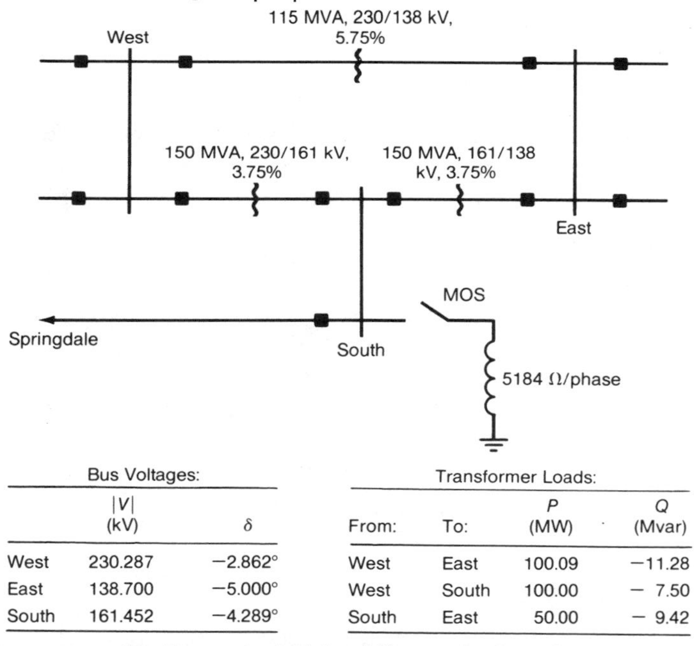

Bus Voltages:

	$\|V\|$ (kV)	δ
West	230.287	−2.862°
East	138.700	−5.000°
South	161.452	−4.289°

Transformer Loads:

From:	To:	P (MW)	Q (Mvar)
West	East	100.09	−11.28
West	South	100.00	− 7.50
South	East	50.00	− 9.42

FIGURE 4.13 System used in Example 4.5; load flow solution shown.

Example 4.5

Figure 4.13 shows a substation and portion of a load flow study, indicating the substation bus voltages. The load schedule used in this study includes loads at East bus (150.09 MW, -20.7 MVAr) and South bus (50 MW, -3.75 MVAr). The South load represents a flow to Springdale. All circuit breakers are closed for the study.

A motor operated air break switch (MOS) at South bus is usually used to outage a shunt reactor. This MOS has an interruption current rating of 19 A. The use of an MOS in this application avoids the use of a more costly circuit breaker, but there is concern that there may be cases in which the MOS cannot be used to outage the reactor. If the MOS cannot be used (because the 19 A rating will be exceeded), it will be necessary to outage the South bus using circuit breakers and subsequently operate the deenergized MOS to outage the reactor. This switching strategy is rather undesirable since the 260 MVA capacity of the substation will be reduced to 115 MVA during the outage of South bus.

In this example, the conditions under which the MOS may be used to outage the reactor are studied.

The reactor is "in" for the study shown and the West bus is voltage regulated at 230.287 kV. The Springdale load is fixed at the indicated value.

Solution

The given loading condition will be referred to as the "base case." In the base case, the reactor current is $|I_r|$,

$$|I_r| = \frac{161.452 \times 10^3/\sqrt{3}}{5184} = 17.9 \text{ A per phase.}$$

This is within the MOS interruption range. The condition that produces higher reactor current is higher bus voltage at South. The maximum tolerable bus voltages is $|V_{South}^{max}|$,

$$|V_{South}^{max}| = (5184)(19)(\sqrt{3})$$

$$= 170.600 \text{ kV.}$$

There are several potential ways in which the bus voltage at South may increase. Only variations in the East bus load are examined here. The base case power factor at East is

$$PF = \frac{150.09}{\sqrt{150.09^2 + (-20.7)^2}} = 0.991 \text{ leading.}$$

A reasonable assessment of bus voltage conditions is obtained by varying the power factor at East (holding the MVA fixed at 151.51). Alternatively, the MVA demand at East may be varied (holding the power factor fixed at 0.991 leading). These considerations require power flow studies. Figure 4.14 shows a one-line diagram of the system with data reduced to a common, 100 MVA per unit base.

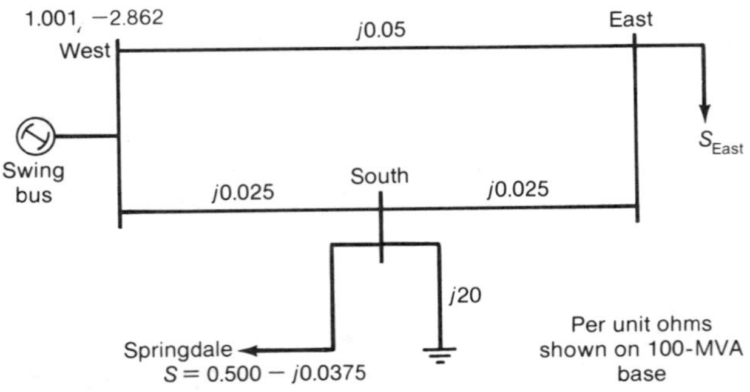

FIGURE 4.14 One-line diagram (100 MVA per unit base) for Example 4.5.

Results of the power flow study are shown in Figure 4.15. Note that the power factor variation test was made over a wide range in the leading configuration since leading power factor loads are most likely to result in high bus voltage. The variation of MVA demand at East was carried up to 255 MVA, which resulted in an overload of the 115 MVA transformer in the West East tie (at 255 MVA demand at East, this transformer delivers about 139 MW, 5 MVAr to East, thus representing about 24 MVA above the transformer rating).

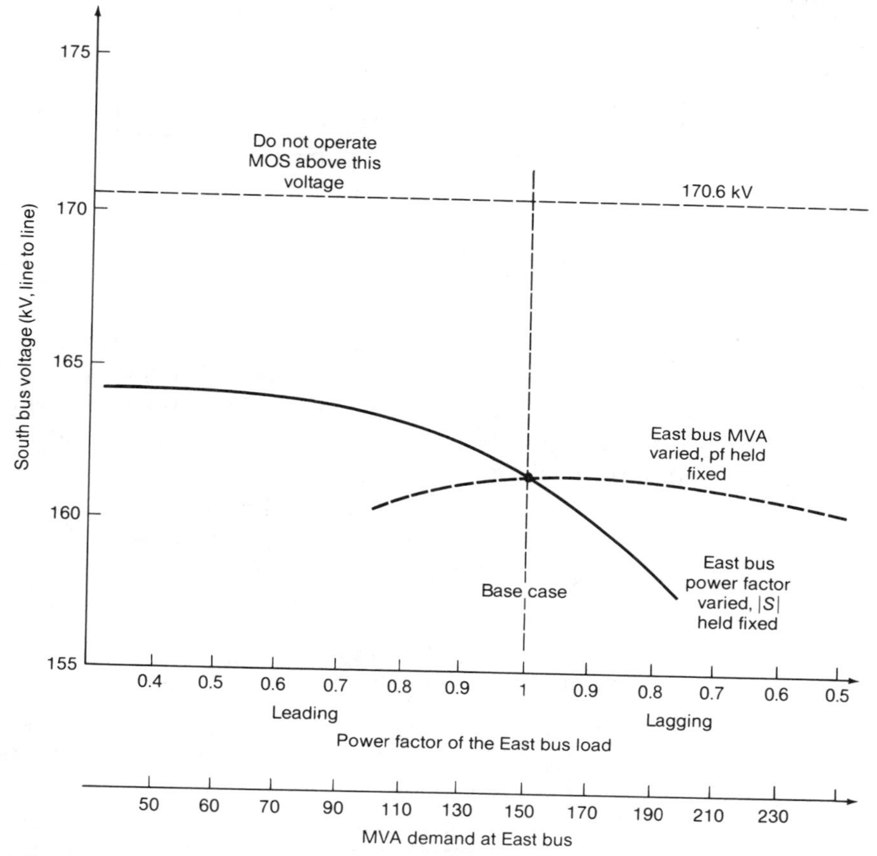

FIGURE 4.15 Power flow results in Example 4.5; bus voltage at South depicted.

It is concluded that over a reasonable loading range at East, the MOS may be used to switch out the shunt reactor.

Example 4.6

In this example, a contingency study is illustrated for a system in which a fixed tap transformer exists. Also, the effects of acceleration factors in a Newton–Raphson power flow study are illustrated.

Figure 4.16 shows a power system with reactances indicated in per unit on a common base. All lines have a rating of 5.05 per unit MVA. The system is to be operated at $|V_{bus}|$ in the range $0.95 \leq |V_{bus}| \leq 1.05$. Consider single and multiple outage contingencies of lines from Imlay to M. Fico and Imlay to Kingville. Also illustrate the use of acceleration factors in the triple line outage contingency.

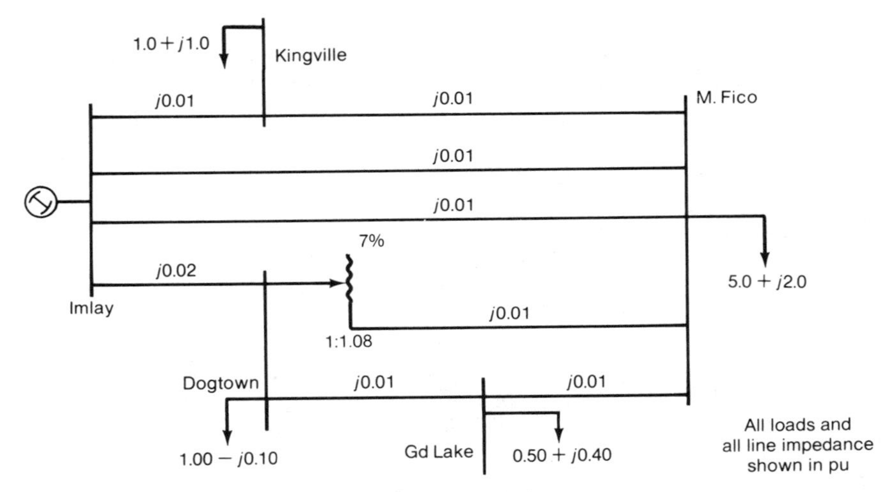

FIGURE 4.16 Sample power system (Example 4.6) with fixed tap transformer.

Solution

The handling of the fixed tap transformers in the Dogtown – M. Fico circuit is accomplished identically to the case of a TCUL transformer except that the tap is held fixed. Figure 4.7 shows the equivalent circuit of this line. The line impedance of Dogtown – M. Fico is lumped with the reflected transformer reactance.

A base case study (all lines in service) is solved to 0.01 per unit mismatch in two Newton–Raphson iterations. Results are as follows:

Imlay	$1.00\,\underline{/0^\circ}$	per unit
Dogtown	$1.00\,\underline{/-1.4^\circ}$	per unit
Gd Lake	$0.99\,\underline{/-1.5^\circ}$	per unit
Kingville	$0.99\,\underline{/-1.0^\circ}$	per unit
M. Fico	$0.99\,\underline{/-1.3^\circ}$	per unit.

Thus all bus voltages are within the required ±5% of nominal. Calculated line flows are as follows:

	P	Q
Imlay to M. Fico	2.32	1.22
Imlay to M. Fico	2.32	1.22
Imlay to Kingville	1.66	1.12
Imlay to Dogtown	1.21	0.01
Dogtown to Gd Lake	0.22	0.80
Dogtown to Imlay	−1.21	0.01
Dogtown to M. Fico	−0.01	0.72
Gd Lake to Dogtown	−0.22	−0.79
Gd Lake to M. Fico	0.28	0.39
Kingville to Imlay	−1.66	−1.08
Kingville to M. Fico	0.66	0.08
M. Fico to Imlay	−2.32	−1.15
M. Fico to Imlay	−2.32	−1.15
M. Fico to Kingville	−0.66	0.08
M. Fico to Gd Lake	0.28	−0.39
M. Fico to Dogtown	0.01	0.77

Note that the heaviest loaded lines are Imlay – M. Fico at $|S| = 2.62$ per unit.

The effect of the off-nominal transformer tap is to force reactive voltamperes from M. Fico to Dogtown. One effect of this reactive flow is to raise the M. Fico bus voltage.

The single line outage case is considered by the outage of Imlay – M. Fico. The bus voltages are still well supported by the considerable interconnection. The lowest voltage occurs at M. Fico, where 98.11% occurs ($\delta = -2.12°$). As may be expected, the remaining Imlay – M. Fico circuit increases in loading (to $3.62 + j1.96$). This represents a load of 4.11 per unit MVA, which is 81.4% of rating. In the single outage case, the fixed tap transformer forces 0.678 per unit reactive power from M. Fico to Dogtown. The solution is attained to 1% mismatch in three iterations.

The outage of the double circuit from Imlay to M. Fico results in further reduction in bus voltage, particularly at M. Fico:

Imlay	$1.00\ \underline{/0°}$
Dogtown	$0.98\ \underline{/-3.36°}$
Gd Lake	$0.97\ \underline{/-4.31°}$
Kingville	$0.97\ \underline{/-2.73°}$
M. Fico	$0.95\ \underline{/-4.97°}$.

Also note the considerable increase in phase angle at M. Fico – this is required to transport power away from Imlay. This double–outage case results in overload of Imlay – Kingville ($4.63+j3.00$ per unit or 9% overload). The fixed tap transformer load in this case is $0.31 - j0.52$ per unit with vars delivered to Dogtown. Thus the double–outage case is unacceptable due to line overload. The fixed tap transformer supports the bus voltage at M. Fico to acceptable levels, however. The double outage case is solved in three iterations.

The triple outage case results in low voltage conditions:

Imlay	$1.00\ \underline{/0.00°}$
Dogtown	$0.88\ \underline{/-9.81°}$
Gd Lake	$0.83\ \underline{/-13.98°}$
Kingville	$0.78\ \underline{/-19.10°}$
M. Fico	$0.79\ \underline{/-18.18°}$.

Also, there are prevalent overloading conditions in the remaining circuits in service:

Imlay to Dogtown	$7.50 + j6.64$
Dogtown to Gd Lake	$5.32 + j4.42$
Dogtown to Imlay	$-7.50 - j4.63$
Dogtown to M. Fico	$1.18 + j0.31$
Gd Lake to Dogtown	$-5.32 - j3.81$
Gd Lake to M. Fico	$4.82 + j3.41$
Kingville to M.Fico	$-1.00 - j1.00$
M. Fico to Kingville	$1.00 + j1.03$
M. Fico to Gd Lake	$-4.82 - j2.90$
M. Fico to Dogtown	$-1.18 - j0.13$

This solution is obtained in four iterations. It is an unacceptable operating regime.

Table 4.1
Use of Acceleration Factors for the Triple Outage Case in Example 4.6

Acceleration Factor α	Number of Iterations [a]
0.714	9
0.769	8
0.833	7
0.909	5
1.000	4
1.111	5
1.250	8
1.429	13

[a] To 1% worst mismatch.

The use of acceleration factors is illustrated on the triple line outage case. Table 4.1 shows results. In this case, the optimal acceleration factor is approximately 1.00. For this example, the logarithm of the number of iterations is very nearly linear in α

$$log\ N \simeq k_1\alpha + k_2 .$$

For $\alpha \leq 1$, k_1 is negative and for $\alpha > 1$, k_1 is positive. This is the expected convergence characteristic.

4.8 MODIFICATIONS OF THE NEWTON-RAPHSON FORMULATION

There are numerous modifications of the Newton–Raphson formulation used for power flow studies. Because of the dimensionality of typical power flow studies, even small improvements have potentially significant impact on central processing time and memory requirements. Perhaps the best known of these alternative formulations is the decoupled power flow study. This is considered in a section of its own in Chapter 5. This chapter concludes with a brief presentation of several modifications of the Newton–Raphson power flow study algorithm.

A Simplified Formula for Jacobian Matrix Entries

Equations (4.38) through (4.45) give the Jacobian matrix entries for a power flow study. In each of the four quadrants of J, the diagonal entries involve a sum over $N - 1$ terms. Each summant is a product of three or four terms. The calculation of $J_\nu(i, i)$, $\nu = 1, 2, 3\ 4$, entails (in total) approximately $14N$ multiplications and $4N$ additions within each iteration. Examination of the ΔP_i and ΔQ_i mismatch formulas indicate a remarkable similarity to the $J_\nu(i, i)$ expressions, and for large N, there is motivation to avoid the sums in $J_\nu(i, i)$. The mismatch formulas are reproduced here for convenience

$$\Delta P_i = -\sum_{j=1}^{N} |Y_{ij}| v_j v_i \cos(-\theta_{ij} - \delta_j + \delta_i) + P_i \tag{4.62}$$

$$\Delta Q_i = -\sum_{j=1}^{N} |Y_{ij}| v_j v_i \sin(-\theta_{ij} - \delta_j + \delta_i) + Q_i \ . \tag{4.63}$$

Rewrite (4.62) and (4.63) as

$$\Delta P_i = P_i - \sum_{\substack{j=1 \\ \neq i}}^{N} |Y_{ij}| v_j v_i \cos(-\theta_{ij} - \delta_j + \delta_i) - |Y_{ii}| v_i^2 \cos(-\theta_{ii}) \tag{4.64}$$

$$\Delta Q_i = Q_i - \sum_{\substack{j=1 \\ \neq i}}^{N} |Y_{ij}| v_j v_i \sin(-\theta_{ij} - \delta_j + \delta_i) - |Y_{ii}| v_i^2 \sin(-\theta_{ii}) \tag{4.65}$$

Substitute the sum term of (4.64) and (4.65) into the expression for $J_\nu(i, i)$ allowing ΔP_i and ΔQ_i to be negligible

$$J_1(i, i) = Q_i - |Y_{ii}| v_i^2 \sin(-\theta_{ii}) \tag{4.66}$$

$$J_2(i, i) = -\frac{P_i}{v_i} - v_i |Y_{ii}| \cos(-\theta_{ii}) \tag{4.67}$$

$$J_3(i, i) = -P_i + |Y_{ii}| v_i^2 \cos(-\theta_{ii}) \tag{4.68}$$

$$J_4(i, i) = \frac{Q_i}{v_i} - v_i |Y_{ii}| \sin(-\theta_{ii}) \ . \tag{4.69}$$

Thus the long sums have been avoided. Note that the approximation

$$\Delta P_i \simeq 0 \qquad \Delta Q_i \simeq 0$$

has been made. Near the solution, this is an excellent approximation.

Replacement of Derivatives with Respect to δ by $1/v$ Times the Derivative

Inspection of the formulas for J_1 and J_3 reveals that there is a convenient factoring of v_i from the sum term. For example, in J_1,

$$J_1(i, i) = v_i \sum_{\substack{j=1 \\ \neq i}}^{N} v_j |Y_{ij}| \sin(\delta_i - \delta_j - \theta_{ij})$$

$$J_1(i, k) = v_i \left[-v_k |Y_{ik}| \sin(\delta_i - \delta_k - \theta_{ik})\right] \ .$$

This suggests calculation and storage of $J_1(i,i)/v_i$ and $J_1(i, k)/v_k$ rather than $J_1(i, i)$ and $J_1(i, k)$. Similarly, in J_3, modified entries are found. The formulas are

$$J_1'(i, i) = \frac{1}{v_i} \frac{\partial P_i}{\partial \delta_i} = \sum_{\substack{j=1 \\ \neq i}}^{N} v_j |Y_{ij}| \sin(\delta_i - \delta_j - \theta_{ij})$$

$$J_1'(i, k) = \frac{1}{v_k}\frac{\partial P_i}{\partial \delta_k} = -v_i |Y_{ij}| \sin(\delta_i - \delta_k - \theta_{ik})$$

$$J_3'(i, i) = \frac{1}{v_i}\frac{\partial Q_i}{\partial \delta_i} = -\sum_{\substack{j=1 \\ \neq i}}^{N} v_j |Y_{ij}| \cos(\delta_i - \delta_j - \theta_{ij})$$

$$J_3'(i, k) = \frac{1}{v_k}\frac{\partial Q_i}{\partial \delta_k} = v_i |Y_{ij}| \cos(\delta_i - \delta_k - \theta_{ik}).$$

These modified Jacobian entries are in common use in commercially available power flow studies. Note that if the procedure to avoid the sums in $J_\nu(i, i)$ (described in this section) is used, (4.66) and (4.68) are conveniently modified to obtain

$$J_1'(i, i) \simeq \frac{Q_i}{v_i} - |Y_{ii}| v_i \sin(-\theta_{ii}) \tag{4.70}$$

$$J_3'(i, i) \simeq \frac{P_i}{v_i} + |Y_{ii}| v_i \cos(-\theta_{ii}) . \tag{4.71}$$

Equations (4.70) and (4.71) bear considerable similarity to (4.67) and (4.69).

Less Frequent Updates of J

As presented, the Newton–Raphson algorithm requires recalculation of J at each iteration. This also implies recalculation of the table of factors of J for each step. As $[\delta \ V]^t$ converges to a solution, J also converges. It has been suggested to refrain from updating J at each iteration, thereby saving considerable time in calculation of J entries and the TOF. This procedure gives particularly good results when $[\delta \ V]^t$ is near the solution vector. A suggested practical implementation is to refrain from updating J when $||(\Delta P \ \ \Delta Q)^t||$ drops below a given threshhold [e.g., $0.2(N - 1)$]; this suggestion is explored in [12].

It may be appropriate to end this section with a remark which the entrepreneur-reader may find intriguing. Like most commercial products, load flow software often contains proprietary procedures which accelerate or otherwise improve convergence. Only by closely examining the Jacobian matrix, updating procedures, and alternative formulations will one discover methods that will potentially improve time and memory requirements. A large network analysis package will have from 2500 to 10,000 executable statements devoted to the power flow study algorithm. This key component of the software often contains features which the developer uses to "sell" the entire package.

Bibliography

[1] J. Ward and H. Hale, "Digital Computer Solution of Power Flow Problems," *Trans. AIEE,* v. 75, pt. III, 1956, pp. 398–404.

[2] J. Henderson, "Automatic Digital Computer Solution of Load Flow Studies," *Trans. AIEE,* v. 73, pt. III-B, 1954, pp. 1696-1702.

[3] H. Brown, G. Carter, H. Happ, and C. Person, "Power Flow Solution by Impedance Matrix Iterative Method," *IEEE Trans. Power Apparatus and Systems,* v. PAS-82, 1963, pp. 1-10.

[4] S. Kuo, *Numerical Methods and Computers,* Addison–Wesley, Reading, Mass., 1965.

[5] J. Meisel, "Application of Fixed Point Techniques to Load Flow Studies," *IEEE Trans. Power Apparatus and Systems,* v. PAS-89, January, 1970, pp. 136–140.

[6] J. Van Ness, and J. Griffin, "Elimination Methods for Load Flow Studies," *Trans. AIEE,* v. 80, pt. III, 1961, pp. 299-304.

[7] A. M. Ostrowski, *Solution of Equations and Systems of Equations,* Academic Press, New York, 1966.

[8] F. Wu, "Theoretical Study of the Convergence of the Fast Decoupled Load Flow," *IEEE Trans. Power Apparatus and Systems,* v. PAS-96, no. 1, January–February 1977, pp. 268–275.

[9] L. Kraft, "Harmonic Resonance in Power Systems," Ph.D. thesis, Purdue University, West Lafayette, Ind., August 1984.

[10] L. Kantorovich, "Functional Analysis and Applied Mathematics," *Uspekhi Matematicheskikh - Nauk,* v. 3, no. 6, 1948, pp. 89–185.

[11] H. Brown, G. Carter, H. Happ, and C. Person, "Power Flow Solution by the Impedance Matrix Methods," *Trans. AIEE,* v. 82, pt. III, v. 82, 1963, p. 1.

[12] Y. Wallach, R. Even, and Y. Yavin, "Improved Methods for Load Flow Calculations," *IEEE Trans. Power Apparatus and Systems,* v. PAS-90, no. 1, January–February 1971, pp. 116–123.

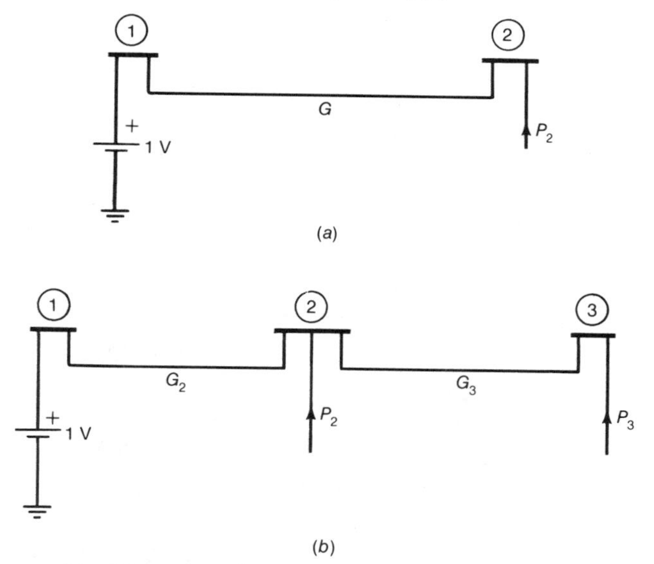

P4.1 Test systems used to illustrate the multiplicity of power flow study solutions.

Exercises

Exercises that do not require programming

4.1 This problem concerns the multiplicity of solutions for the power flow problem. This is examined for two very simple example networks.

a. Consider the two bus dc system in Figure P4.1a. The swing bus is bus 1, where 1.0 Volt per unit occurs. Write an expression for the mismatch power at bus 2 and set this expression equal to zero. Introduce a variable $P' = P/G$ (note that G is the line conductance) and solve for V_2 in terms of P'. Note that the mismatch expression is quadratic in V_2 and there will be two solutions for V_2. Define V_{2OC}, the open circuit solution, as the larger of the two, and V_{2SC}, the short circuit solution as the smaller.

b. Plot V_{2OC} and V_{2SC} versus P'. Discard imaginary solutions for these voltages, as they indicate that P violated the maximum power transfer theorem at bus 2. Note that if negative voltages are rejected because of usual operating conditions, the only region of interest for open circuit solutions is $P' \geq -1/4$. Under the same assumption, the only region of interest for short circuit solution is $-1/4 \leq P' \leq 0$.

c. A mathematical formulation of any problem in which there are multiple solutions is of concern since it is important to distinguish the most reasonable solution. If the solutions are near each other, this distinction is a problem. If $G = 100$ and the region of interest in P is $-2 < P < 0$, are the values of V_{2OC} and V_{2SC} close to each other? Which of the two is the more reasonable solution?

d. Figure 4.1b shows a two bus extension of this study. Write two simultaneous algebraic expressions for the mismatch powers at buses 2 and 3.

e. Eliminate V_2 from these expressions. The result is a quartic expression in V_3. Define

$$P_2' = \frac{P_2}{G_2} \qquad P_3' = \frac{P_3}{G_3}$$

and simplify the quartic expression.

f. For the very special case $G_2 = G_3$ and $P_2 = P_3$, simplify the quartic expression in V_3. Solve the equation for a few points in the range $-0.08 < P < 0.2$. Identify open and short circuit solutions and plot them versus P. Discard imaginary solutions.

g. For the region $-0.02 < P < 0$, are the values of V_{3OC} and V_{3SC} close enough to each other to cause problems of distinction of the most reasonable solution?

4.2 This problem concerns Brown's Z_{bus} alternative power flow formulation, which is intended to result in accelerated solutions. The basis of the approach is that heavily loaded buses frequently cause slowed solutions, but much of the bus load may be modeled as an impedance. Fixed impedances are easily handled as part of the network equations.

Figure P4.2a shows a load model at bus i in which a parallel ground tie and current source are used. The net current injection by this load model is $+I_i$, shown in the figure. Let the impedance of the ground tie be such that the specified load complex power is consumed when $|v_i| = 1.0$; thus

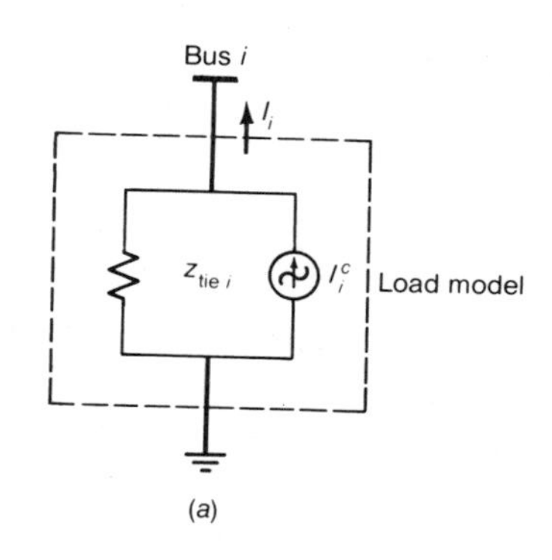

(a)

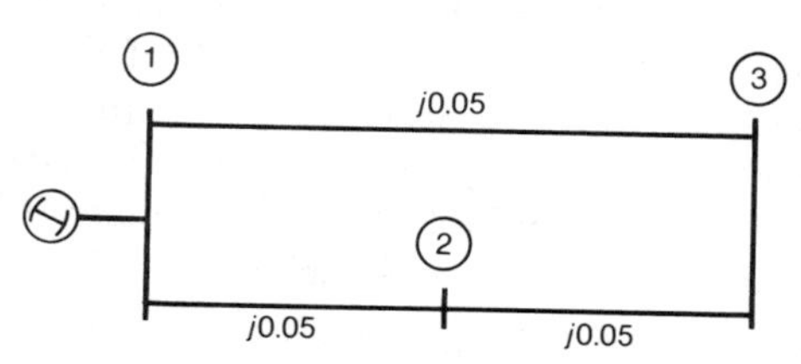

(b)

FIGURE P4.2 Brown's $\mathbf{Z_{bus}}$ load flow study.

$$z_{tie\ i} = \frac{|v_i|^2}{-S_i^*} = \frac{1}{-S_i^*} .$$

The parallel current source is a correction current to allow for the fact that the solution $|v_i|$ differs from unity.

a. Show that the correction current is given by

$$I_i^c = \frac{S_i^*}{v_i^*} + \frac{v_i}{z_{tie\ i}} .$$

(Note: S_i is the specified injected power at bus i.)

b. What is the correction current when $|v_i| = 1.00$?

c. Let the three bus system in Figure P4.2b be studied using this modified formulation. Busses 2 and 3 are PQ load buses,

$$S_2 = -1 + j0 \qquad S_3 = -1 - j1,$$

and the per unit line impedances are each $j0.05$. The swing bus voltage is $1+j0$ per unit. Calculate the required z_{tie} impedances and form Y_{bus} using these .impedances in the network.

d. Write the equations to be solved to study the system in part (c). The Gauss–Seidel method is to be used.

4.3 Consider the simultaneous set of nonlinear algebraic equations

$$X^2 - 2X - 1 - Y = 0$$

$$X + 3 - Y = 0.$$

The solutions to these equations are $(X, Y) = (4, 7)$ and (-1, 2). The application of the Newton–Raphson method will be illustrated for this example.

a. Find the Jacobian matrix for these equations.

b. Solve the equations using the Newton–Raphson method and initial vector $(-1.5 \;\; -1.5)^t$.

c. Solve the equations using the initial vector $(5 \;\; 5)^t$.

d. From parts (b) and (c), it is clear that the solution determined by the iterative procedure depends on the initial vector. Also, whether or not the procedure converges depends on the initial vector. Let R_a denote the set of points (initial vectors) for which convergence to (-1, 2) occurs, R_b be the points for which convergence to (4, 7) occurs, and R_c be the region for which no convergence occurs. Plot Y versus X for the two equations. On the same graph, experimentally find and plot the regions R_a, R_b, R_c. (Note: A programmable calculator can be used to find the convergence of a particular given factor. A few trials will yield some obvious conclusions.)

e. Consider the initialization (0, 0). At the end of one step in the iterative process one finds

$$\begin{pmatrix} X \\ Y \end{pmatrix}^{(1)} = \begin{pmatrix} -4/3 \\ 5/3 \end{pmatrix}$$

and the error is E,

$$E = \begin{pmatrix} X \\ Y \end{pmatrix} - \begin{pmatrix} -1 \\ 2 \end{pmatrix}$$

$$E^{(1)} = \begin{pmatrix} 1/3 \\ -1/3 \end{pmatrix}$$

and $||E|| \equiv \sqrt{E^t E}$ is $\sqrt{5}$. Carry through four additional iterations to find $E^{(2)}, \ldots, E^{(5)}$. Plot $||E^{(0)}||$ through $||E^{(5)}||$ on a logarithmic scale versus iteration number on a linear scale. Fit a straight line through the points and estimate the slope of this line. If perfect quadratic convergence were present, one would expect a slope of -2 as the iterates come close to the solution.

4.4 Consider an N bus dc power system in which each bus is connected to each other bus by a line of conductance g. Also, let bus 1 be the swing bus and buses 2 through N be load buses at which the specified injected power is the same value, P. The swing bus voltage is $v_1 = 1$.

a. Show that

$$v_i = \frac{1 \pm \sqrt{1 + 4P/g}}{2} \qquad i = 2, 3, \ldots, N .$$

b. Show that for the open circuit solution (i.e., upper sign used for v_i)

$$I_i = \frac{g}{2}\left(\sqrt{1 + \frac{4P}{g}} - 1\right) \qquad i = 2, 3, \ldots, N .$$

(Note: To obtain these results, consider a Y_{bus} formulation of the dc power flow problem.)

c. For the case that P/g is small and negative (i.e., lightly loaded system), find a simplified approximate expression for v_i and I_i which is free of square roots. Consider only the "open circuit solution."

4.5 Starting with the mismatch expressions for ΔP at bus i, demonstrate that the J_2 formulas (i.e., $\partial \Delta P / \partial |V|$ expressions) are correct as given in the text.

4.6 The usual formulation for the Newton–Raphson power flow study is in terms of polar bus voltages. That is, the unknown vector is $[\delta \mid |v|]^t$. Consider now a rectangular formulation in which each bus voltage is broken into a real and an imaginary part,

$$v_i = v_{Ri} + jv_{Ii} .$$

Then the mismatch expressions for ΔP and ΔQ must be rewritten in terms of $[V_R \quad V_I]^t$, where V_R and V_I are the real and imaginary bus voltage components.

a. Find the expressions for ΔP and ΔQ in terms of $[V_R \quad V_I]^t$.

b. The rectangular formulation Jacobian matrix is

$$J = \begin{bmatrix} J_\alpha & J_\beta \\ J_\gamma & J_\delta \end{bmatrix}$$

where

$$J_\alpha = \frac{\partial \Delta P}{\partial V_R} \qquad J_\beta = \frac{\partial \Delta P}{\partial V_I}$$

$$J_\gamma = \frac{\partial \Delta Q}{\partial V_R} \quad J_\delta = \frac{\partial \Delta Q}{\partial V_I} .$$

Illustrate the calculation of these Jacobian entries by finding formulas for J_α and J_γ.

c. In the polar formulation, J_2 and J_3 are submatrices that have entries which are small in comparison with entries of J_1 and J_4. Can the same be said of J_β and J_γ in the rectangular formulation.

4.7 Consider the definition of rate of convergence, R_c, of any iterative process to a tolerance ϵ as

$$R_c = \frac{ln\ (||F(X^{(0)})||_\infty / \epsilon)}{n_\epsilon},$$

where n_ϵ is the required number of iterations. Also note that $\rho(\Xi)$ denotes the spectral radius of matrix Ξ,

$$\rho(\Xi) = \max_i (|\lambda_i|),$$

where λ_i are the eigenvalues of Ξ. In this exercise, the property

$$R_c = -ln\ (\rho(\Xi))$$

will be examined for a Newton–Raphson process in which Ξ is the inverse Jacobian matrix.

a. For the simultaneous set of equations

$$F(X) = AX - b = 0$$

using

$$A = \begin{pmatrix} 1.5 & 0.1 \\ 0.2 & 1.6 \end{pmatrix},$$

estimate the rate of convergence and number of iterations to obtain a solution to within 10^{-3} in each row of F. The Newton–Raphson method is to be used.

b. Repeat part (a), but introduce an acceleration factor into the iterative process. If the acceleration factor $\alpha = 0.7$ is used, find the required number of iterations to obtain a solution. Comment on your result in the general case of A matrices which are not constant. Is there an optimal value of α?

c. Consider now an A matrix which is nearly decoupled,

$$A = \begin{bmatrix} A_1 & A_2 \\ A_3 & A_4 \end{bmatrix}$$

$$A_2 \simeq 0 \qquad A_3 \simeq 0 .$$

Thus the upper rows of F are closely coupled to the upper rows of X, and the lower rows exhibit similar coupling. Obtain an estimate of the convergence rate for the solution of

$$F(X) = AX - b = 0$$

using Newton's method. The spectral radii of A_1 and A_4, are ρ_1 and ρ_4 respectively.

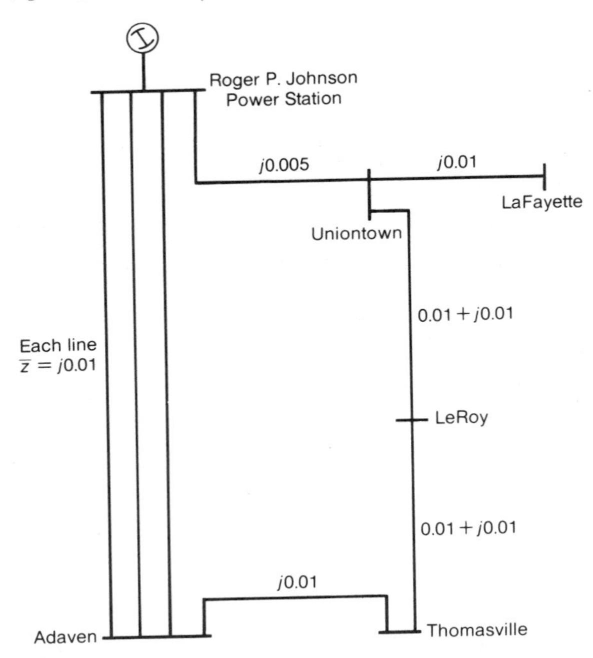

FIGURE P4.8 **Six bus power system.**

Exercises that require computer programming

4.8 Consider the power system shown in Figure P4.8. Per unit reactances are shown in the figure on a 100 MVA, 138 kV base. The load schedule is as follows:

Bus	Load	
Uniontown	75 MW	12 MVAr
LaFayette	10 MW	1 MVAr
LeRoy	5 MW	-1 MVAr
Thomasville	10 MW	-10 MVAr
Adaven	100 MW	80 MVAr

and the R. P. Johnson power station bus has rated voltage (138 kV) and is taken as the swing bus.

a. Solve the system in a power flow study using the Gauss–Seidel method. State the mismatch criterion used. Your output should include the bus voltages and line flows. Identify any buses with voltage magnitudes "out of range." For the latter, consider that a voltage magnitude outside ±7% as "out of range."

b. It is desired to outage one or more circuits, R. P. Johnson–Adaven, for maintenance. Repeat part (a) and conclude whether it is possible to outage these circuits yet retain the proper bus voltages.

c. In part (b), the heavy load at Adaven was "rerouted" by the outage. Examine the load flow study results and explain the line loading from the swing bus to Unionville as the number of outaged lines increases.

4.9 Write a Newton–Raphson power flow study program and verify the Y_{bus}, θ, and J matrices in Example 4.4. Solve the system to 0.01 maximum mismatch, and compare the bus voltages to the stated

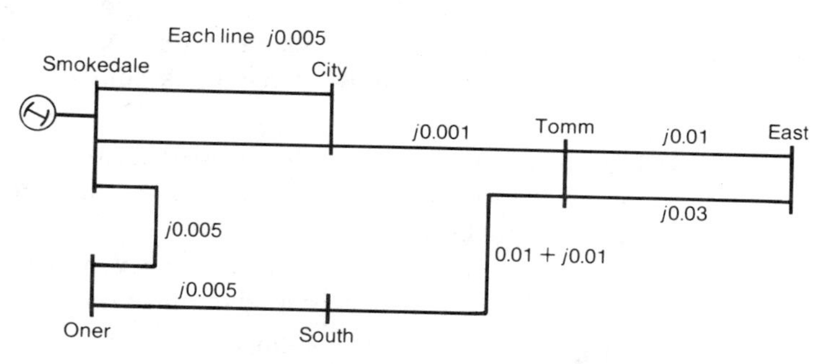

FIGURE P4.10 Smokesale power system.

"exact solution."

4.10 Consider the power system shown in Figure P4.10. The Smokedale bus is voltage regulated at 140 kV in this 138 kV system. Line impedances are shown in per unit on a 100 MVA base. The bus loads are as follows:

Bus	P_{load}	Q_{load}
City	100 MW	21 MVAr
Tomm	12 MW	0 MVAr
East	25 MW	-20 MVAr
Oner	16 MW	12 MVAr
South	1 MW	1 MVAr

a. Solve the system as stated using the Newton–Raphson method. Your output should include bus voltages and line flows. Use a 1 MVA maximum mismatch tolerance.

b. Vary the reactive power demand at the South bus and plot the line–to–line bus voltage versus the reactive demand. Explain this curve.

4.11 In this problem, the system and load schedule of Exercise 4.10 will be restudied, but the South bus will no longer be a *PQ* load bus. Consider, instead, a generator sited at South. Let the generator be fixed at 20 MW and let the voltage regulator be adjusted to give fixed 138 kV voltage of the bus.

a. Solve the resulting network using South as a *PV* bus.

b. Considering the bus voltage response to reactive power injection at the South bus, conjecture as to the result of raising the voltage regulator set point at this bus. In other words, if $|V|$ is increased at South, what will happen to the reactive power required of the generator? Verify your conjecture with test cases.

4.12 In this problem, the system described in Exercise 4.10 will be solved again using the Newton–Raphson method, but the system will be modified to accommodate a new synchronous condenser at the Farad bus, which is located on a radial line from the City bus. The line has 1% reactance. A synchronous condenser is a synchronous motor in which the active power is very low due to no shaft load and the field is generally overexcited to produce high internal $|E_f|$. The result is a machine which is a source of reactive power. Usually, the field excitation is varied by a control system to hold the bus voltage constant.

a. Using the excitation controller described and considering zero input power to the machine, describe the mathematical model used for a synchronous condenser bus.

b. Using a bus voltage set point of 140 kV at Farad, solve the system with the load schedule given in Exercise 4.10.

c. The Farad synchronous condenser has a rating of 30 MVA (in the overexcited mode of operation). How does this limit affect your solution?

d. The Jacobian matrix entry in the $\partial Q/\partial|V|$ submatrix corresponding to the Farad bus (diagonal position) may be used to estimate the required reactive power injection to obtain an additional 1 kV (line–to–line) at the Farad bus (i.e., Farad will be operated at 141 kV rather than 140 kV). Estimate the change in injected reactive power at Farad and comment on the 30 MVA machine rating.

e. A typical 30 MVA synchronous condenser might require as much as 1 MW to rotate the machine and overcome losses. One way to model this active power demand is to use a fictitious ground tie at the synchronous condenser bus. With a 140 kV bus voltage set point, what per unit ground tie is required to model the 1 MW loss?

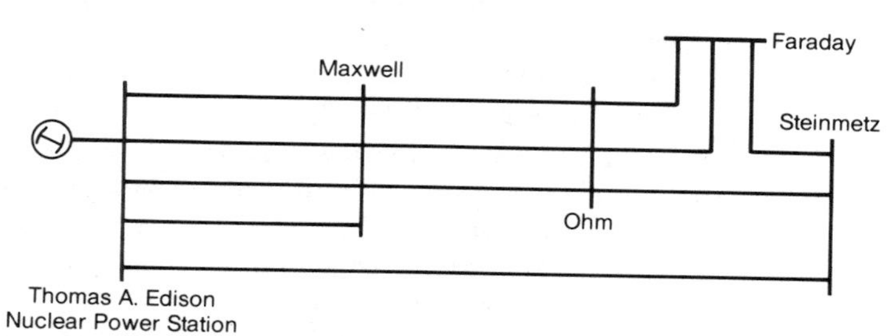

Bus loads (MVA):

Bus	Load *P*	Load *Q*
Maxwell	500 MW	300 Mvar
Ohm	220	70
Steinmetz	100	5
Faraday	Interchange with Induction Power and Light (IPL) Co.	

FIGURE P4.13 138 kV power system and loads.

4.13 Figure P4.13 shows a 138 kV power system and loads. The per unit line impedances on a 100 MVA base are:

Edison–Maxwell	$\bar{z} = j0.005$	$\frac{j\bar{B}}{2} = j0.05$	each circuit
Maxwell–Ohm	$\bar{z} = j0.010$	$\frac{j\bar{B}}{2} = j0.06$	each circuit
Ohm–Faraday	$\bar{z} = j0.0005 + j0.010$	$\frac{j\bar{B}}{2} = j0.001$	each circuit
Edison–Steinmetz	$\bar{z} = j0.09$	$\frac{j\bar{B}}{2} = j0.1$	
Faraday–Steinmetz	$\bar{z} = 0.005 + j0.010$	$\frac{j\bar{B}}{2} = j0.001$	

where $\frac{j\bar{B}}{2}$ is the line charging susceptance lumped equivalent. This susceptance should be considered as a ground tie on each bus terminal. The Edison *NPS* is the swing bus and is voltage regulated at 139.5 kV. The Faraday bus is an intertie bus with the Induction Power and Light Co. This task relates to the study of interchange power at Faraday. Company policies are to hold all bus voltages at 138 ± 4 kV. All system lines are rated 333 MVA except for each circuit of Faraday–Ohm, which is 370 MVA.

a. The Faraday intertie is nominally at 280 MVA at 90% power factor lagging. Run a base case Newton–Raphson power flow study and determine system compliance with company policies.

b. The load schedule shown is the present summer peak value.

If the load at the Maxwell bus is increasing, find the maximum load that may be tolerated at Maxwell considering the voltage and line loading limitations. Also, note that the Edison nuclear power station is a 900 MVA station. Is the system *transmission* or *generation limited?*

c. Under some circumstances, the concept of *interchange capacity* is important. A loose definition of this term is the maximum MVA capability of a power system to accept interchange complex power. Obviously, the interchange capacity of a system depends on the power factor of the interchange power, the load/generation schedule, and other system parameters. The concept is most often used in very large systems; in such systems, the interchange capacity tends to be weakly dependent on electrical system parameters. With the Maxwell bus loaded as given in the present peak load schedule, find the maximum interchange MVA which can be accepted at Faraday. Consider fixed 90% lagging power factor. Use the bus voltage/line load limits given and also use the 900 MVA Edison limit. Is the interchange capacity *transmission* or *generation limited?*

CHAPTER 5

Approximate, Fast, and Special Purpose Power Flow Studies

5.1 INTRODUCTION

The power flow study is used to analyze a variety of system planning and operation questions. Very large studies have been run to study most of North America at transmission voltages; such studies encompass over 10,000 buses and 12,000 lines. Although the number of iterations from a flat start in a Newton–Raphson power flow study is insensitive to the number of system buses, the time and memory requirements of each iteration are quite dependent on the number of system buses. The solution for the unknown bus voltages and angles requires a triangular factorization and forward-backward substitution. Even using sparsity programming, this process requires central processor time that varies as the cube of the number of buses. Also, the memory requirement varies as the square of the number of buses. Although it is true that the greater sparsity for very large systems greatly reduces the time and memory requirements, there is nonetheless a significant motivation to seek approximations that will accelerate the overall computer time requirements and further reduce memory requirement. The two best known and most widely used fast power flow methods rely on somewhat different phenomena: The decoupled power flow exploits the close dependence of active power flow and bus voltage phase angle and reactive power flow and bus voltage magnitude. The methods of distribution factors relate to the approximately linear behavior of line flows with bus injections. These topics are considered in some detail in this chapter.

Also considered in this chapter are a selection of special purpose power flow studies. These are specialized techniques that employ considerable modeling detail in one aspect of the power flow study. Although such special purpose power flow studies have not generally enjoyed widespread application, in some instances their application may yield valuable information.

5.2

THE DECOUPLED POWER FLOW STUDY

The close relationship between active power flow and bus voltage phase angle is evident in systems with modest loads and generally low R/X ratios in the transmission system. Similarly, the reactive power flows are highly dependent on bus voltages. These observations are reinforced by examining the $\partial P/\partial \delta$ and $\partial Q/\partial V$ sections of the power flow Jacobian matrix; these submatrices are dominant in J. This dominance suggests the approximation in which the $\partial P/\partial V$ and $\partial Q/\partial \delta$ submatrices are assumed small enough to be ignored

$$\begin{bmatrix} \Delta \delta \\ \Delta V \end{bmatrix} \simeq - \begin{bmatrix} J_1 & 0 \\ 0 & J_4 \end{bmatrix} \begin{bmatrix} \Delta P \\ \Delta Q \end{bmatrix} . \tag{5.1}$$

Thus the $\Delta\delta$–ΔP equations may be decoupled from the ΔV–ΔQ equations,

$$\Delta \delta \simeq - J_1 \Delta P \tag{5.2}$$

$$\Delta V \simeq - J_4 \Delta Q . \tag{5.3}$$

Although the equations are still coupled in the sense that J_4 depends on δ and J_1 depends on V, it is possible to accelerate the calculation of $[\Delta\delta \ \Delta V]^t$ since the dimension of J_1 and J_4 are of the order of half of the dimension of the entire Jacobian matrix. This decoupling process results in the decoupled power flow study formulation. It is also known by the names of Stott and Alsac, who were among the first to formalize the formulation [1–3]. In the absence of PV and voltage regulated buses, the dimensions of J_1 and J_4 will be $\frac{1}{2}N_J$, where N_J is the dimension of the full Jacobian matrix. Proceed to examine the expected solution times in the full and decoupled cases

$$\textit{Full Jacobian Matrix} \quad O(t) = N_J^3 \tag{5.4}$$

$$\textit{Decoupled Jacobian Matrix} \quad O(t) = 2(\frac{1}{2}N_J)^3 \tag{5.5}$$

$$= \frac{1}{4}N_J^3 .$$

Thus one would expect a considerable savings in time requirement per iteration in the decoupled case. This simplified approach to the analysis of the order of the time required ($O(t)$) does not take into account the effect of the approximation $\partial Q/\partial \delta \simeq \partial P/\partial V \simeq 0$. This approximation will introduce an error into the $[\Delta\delta \ \Delta V]^t$ corrections, but the Newton–Raphson algorithm is so robust that this error rarely causes more than a single added iteration required to reach the required mismatch tolerance. It is important to note that the decoupling assumptions do *not* result in added error in the solution voltages and angles. The mismatch tolerance controls the solution error; for a given mismatch tolerance, the decoupled load flow solution is as accurate as a full solution.

The presence of TCUL buses are considered in the decoupled power flow study as

$$\frac{\partial P}{\partial t} \simeq 0$$

with the $\partial Q/\partial t$ terms retained. Phase shifters are considered as

$$\frac{\partial Q}{\partial \phi} \simeq 0$$

with the $\partial P/\partial \phi$ terms retained. Busses of the PV type do not require special consideration except to note that the J_1 and J_4 submatrices may be of different dimension.

Convergence of the Decoupled Power Flow Study

The convergence properties of the decoupled power flow study are a separate issue from the accuracy of the method. As stated above, the accuracy of the method is not inferior compared to the full equations, provided that the same mismatch criteria (and formulas) are applied in each case. The convergence, however, may be affected because certain properties of the $\partial P/\partial \delta$ and $\partial Q/\partial V$ matrices may not be the same as those of the full Jacobian matrix [4]. When the Jacobian matrix and the $\partial P/\partial \delta$ and $\partial Q/\partial V$ submatrices are nearly constant, the approximate rate of convergence of the coupled equations is

$$R_c = -ln\ (\rho(\Xi)),$$

where $\rho(\Xi)$ is the spectral radius of the Jacobian inverse. For the perfectly decoupled case, the spectral radius of Ξ is the larger of the spectral radii of $(\partial P/\partial \delta)^{-1}$ and $(\partial Q/\partial V)^{-1}$,

$$\rho(\Xi) = max\ [\rho((\partial P/\partial \delta)^{-1}),\ \rho((\partial Q/\partial V)^{-1})\ . \tag{5.6}$$

In cases of high line loading, high reactive power line dissipation, low bus voltages, and low load power factor, the $\partial P/\partial V$ and $\partial Q/\partial \delta$ submatrices may be significant. In such cases, the spectral radius of Ξ is not given by (5.6) because the eigenvalues of Ξ depend on potentially large entries in $\partial P/\partial V$ and $\partial Q/\partial \delta$. It is possible to degrade convergence by decoupling in these cases, although in the majority of practical cases, the advantages indicated in (5.4) and (5.5) outweigh any reduction in R_c.

This topic is discussed in detail - and numerical examples are given in [4–6].

Connection Between the Decoupled Power Flow and Properties of the Bus Susceptance Matrix

In the $\partial P/\partial \delta$ submatrix of the Jacobian [Eqs. (4.38 and 4.39)], one finds that for the case of a purely reactive transmission system,

$$\frac{\partial P_i}{\partial \delta_i} = \sum_{\substack{j=1 \\ \neq i}}^{N} v_i v_j |B_{ij}|\ sin\ (\delta_i - \delta_j - \theta_{ij})$$

$$\frac{\partial P_i}{\partial \delta_k} = -v_i v_k \left| B_{ik} \right| \sin (\delta_i - \delta_k - \theta_{ik}),$$

where the notation B is used instead of Y_{bus} to emphasize that Y_{bus} is completely imaginary. Note that $|B|$ is the bus susceptance matrix and θ is the angle of the corresponding Y_{bus} matrix (i.e., a matrix containing $\pm \pi/2$ entries). If two further approximations are made, namely unity bus voltage and that the cosine of a small angle is unity, then

$$\frac{\partial P_i}{\partial \delta_i} \simeq \sum_{\substack{j=1 \\ \neq i}}^{N} \left| B_{ij} \right|$$

$$\frac{\partial P_i}{\partial \delta_k} \simeq -\left| B_{ik} \right| .$$

In the absence of ground ties, these $\partial P / \partial \delta$ entries are simply $-B$. Therefore,

$$\Delta P \simeq B \, \Delta \delta .$$

This is the same observation made in (4.59). A similar conclusion is reached using the same assumptions in the $\partial Q / \partial V$ submatrix. The result is (4.61). Thus the first iteration of the decoupled power flow study is similar to the bus susceptance matrix $[\delta \;\; V]^t$ initialization procedure presented in Chapter 4.

5.3 DISTRIBUTION FACTORS

In many instances, power flow studies are performed to determine line loading. Also, power flow studies are frequently performed to determine the effects of line outages (contingencies). These two facts lead to the investigation of the sensitivities of line loading with respect to bus demands and line loading with respect to the power level in lines to be outaged. These sensitivities are termed ***distribution factors***.

The power transfer distribution factor (PTDF) relating the loading in the line from bus i to bus j with respect to injected bus power S_k is denoted $\rho_{ij,k}$,

$$\rho_{ij,k} = \frac{\partial \bar{S}_{ij}}{\partial S_k} .$$

The bar above $\bar{S}_{ij}$ denotes a line flow. For this application, the power is considered as metered at bus j. The notion of bus voltages and injection currents are readily introduced into the definition of $\rho_{ij,k}$,

$$\rho_{ij,k} = \frac{\partial ((v_i - v_j)/\bar{z}_{ij})^* v_j}{\partial (v_k i_k^*)} . \tag{5.7}$$

The notation $\bar{z}_{ij}$ refers to the impedance of line ij. For the case of near unity bus voltage,

$$\rho_{ij,k} = \left[\frac{\partial v_i}{\partial i_k}\ \frac{1}{\bar{z}_{ij}}\right]^* - \left[\frac{\partial v_j}{\partial i_k}\ \frac{1}{\bar{z}_{ij}}\right]^* .$$

Thus

$$\rho_{ij,k} = \left[\frac{(Z_{bus})_{ik} - (Z_{bus})_{jk}}{\bar{z}_{ij}}\right]^* . \tag{5.8}$$

The bus voltage subtraction in the numerator of (5.7) is the difference between two very similar quantities. For this reason, the Z_{bus} entries in (5.8) must be from the bus impedance matrix referenced to the swing bus. This is borne out by the fact that $\rho_{ij,k}$ must be zero when k is the swing bus; this will be the case if $Z_{bus}^{(swing)}$ is used in (5.8) since

$$(Z_{bus}^{(swing)})_{i\ swing} \equiv 0$$

for all i. Note that

$$\rho_{ij,k} = -\rho_{ji,k} . \tag{5.9}$$

This is readily verified by examination of (5.8).

References [7–9] further document power transfer distribution factors and their applications.

The Use of Power Transfer Distribution Factors in a Power Flow Study

Since $\rho_{ik,k}$ is the sensitivity of line flow in ij with respect to injection in k, one may relate line flows to bus injections under two different loading schedules. Denote these loading schedules as A and B. Then

$$\Delta\bar{S}_{ij} = \bar{S}_{ij}^A - \bar{S}_{ij}^B \simeq \rho_{ij,k}(S_k^A - S_k^B) . \tag{5.10}$$

Thus a quick, approximate method to calculate the line flow under schedule B is available

$$\bar{S}_{ij}^B \simeq \bar{S}_{ij}^A + \rho_{ij,k}(S_k^B - S_k^A) . \tag{5.11}$$

Equation (5.11) does require a power flow study to obtain $\bar{S}_{ij}^A$. The value of $\bar{S}_{ij}^B$ may be viewed as the line flow in a "changed case" and the distribution factor term in (5.11) is the amount of the change above the "base case". Figure 5.1 is a pictorial interpretation.

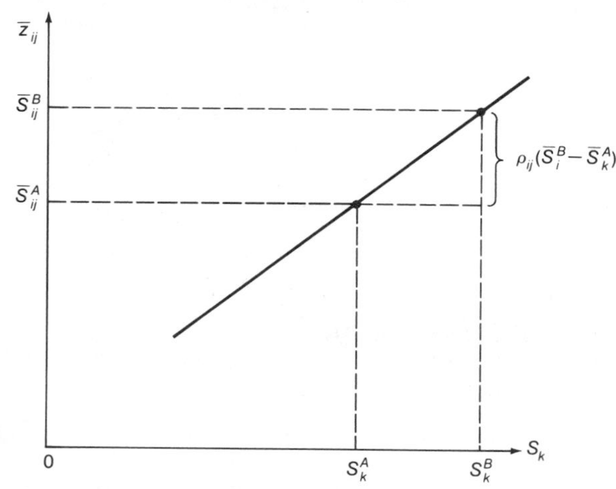

FIGURE 5.1 Pictorial of linear power flow study results.

Usually, power flow studies are used to determine many line flows. Also, if a "base case" (schedule A case) is available, often the changed case involves changes in many bus loads. Therefore, it is appropriate to reexamine (5.11) and extend this expression to include multiple lines and multiple buses. Since the formulation in (5.11) is linear, superposition is applicable. Therefore, multiple bus injections are readily handled

$$\bar{S}_{ij}^B = \bar{S}_{ij}^A + \sum_k \rho_{ij,k}(S_k^B - S_k^A),$$

where $\sum_k$ denotes summing over all buses which have changed bus injection. Several (or all) system lines are readily handled using vector notation

$$\bar{S}^B = \bar{S}^A + \begin{bmatrix} \sum_k \rho_{1,k}(S_k^B - S_k^A) \\ \sum_k \rho_{2,k}(S_k^B - S_k^A) \\ \vdots \\ \sum_k \rho_{NL,k}(S_k^B - S_k^A) \end{bmatrix}$$

$$= \bar{S}^A + \rho(S_{bus}^B - S_{bus}^A), \tag{5.12}$$

where $\bar{S}$ is a vector of NL line flows and the distribution factors $\rho_{1,k}$, $\rho_{2,k}$, ... represent the distribution factors of lines 1, 2, ... These distribution factors arranged in a rectangular array to form ρ,

$$\rho = \begin{bmatrix} \rho_{1,1} & \rho_{1,2} & \cdots & \rho_{1,NB} \\ \cdots & & \cdots & \\ \rho_{NL,1} & \rho_{NL,2} & \cdots & \rho_{NL,NB} \end{bmatrix},$$

where NB is the number of buses and this is the dimension of vectors S_{bus}^A and S_{bus}^B in (5.12).

Example 5.1

In this example, the simplicity of the distribution factor method is illustrated in a novel example. Figure 5.2 shows a one line diagram of a six bus, three phase power system. The bus demands for this sample system are

bus	load (pu)
2	$2.325 - j1.075$
3	$4.725 - j0.400$
4	$7.075 - j1.325$
5	$3.820 + j1.560$
6	$0.918 - j0.982$

Of course, for balanced operation, these loads appear in each phase at the buses indicated. Bus 1 is the swing bus. The base case power flow solution is indicated in Figure 5.2.

Find the power flow solution for the case of load phase B at bus 5 open.

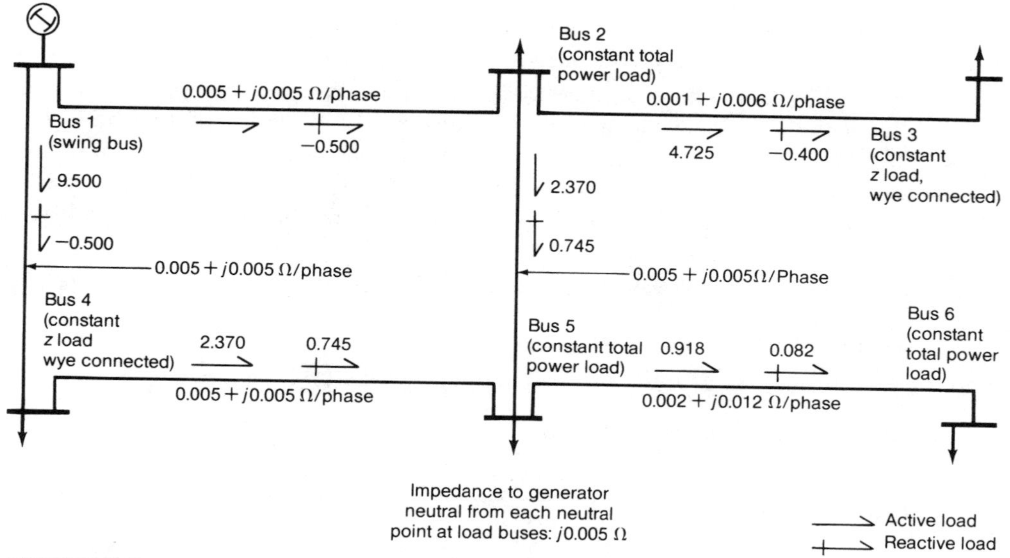

FIGURE 5.2 Six-bus example: single-line diagram.

Solution

In this example, there are six three-phase buses. There are actually 18 circuit nodes. It is assumed that the swing bus is a fixed voltage bus in each phase, and therefore there are 15 nonzero axes in the three phase Z_{bus} referenced to the swing bus. The only ΔS_{bus} that is considered is

$$\Delta S_{bus\ 5B} = +3.82 + j1.56 \quad \text{per unit}$$

(injection notation). The power flow change in line i is $\rho_{i,5B} \Delta S_{bus\ 5B}$. The result is shown in Figure 5.3. In comparison with a three phase Newton–Raphson power flow study (with 0.01 mismatch tolerance), approximately 3% line flow discrepancy occurs – primarily near buses 2 and 4.

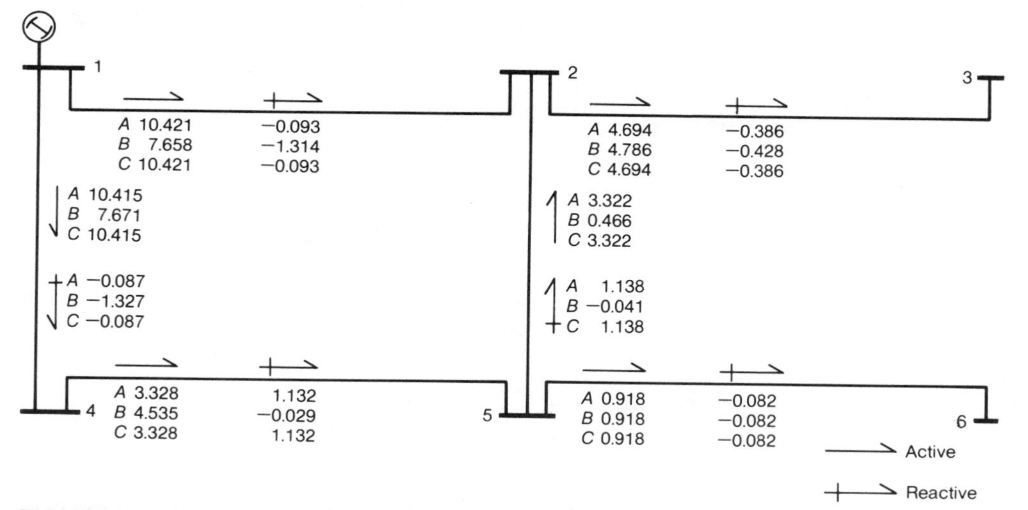

FIGURE 5.3 Three-phase line flows after opening phase *B* of load at bus 5.

Properties of Power Transfer Distribution Factors

Several interesting properties of distribution factors are directly attainable from the formula (5.8). It is interesting to examine how ρ varies with a line outage or line addition. In the case of line addition, $Z_{bus}^{(swing)}$ modifies as given by the Z_{bus} building algorithm rules; only the case of loop closure line additions is considered, since radial additions will not cause any old entries in Z_{bus} to change. For the case of a line addition from i to j,

$$Z_{bus}^{(swing)\,(new)} = Z_{bus}^{(swing)\,(old)} - \Delta_{ij}^{t}(z_{ii} - 2z_{ij} + z_{jj} + \bar{z}_{ij})^{-1}\Delta_{ij}, \qquad (5.13)$$

where

$$\Delta_{ij} = col_i(Z_{bus}^{(swing)\,(old)}) = col_j(Z_{bus}^{(swing)\,(old)}) .$$

Applying (5.13) to $\rho_{\ell m,n}$ yields

$$\rho_{\ell m,n}^{add\ ij} = \rho_{\ell m,n} - \frac{(z_{\ell i} - z_{\ell j})^*(z_{ni} - z_{nj})^* - (z_{mi} - z_{mj})^*(z_{ni} - z_{nj})^*}{\bar{z}_{\ell m}^*(z_{ii} - 2z_{ij} + j_{jj} + \bar{z}_{ij})^*}$$

$$\rho_{\ell m,n}^{add\ ij} = \rho_{\ell m,n} - \frac{(\rho_{\ell m,i} - \rho_{\ell n,j})(z_{ni} - z_{nj})^*}{(z_{ii} - 2z_{ij} + z_{jj} + \bar{z}_{ij})^*} \, . \qquad (5.14)$$

Table 5.1
Properties of the Power Transfer Distribution Factor

$$\rho_{\ell m,n} = \frac{\partial \bar{S}_{\ell m}}{\partial S_n} = \frac{z_{\ell n}^* - z_{mn}^*}{\bar{z}_{\ell m}^*}$$

$$\rho_{\ell m,n}^{add\ ij} = \rho_{\ell m,n} - \frac{(\rho_{\ell m,i} - \rho_{\ell m,j})(z_{ni} - z_{nj})^*}{(z_{ii} - 2z_{ij} + z_{jj} + \bar{z}_{ij})^*}$$

$$\rho_{\ell m,n}^{out\ ij} = \rho_{\ell m,n} - \frac{\bar{z}_{ij}^*(\rho_{ij,\ell} - \rho_{ij,m})\rho_{ij,n}}{\bar{z}_{\ell m}^*(\rho_{ij,i} - \rho_{ij,j} - 1)}$$

$$\rho_{k\ell,m} = - \rho_{\ell k,m}$$

$$\rho_{kk,m} = 0$$

$$\rho_{k\ell,m}^{out\ ij} = \rho_{k\ell,m}^{out\ ji} \qquad \rho_{k\ell,m}^{add\ ij} = \rho_{k\ell,m}^{add\ ji}$$

$$\rho_{k\ell,m}^{out\ k\ell} = 0$$

$$\rho_{\ell m,\ell} = \left[\frac{\rho_{jk,\ell}^{out\ \ell m} - \rho_{jk,\ell}}{\rho_{jk,m}^{out\ \ell m} - \rho_{jk,m}} \right] \cdot \rho_{\ell m,m}$$

The superscript "add ij" denotes addition of the line i to j with line impedance $\bar{z}_{ij}$:

A similar expression is obtainable for $\rho_{\ell m,n}$ after the outage of line ij,

$$\rho_{\ell m,n}^{out\ ij} = \rho_{\ell m,n} - \frac{\bar{z}_{ij}^*(\rho_{ij,\ell} - \rho_{ij,m})\rho_{ij,n}}{\bar{z}_{\ell m}^*(\rho_{ij,i} - \rho_{ij,j} - 1)} . \tag{5.15}$$

Equation (5.15) holds for $(ij) \neq (\ell m)$. When (ij) and (ℓm) are one and the same line, $\bar{z}_{\ell m}$ goes to infinity and $\rho_{\ell m,n}^{out\ ij}$ is zero.

Table 5.1 gives a few properties of power transfer distribution factors.

Line Outage Distribution Factors

It is possible to define a different type of distribution factor which relates the change in line loading not to bus injection but to the load in a line that is outaged. The line outage distribution factor, $\tau_{jk,\ell m}$, is defined as the ratio of the incremental load in line jk, $\Delta\bar{S}_{jk}$, to the load in line ℓm for an outage of line ℓm

$$\tau_{jk,\ell m} = \left. \frac{\Delta\bar{S}_{jk}}{\bar{S}_{\ell m}} \right|_{\ell m \text{ outaged}} . \tag{5.16}$$

To obtain τ directly from Z_{bus}, a technique known as the compensation theorem is used. With reference to Figure 5.4, S_m is the injection at bus m; this injected complex power is adjusted to reduce $\bar{S}_{\ell m}$ to zero. At this point, line ℓm is outaged resulting in a network with power transfer distribution factors $\rho^{out\ \ell m}$. The outage of ℓm does not change the line loading in jk since ℓm is unloaded at the instance of outage. The load in line jk, $\bar{S}_{jk}$, must then be corrected by removing the fictitious injection S_m which had been introduced to render $\bar{S}_{\ell m} = 0$. The corrected value of $\bar{S}_{jk}$ is used in (5.16) to find $\tau_{jk,\ell m}$.

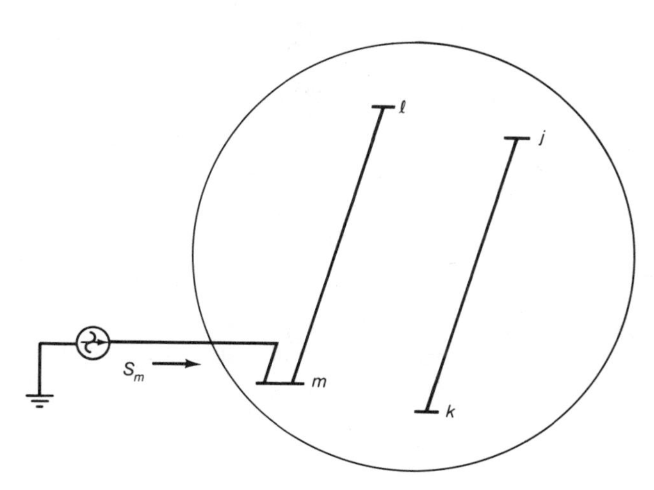

FIGURE 5.4 Application of the compensation theorem to outage a line.

The compensation process details are as follows: first introduce S_m such that

$$\Delta \bar{S}_{\ell m} = \rho_{\ell m,m} S_m ,$$

is such as to reduce $\bar{S}_{\ell m}$ to zero. Thus

$$S_m = \frac{-\bar{S}_{\ell m}}{\rho_{\ell m,m}} .$$

The incremental loading in jk due to this injection is

$$\Delta \bar{S}_{jk} = \rho_{jk,m} S_m = \frac{-\rho_{jk,m} \bar{S}_{\ell m}}{\rho_{\ell m,m}} .$$

The correction of S_m back to zero results in a correction of

$$\frac{\rho_{jk,m}^{out\ \ell m}(-\Delta \bar{S}_{\ell m})}{\rho_{\ell m,m}} .$$

The result is

$$\Delta \bar{S}_{jk} = \frac{\bar{S}_{\ell m}}{\rho_{\ell m,m}} (\rho_{jk,m}^{out\ \ell m} - \rho_{jk,m}) .$$

This is used in (5.16)

$$\tau_{jk,\ell m} = \frac{1}{\rho_{\ell m,m}} (\rho_{jk,m}^{out\ \ell m} - \rho_{jk,m}).$$

Table 5.2
Properties of the Line Outage Distribution Factor

$$\tau_{jk,\ell m} = \left. \frac{\Delta \bar{S}_{jk}}{S_{\ell m}} \right|_{out\ \ell m} = \frac{1}{\rho_{\ell m,m}} (\rho_{jk,m}^{out\ \ell m} - \rho_{jk,m})$$

$$\tau_{jk,\ell m} = \frac{(\rho_{\ell m,k} - \rho_{\ell m,j}) \bar{z}_{\ell m}^{*}}{(\rho_{\ell m,\ell} - \rho_{\ell m,m} - 1) \bar{z}_{jk}^{*}}$$

$$\tau_{jk,jk} = -1$$

$$\tau_{jk,\ell m} = -\tau_{kj,\ell m} \qquad \tau_{jk,\ell m} = -\tau_{jk,m\ell}$$

Table 5.2 summarizes a few properties of the line outage distribution factor, τ. Additional information on this may be topic is found in [10].

The line outage distribution factor, τ, is used in much the same way as ρ. The outage of line ℓm is considered to cause a redistribution of power such that

$$\bar{S}_{jk}^{B} \simeq \bar{S}_{jk}^{A} + \tau_{jk,\ell m} \bar{S}_{\ell m}^{A}, \qquad (5.17)$$

where case A is the "all lines in" case and case B is the case in which line

ℓm is outaged. Linear system behavior was assumed in calculating τ, and for this reason, (5.17) is approximate. At least in theory, multiple line outages may be handled using superposition of several terms in (5.17)

$$\overline{S}_{jk}^{B} \simeq \overline{S}_{jk}^{A} + \sum_{\ell m} \tau_{jk,\ell m} \overline{S}_{\ell m}^{A}, \tag{5.18}$$

where $\sum_{\ell m}$ refers to summation over the simultaneous outage of several lines denoted ℓm. The accuracy of (5.17) degrades when $\overline{S}_{\ell m}^{A}$ in large (i.e., the outage produces a large disturbance). Since the outage of several lines is likely to produce a greater disturbance, (5.18) should be used with great caution.

Example 5.2

In this example, the use of line outage distribution factors is illustrated for the five bus sample system in Figure 5.5. For the system and load schedule shown in the figure, obtain the line flows $\overline{S}_{24}$ and $\overline{S}_{32}$ for a changed case in which line 3–4 is outaged. The line flows in Figure 5.5 were obtained by load flow analysis.

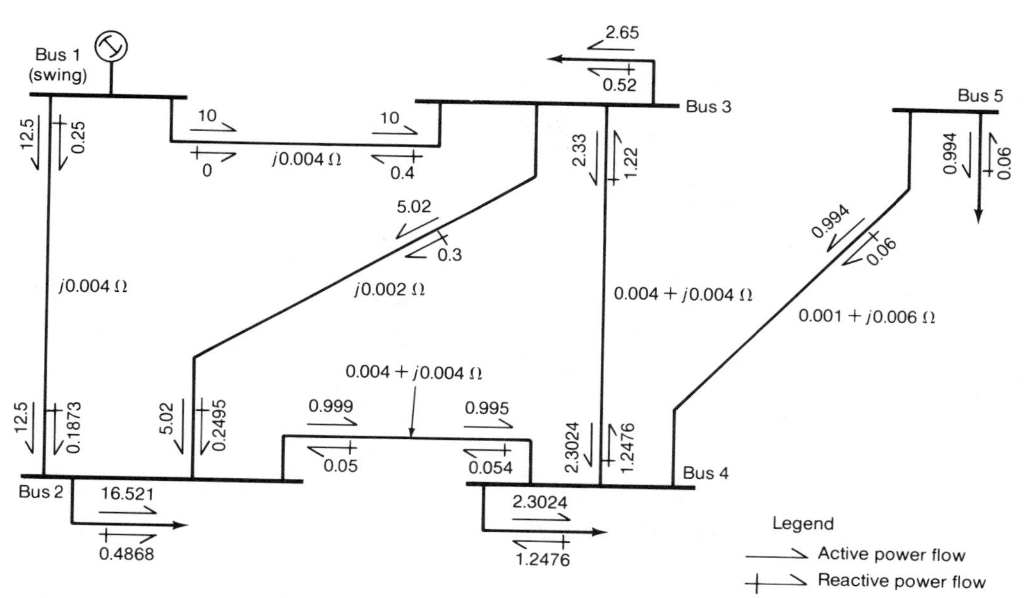

FIGURE 5.5 Five-bus system used in Exampe 5.2.

Solution

The line outage distribution factor method gives the required line flows as

$$\overline{S}_{24}^{B} = \overline{S}_{24}^{A} + \tau_{24,34}\, \overline{S}_{34}^{A}$$

$$\overline{S}_{32}^{B} = \overline{S}_{32}^{A} + \tau_{32,34}\, \overline{S}_{34}^{A}\,.$$

The superscript A denotes the base case (shown in Figure 5.5), and B denotes the line outaged case. The bus impedance matrix referenced to the swing bus is

$$Z_{bus}^{(1)} = \begin{bmatrix} 0.03279+j2.3543 & -0.03279+j1.6457 & j2 & j2 \\ -0.03279+j1.6457 & 0.03279+j2.3543 & j2 & j2 \\ j2 & j2 & 2+j4 & 2+j4 \\ j2 & j2 & 2+j4 & 3+j10 \end{bmatrix} \cdot 10^{-3}.$$

The axes of this matrix correspond to buses 2, 3, 4, and 5, respectively. The required line outage distribution factors are found as follows

$$\tau_{24,34} = \frac{(\rho_{34,4} - \rho_{34,2})\bar{z}_{34}^*}{(\rho_{34,3} - \rho_{34,4} - 1)\bar{z}_{24}^*} = 1.0 + j0.0$$

$$\tau_{32,34} = \frac{(\rho_{34,2} - \rho_{34,3})\bar{z}_{34}^*}{(\rho_{34,3} - \rho_{34,4} - 1)\bar{z}_{32}^*} = 0.784 + j0.004 \; .$$

Finally, the required complex powers are

$$\bar{S}_{24}^B = 1 + j0 + (1 + j0)(2.375 - j1.125)$$

$$= 3.375 - j1.125$$

$$\bar{S}_{32}^B = 5 + j0.5 + (0.784 + j0.004)(2.375 - j1.125)$$

$$= 6.863 - j0.371 \; .$$

If the interpretation is made that $\bar{S}_{24}^B$ is the complex power leaving bus 2 in line 24, the magnitude of the mismatch power, expressed as a percent of the load power at bus 4, is 0.07%. Similarly, at bus 5 the percent mismatch is about 0.61%. At other system buses, the mismatch is still less. The "correct" power flows used to make these comparison calculations were obtained from a conventional load flow study.

5.4

TRANSPORTATION METHODS

The generalized problem of calculating the flow of some commodity from its source to its sink is termed a *transportation problem.* In the case of a commodity such as milk, which does not incur loss and which does not satisfy formulas that interrelate the quantity and flow rate, a satisfactory transportation method of analyzing the source to sink flow is to guess the amount of milk carried by each truck, train, and so on, and adjust these guesses so that the demands are satisfied. There may be many solutions to a given problem. Another example of a transportation formulated problem is that of passengers boarding a subway train system. The given data include the number of incoming passengers and the number of disembarking passengers at each station. Also, the frequency and routing of trains on the system must be known. The quantities to be calculated include the passenger count on each train. Obviously, there may be many solutions, depending on the itinerary of each passenger (which route they take, whether they delay at a station, etc.). As the number of trains and the complexity of the network increases, and if one accounts for the capacity of each train, it is expected that there will be some commonality to the many solutions. A representative or feasible solution, even with its inaccuracies and differences from the actual solution, may be adequate for many applications.

An electric power network differs from a milk delivery system or a passenger train system in that the Kirchhoff laws must be satisfied. But we have just seen that an approximate method such as the use of distribution factors (which yields a solution that does not satisfy the Kirchhoff laws) may give a solution which is satisfactory for certain applications. Other than the distribution factor method, transportation solutions for power networks are not in common use, but they have been proposed for certain research applications particularly for the approximate solution of very large networks. A transportation analysis of a large power network is the calculation of feasible line flows without the calculation of detailed bus voltages and currents and without the explicit use of the Kirchhoff laws.

A feasible power flow solution must satisfy the zero mismatch criterion at each system bus. If the injected power were known at all system buses, there will typically be an insufficient number of equations compared with the unknown line flows (reactive power flow is ignored). Although there are an infinite number of solutions to such a set of equations, it may be difficult to obtain one of them in a power engineering application due to the complexity of the network. Although trial and error is a possibility, this method does not lend itself to computer analysis in this problem. Let n_b and n_ℓ refer to the number of system buses and lines, respectively. If the system is totally radial (i.e., a long radial feeder with one bus at each end and $n_b - 2$ buses in line), there will be n_b buses and $n_\ell = n_b - 1$ lines. This radial configuration is the *only* configuration for which

$$n_\ell < n_b \ .$$

For this simple radial case, using S_{bus} and $\overline{S}$ to denote the bus injection and line flow vectors respectively, it is an easy matter to demonstrate that

$$L^t \overline{S} = S_{bus}, \tag{5.19}$$

where L is the line–bus incidence matrix. To use (5.19), all transmission losses are ignored, and the elements of S_{bus} are the loads and generation written in injection notation (the sum of these elements is zero). Since (5.19) is n_b equations in n_ℓ unknowns, there are more equations than unknowns. This particular set of equations is consistent, and there is a single solution which is readily obtained by solving the first n_b rows of (5.19).

Virtually no power system is purely radial. It is only in the purely radial case that

$$n_\ell < n_b \ .$$

One concludes that for the general case of interest,

$$n_\ell \geq n_b \ .$$

The remainder of this section relates to the solution of (5.19) to obtain a feasible solution for this general case.

A Technique to Obtain a "Transportation" Feasible Solution

Consider a transportation feasible solution that does not consider the Kirchhoff laws or losses: Figure 5.6 shows such a feasible solution in which buses 1 and 2 are sources, and 3, 4, and 5 represent load buses whose total demand is 10 MW. A further simplification is made in that the scheduled

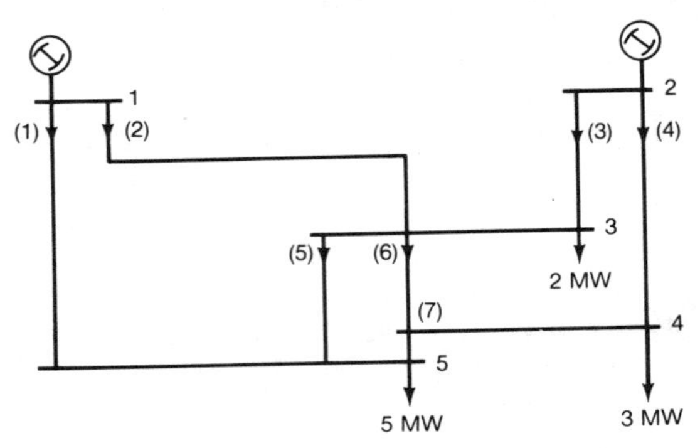

FIGURE 5.6 Transportation power flow study.

power at the sources is known – this may be known through the application of participation factors. In this case, let the 10 MW be distributed as 6 MW and 4 MW at buses 1 and 2. Thus, using injected notation,

$$S_{bus} = \begin{bmatrix} 6 \\ 4 \\ -2 \\ -3 \\ -5 \end{bmatrix} .$$

In the absence of losses,

$$\sum_{i=1}^{N} (S_{bus})_i = 0 . \tag{5.20}$$

Although there is no unique way in which the transmission line loads will deliver the given load schedule (Figure 5.7 shows two such possibilities), it

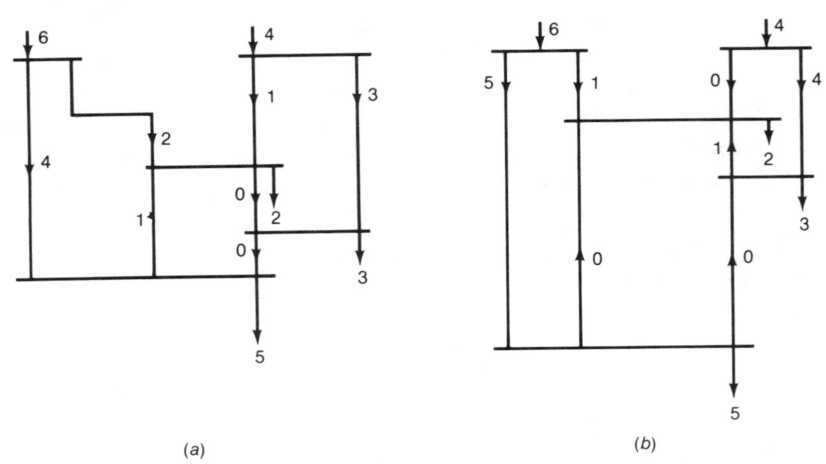

FIGURE 5.7 Alternative solutions for power flow study.

is desirable to develop an algorithm that will generate *one* such transmission loading configuration. Also, it would be preferable to find a *reasonable* solution; the concept of *reasonability* will be formalized momentarily. When the number of lines is greater than the number of buses (i.e., $n_\ell \geq n_b$), there are, in general, an infinite number of ways to solve

$$L^t \bar{S} = S_{bus}$$

for $\bar{S}$ given the bus injection vector, S_{bus} and the line–bus incidence matrix, L. As a reminder, L in (5.19) consists of 1's in position ij where i corresponds to a line "starting" at bus j; in position ij, a -1 occurs when line i "ends" at bus j. Otherwise, the n_ℓ by n_b matrix L contains zeros.

Equation (5.19) is n_b equations in n_ℓ unknowns. The minimum square error (mse) solution entails the use of the Moore–Penrose pseudoinverse (see Appendix B),

$$\bar{S} = (L^t)^+ S_{bus} \ . \tag{5.21}$$

The notation $(\cdot)^+$ denotes the pseudoinverse. Unfortunately, (5.21) may not give an acceptable solution for $\bar{S}$ since the mse solution may be of insufficient accuracy to be of use. This difficulty may be obviated in several ways. Impedance and/or load data may be used to adjust $\bar{S}$. This article will indicate only one of these alternatives: load data are used to obtain a better solution for $\bar{S}$.

Consider a case in which certain line flows $\bar{S}_b$ are known (from telemetered measurements) and the remaining entries of $\bar{S}$, namely $\bar{S}_a$, are unknown

$$\bar{S} = \begin{bmatrix} \bar{S}_a \\ \bar{S}_b \end{bmatrix} .$$

Then (5.19) is

$$L^t \begin{bmatrix} \bar{S}_a \\ \bar{S}_b \end{bmatrix} = S_{bus} \ . \tag{5.22}$$

Partition L^t into $[A \ \ B]$ such that A corresponds to the unknown lines represented in $\bar{S}_a$ and B corresponds to $\bar{S}_b$. Therefore,

$$A\bar{S}_a + B\bar{S}_b = S_{bus},$$

from which

$$A\bar{S}_a = S_{bus} - B\bar{S}_b \ . \tag{5.23}$$

In (5.23), let n_u be the number of unknown line flows and n_k be the number of line telemetered measurements

$$n_u + n_k = n_\ell .$$

Therefore, A is n_b by n_u and B is n_b by n_k. The mse solution of (5.32) is

$$\bar{S}_a = A^+(S_{bus} - B\bar{S}_b) \ . \tag{5.24}$$

Even if A were nonsingular, (5.24) may not yield an authentic power flow solution since losses are ignored. For some applications, however, an mse solution may suffice.

The application of pseudoinverses of the load flow problem has been analyzed in some detail by Johnson [22].

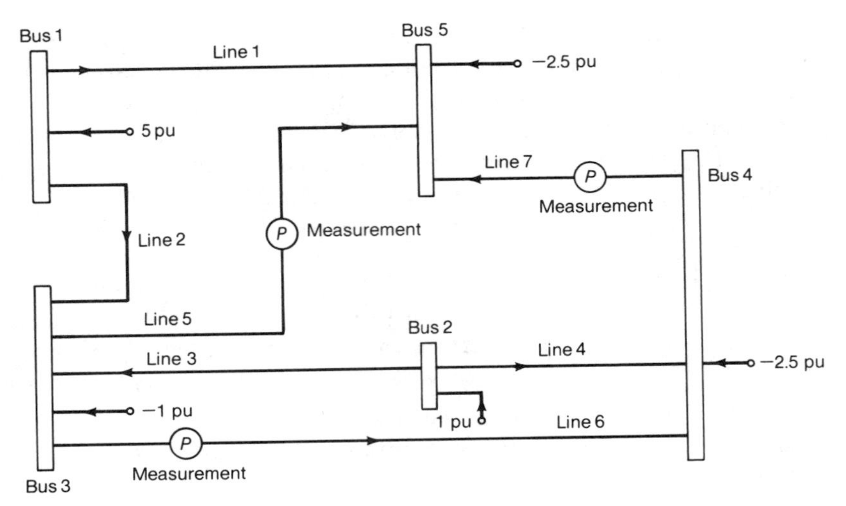

FIGURE 5.8 Sample system used for a transportation power flow study (Example 5.3).

Example 5.3

For the system in Figure 5.8, calculate a feasible flow diagram. For this system, losses are ignored and only active power flow is considered by virtue of the fact that a bulk transmission system is considered. The perunitized bus injection data are

$$S_{bus} = [5.0 \;\; 1.0 \;\; -1.0 \;\; -2.5 \;\; -2.5]^t$$

and measurements in lines 5, 6, and 7 reveal that

$$\bar{S}_5 = -0.5 \qquad \bar{S}_6 = 0.0 \qquad \bar{S}_7 = -0.5.$$

Solution

The L matrix is

$$L = \begin{bmatrix} 1 & 0 & 0 & 0 & -1 \\ 1 & 0 & -1 & 0 & 0 \\ 0 & 1 & -1 & 0 & 0 \\ 0 & 1 & 0 & -1 & 0 \\ 0 & 0 & 1 & 0 & -1 \\ 0 & 0 & 1 & -1 & 0 \\ 0 & 0 & 0 & 1 & -1 \end{bmatrix} \tag{5.25}$$

where the lines and buses are numbered as in Figure 5.8. Note that lines are directed in this pictorial. Equation (5.24) is

$$\bar{S}_a = A^+ \left(\begin{bmatrix} 5.0 \\ 1.0 \\ -1.0 \\ -2.5 \\ -2.5 \end{bmatrix} - B \begin{bmatrix} -0.5 \\ 0.0 \\ -0.5 \end{bmatrix} \right), \tag{5.26}$$

where A is the 5 by 4 left submatrix of L^t and B is the 5 by 3 right submatrix,

$$A = \begin{bmatrix} 1 & 1 & 0 & 0 \\ 0 & 0 & 1 & 1 \\ 0 & -1 & -1 & 0 \\ 0 & 0 & 0 & -1 \\ -1 & 0 & 0 & 0 \end{bmatrix} \qquad B = \begin{bmatrix} 0 & 0 & 0 \\ 0 & 0 & 0 \\ 1 & 1 & 0 \\ 0 & -1 & 1 \\ -1 & 0 & -1 \end{bmatrix}.$$

The pseudoinverse of A in this case is found using

$$A^+ = (A^tA)^+A^t \ . \tag{5.27}$$

In this case, A^tA is nonsingular, and

$$(A^tA)^+ = (A^tA)^{-1} = \begin{bmatrix} 2 & 1 & 0 & 0 \\ 1 & 2 & 1 & 0 \\ 0 & 1 & 2 & 1 \\ 0 & 0 & 1 & 2 \end{bmatrix}^{-1} = \begin{bmatrix} 0.8 & -0.6 & 0.4 & -0.2 \\ -0.6 & 1.2 & -0.8 & 0.4 \\ 0.4 & -0.8 & 1.2 & -0.6 \\ -0.2 & 0.4 & -0.6 & 0.8 \end{bmatrix}.$$

Substitution of $(A^tA)^+$ into (5.27) reveals that

$$A^+ = \begin{bmatrix} 0.2 & 0.2 & 0.2 & -0.2 & -0.8 \\ 0.6 & -0.4 & -0.4 & -0.4 & 0.6 \\ -0.4 & 0.6 & -0.4 & 0.6 & -0.4 \\ 0.2 & 0.2 & 0.2 & -0.8 & 0.2 \end{bmatrix}. \tag{5.28}$$

and further substitution into (5.24) gives

$$\overline{S}_a^t = [3.5 \;\; 1.5 \;\; -1 \;\; 2] \ .$$

Without impedance data, it is not possible to "check" this result versus load flow results. Of course, bus power mismatch is zero in all cases (e.g., at bus 4, the active power inflows to the bus from lines 4, 6, and 7, respectively, are 2.0, 0.0, and 0.5 and the bus injection is -2.5; this gives zero mismatch). The Kirchhoff laws and Ohm's law, however, are not generally satisfied simultaneously by transportation solutions.

5.5

SPECIAL PURPOSE POWER FLOW STUDIES

The Newton–Raphson power flow study with conventional load and transmission models comprises the majority of power flow studies done by electric utility companies. There are a few additional special purpose power flow studies which are occasionally needed – these are presented here for completeness. The general topic of optimal power flow studies (i.e., optimal dispatch) is relegated to Chapter 6 and statistical modeling in power flow studies is considered in Chapter 9 because topics in those chapters closely relate to these special purpose power flows. For planning and operation of electric power systems with high voltage dc transmission subsystems, power flow study packages with special capabilities are required. The formulation of these special capabilities is covered in detail by Arrillaga, Arnold, and Harker [23].

Harmonic Power Flow Studies

Most power system studies are sinusoidal steady state studies in which all voltages and currents are sinusoids of the same, power frequency. In the presence of a nonlinear load or transmission element (e.g., a rectifier load), nonsinusoidal currents and voltages will occur. These nonsinusoidal signals are periodic, and usually possess Fourier series,

$$v(t) = \sqrt{2} \sum_{i=1}^{\infty} a_i \cos (i\omega_o + \phi_i) \tag{5.29}$$

$$i(t) = \sqrt{2} \sum_{i=1}^{\infty} b_i \cos (i\omega_o + \alpha_i) . \tag{5.30}$$

In these expansions, dc components are assumed to be zero and magnitudes a_i and b_i are expressed in rms values. In three phase power systems, (5.29) and (5.30) represent phase A signals. For the case in which the power frequency (ω_o) is in positive sequence, the signals at frequency $i\omega_o$ have phase angle $-(\pi/3)\, i$ in phase B and $+(\pi/3)\, i$ in phase C. Thus the signals at frequency $i\omega_o$ are positive sequence for i = 1, 4, 7, 10, ...; the signals at $i\omega_o$ will be negative sequence for i = 2, 5, 8, 11, ...; and zero sequence will result at i = 3, 6, 9, ...

The resolution of bus voltages and injection currents into Fourier series results in a quantity in addition to P and Q which occurs in the complex voltamperes, S. For terms of like frequency in (5.29) and (5.30), active power results by summing the average power at each frequency

$$P_{av} = \sum_{i=1}^{\infty} a_i b_i \cos (\phi_i - \alpha_i) ,$$

and the reactive voltamperes is defined similarly for this nonsinusoidal case,

$$Q = \sum_{i=1}^{\infty} a_i b_i \sin (\phi_i - \alpha_i) .$$

If $v(t)$ and $i(t)$ were sinusoidal of the same frequency, the familiar relationship

$$|S| = |V||I| = \sqrt{P^2 + Q^2}$$

is readily shown. For the nonsinusoidal case, product terms of dissimilar frequencies in $|V||I|$ result in an additional term, D, termed ***distortion voltamperes***. The distortion voltamperes is defined such that

$$|S| = \sqrt{P^2 + Q^2 + D^2} .$$

Note that only active power, P, is conserved; reactive and distortion voltamperes may be converted to each other and therefore Q and D themselves are not conserved. These remarks are a consequence of the physical interpretation of active power as average power and the definitions of Q and D. Further comments are given by Shephard and Zand [18].

A harmonic power flow study is a system study that analyzes power systems with nonlinear loads. The input data to harmonic power flow studies are the usual voltampere data (line impedances, transformer data, etc.), load data, and specifications of the nonlinear loads. The output data are the levels of harmonic voltages and currents, their sequence, and harmonic distortion data.

The harmonic power flow study is used in cases where nonlinear loads (or sources) cause concern over potentially high harmonic currents and voltages. Typical applications are in designing harmonic filters, assessing communications/relaying interference, and sizing shunt capacitors to avoid voltage resonant conditions. A particularly important application ties in the analysis of transmission systems in which HVDC subsystems are present.

Methods of harmonic power flow analysis vary and may be broadly classified according to complexity of modeling of the nonlinear elements. In cases where bus voltages have less than 3% total harmonic distortion, rather simple modeling of the nonlinear bus load is usually justified. For the case of six and twelve pulse rectifier loads, approximately $1/i$ attenuation of the ith harmonic occurs (for $i \geq 5$ in the case of six pulse converters, or for $i \geq 11$ for twelve pulse converters). The $1/i$ attenuation is a consequence of the assumption of nearly square wave shape; however, the effects of commutation in the rectifier cause rounding of the edges of the square wave. This rounding is, in fact, attenuation of higher frequencies. Thus $1/i$ attenuation is a conservative estimate since actual harmonic levels should be somewhat lower. Thus the harmonic content of the load current is known without complex analysis, and resulting line flows are found by superposition. The behavior of the transmission network is essentially that of $\overline{z} = \overline{r} + ji\overline{x}$, where $\overline{x}$ is the line reactance at power frequency (i.e., at $\omega = \omega_o$).

The precalculation and superposition approach is not a true power flow formulation since the variation of the load harmonic current components with bus harmonic voltage components is not considered. If it is desired to perform a harmonic power flow study, it is necessary to reformulate the conventional power flow problem to reflect the changed mismatch requirements. Xia and Heydt [11–14] reformulated the Newton–Raphson power flow to accommodate line commutated converters. Network voltages and currents are represented by Fourier series. Harmonic bus voltage magnitudes and angles are unknowns for which additional equations are needed. The additional equations are based on Kirchhoff's current law for each harmonic and on conservation of apparent voltamperes at certain buses. The assumption of a balanced bilateral system permits the exclusion of even numbered harmonics. The restriction to delta connected nonlinear devices eliminates zero sequence currents. Triple harmonics ($3n$, $n = 1, 2, 3, \ldots$) in a balanced system are strictly zero sequence, and since zero sequence currents cannot flow from the network into a delta or ungrounded wye connected device, all triple harmonics are usually excluded from the formulation. Power system components, loads, and generators which produce harmonics in an otherwise pure fundamental frequency network are termed nonlinear. Standard PQ or PV power flow buses with no converters or other nonlinear devices are termed *linear*. All other buses are termed *nonlinear*. Total real power is specified at all buses except the swing bus. Total reactive power is specified at all PQ linear buses (except the swing bus), and the total apparent voltamperes or total reactive voltamperes, whichever is known, is specified at all nonlinear buses. The unknowns in the harmonic power flow are

Fundamental bus voltage magnitude and angle at all but the swing bus	$2(n-1)$
Real and reactive power at the swing bus	2
Harmonic bus voltage magnitude and angle at all buses	$2nh$
Total reactive voltamperes at each nonlinear bus	m
Two state variables (α, β) describing each nonlinear bus	$2m$
Total	$2n(1+h) + 3m$

where m is the number of nonlinear buses and h is the number of harmonics considered (excluding the fundamental). The equations are

Active and reactive power at all but the swing bus	$2(n-1)$
Fundamental voltage magnitude and angle at the swing bus	2
Fundamental real and imaginary current balance at each nonlinear bus	$2m$
Harmonic real and imaginary current balance at each bus	$2nh$
Apparent voltampere balance at each nonlinear bus	m
Total	$2n(1+h) + 3m$

It should be noted that if reactive voltamperes are specified at a nonlinear bus, apparent voltampere balance is not performed at that bus.

To reduce the complexity of the notation, let the swing bus be numbered *one*, the linear buses be numbered 1 through $(m-1)$, consecutively, and the nonlinear buses be numbered m through n, consecutively. The matrix formulation of the problem is now

$$\begin{bmatrix} \Delta w \\ \Delta I^{(5)} \\ \vdots \\ \Delta I^{(L)} \\ \Delta I^{(1)} \end{bmatrix} = \begin{bmatrix} J^{(1)} & J^{(5)} & \cdots & J^{(L)} & 0 \\ YG^{(5,1)} & YG^{(5,5)} & & YG^{(5,L)} & H^{(5)} \\ \cdot & \cdot & \vdots & \cdot & \cdot \\ YG^{(L,1)} & YG^{(L,5)} & & YG^{(L,L)} & H^{(L)} \\ YG^{(1,1)} & YG^{(1,5)} & \cdots & YG^{(1,L)} & H^{(1)} \end{bmatrix} \begin{bmatrix} \Delta V^{(1)} \\ \Delta V^{(5)} \\ \vdots \\ \Delta V^{(L)} \\ \Delta\Phi \end{bmatrix}, \tag{5.31}$$

where

$$\Delta W = [P_1 - f_{1,r},\ Q_1 - f_{1,i},\ \ldots,\ P_n - f_{n,r},\ Q_n - f_{n,i}]^t$$

=active and reactive power balance at all buses

$$\Delta I^{(1)} = [I_{m,r}^{(1)} + g_{m,r}^{(1)},\ I_{m,i}^{(1)} + g_{m,i}^{(1)},\ \ldots,\ I_{n,r}^{(1)} + g_{n,r}^{(1)},\ I_{n,i}^{(1)} + g_{n,i}^{(1)}]^t$$

=fundamental active and reactive current balance at each nonlinear bus

$$\Delta I^{(k)} = [I_{1,r}^{(k)}, I_{1,i}^{(k)}, \ldots, I_{m-1,r}^{(k)}, I_{m-1,i}^{(k)}, I_{m,r}^{(k)} + g_{m,r}^{(k)}, I_{m,i}^{(k)} + g_{m,i}^{(k)}, \ldots, I_{n,r}^{(k)} + g_{n,r}^{(k)}, I_{n,i}^{(k)} + g_{n,i}^{(k)}]^t \quad k=5, 7, 11, \ldots, L$$

=kth harmonic active and reactive current balance at each bus

$$\Delta V^{(k)} = [V_1^{(k)}, \Delta\Theta_1^{(k)}, \Delta V_1^{(k)}, \ldots, V_n^{(k)}, \Delta\Theta_n^{(k)}, \Delta V_n^{(k)}]^t \quad k = 1, 5, 7, 11, \ldots, L$$

=kth harmonic voltage magnitude and angle at all buses

$$\Delta\Phi = [\Delta\alpha_m, \Delta\beta_m, \ldots, \Delta\alpha_n, \Delta\beta_n]^t$$

=state variable mismatch for each nonlinear device

$J^{(1)}$=conventional fundamental Jacobian matrix

=partial derivatives of P, Q with respect to $V_{(k)}$, $\Theta^{(k)}$ formed in the conventional way, where $\theta^{(k)}$ is used to denote bus harmonic voltage phase angles

$$YG^{(k,j)} = \begin{cases} Y^{(k,k)} + G^{(k,k)} & k = j \\ G^{(k,j)} & k \neq j \end{cases} \tag{5.32}$$

$G^{(k,j)}$ = partial derivatives of kth harmonic device currents with respect to jth harmonic applied voltages as determined by the nonlinear devices

$$A_{m,r} = \frac{\partial g_{m,r}^{(k)}}{V_m^{(j)} \partial\theta_m^{(j)}} \qquad B_{m,r} = \frac{\partial g_{m,r}^{(k)}}{\partial V_m^{(j)}}$$

$$G^{(k,j)} = \begin{bmatrix} O_{2n,2n} & O_{2n,2m} \\ O_{2m,2n} & diag\begin{bmatrix} A_{m,r} & B_{m,r} \\ A_{m,i} & B_{m,i} \end{bmatrix} \end{bmatrix} \tag{5.33}$$

where the diagonal matrix indication in Eq. (5.33) denotes m submatrices. In (5.31), H has the form

$$H^{(k)} = diag \begin{bmatrix} \dfrac{\partial g_{t,r}^{(k)}}{\partial\alpha_t} & \dfrac{\partial g_{t,r}^{(k)}}{\partial\beta_t} \\ \dfrac{\partial g_{t,i}^{(k)}}{\partial\alpha_t} & \dfrac{\partial g_{t,i}^{(k)}}{\partial\beta_t} \end{bmatrix} \quad t = m, \ldots, n;\ k = 1, 5, 7, 11, \ldots, L \tag{5.34}$$

=partial derivatives of nonlinear device currents with respect to nonlinear device state variables, α and β

$O_{j,k}$=j by k dimensioned array of zeros

$Y^{(k,k)}$=partial derivatives of kth harmonic injection currents with respect to kth harmonic bus voltages as derived from the system admittance matrix

$f_{t,r}, f_{t,i}$ =real and reactive injected power at bus t as calculated from network variables

$g_{t,r}^{(k)}, g_{t,i}^{(k)}$=real and imaginary injection current for harmonic k at bus t as calculated by nonlinear device variables

L = highest harmonic considered

n = number of buses

m = number of nonlinear buses.

Typically, the fundamental frequency power flow study converges in two to four iterations, whereas the harmonic frequency power flow study converges in 10 to 30 iterations. Matrix G is typically over 98% sparse for $n+m \geq 20$. References [14–19] document the Xia–Heydt algorithm and a few alternative methods.

Three Phase Power Flow Studies

The term "three phase power flow study" is applied to studies in which three phase detail is retained and the simplification of a single line diagram is not used. This type of study is occasionally required when imbalanced three phase (3ϕ) operation or conditions occur. Imbalanced operation refers to the case of single phase (1ϕ) loads on the three phase system, use of an imbalanced transmission circuits, or energization of the 3ϕ bus with an imbalanced set of 3ϕ voltages. Imbalanced transmission configuration occurs in the case of single phase taps off the three phase substation bus, cases of a single conductor outage in a 3ϕ circuit (thus resulting in an effective 1ϕ subnetwork), untransposed transmission circuits, or similar configurations that result in dissimilar voltampere characteristics among the three phases. Although virtually all transmission and subtransmission circuits are 3ϕ and many distribution circuits are 3ϕ, the three phase load flow study is not often needed at these voltage levels since systems are very nearly balanced such as to allow the usual assumptions which are, in effect, the analysis of one phase of the system. This is also equivalent to analysis of only the positive sequence network. There are two instances in which a full 3ϕ analysis is occasionally required: the case of untransposed EHV transmission systems and, at the other end of the power level spectrum, the case of mixed 1ϕ and 3ϕ loads and networks at the distribution voltage level.

To retain 3ϕ detail, it is necessary to expand the V_{bus} and I_{bus} vectors by replacing each 3ϕ bus voltage with its three individual phase-to-ground voltages and each 3ϕ bus injection current with three line currents. Let the notation $V_{bus}^{3\phi}$ and $I_{bus}^{3\phi}$ denote these vectors, which have three times the dimension of the corresponding V_{bus} and I_{bus} vectors. Then

$$V_{bus}^{3\phi} = Z_{bus}^{3\phi} I_{bus}^{3\phi}$$

and

$$I_{bus}^{3\phi} = Y_{bus}^{3\phi} V_{bus}^{3\phi} .$$

The rules for the formulation of the 3ϕ impedance and admittance matrices are the same as those for the balanced 3ϕ analogs.

For a 3ϕ Newton–Raphson power flow study, the $|V|$ and δ subvectors increase in dimension by a factor of 3 to reflect the magnitude and angle of the three phase-to-ground voltages. In the absence of PV buses, the Jacobian matrix becomes $6(N-1)$ by $6(N-1)$ and there are three swing buses, corresponding to phases A, B, and C of the 3ϕ swing bus. The formulas for forming $J^{3\phi}$ are identical to those for the balanced case. The 3ϕ bus admittance matrix, $Y_{bus}^{3\phi}$, is needed to form $J^{3\phi}$.

When the network itself is balanced, it is possible to decouple $Y_{bus}^{3\phi}$ into three parts in such a way that each component is totally ***disconnected*** from the other two. One such transformation that will perform this decoupling is the symmetrical component transformation. This approach is considered in some detail in Section 7.4. The decoupling process results in resolution of the three phase power flow problem into three separate single phase power flow studies. Each of these problems is solved separately and Jacobian matrices are constructed using symmetrical components of the network data. This approach is valid only when the network consists entirely of fully transposed three phase circuits. The loads may, nonetheless, be single phase or imbalanced 3ϕ. The symmetrical component solutions are used with an inverse transformation to obtain phase voltages and currents.

References [19,20,21] further discuss three phase power flow studies.

Distribution Power Flow Studies

Most power flow studies are performed on transmission networks because it is at these high power and voltage levels that important decisions on sizing and loading of components must be made. Also, at transmission voltage levels, operating decisions are very important since they may effect thousands of service buses and they may result in flows of hundreds of megawatts. Nonetheless, distribution circuits are occasionally analyzed in some detail using a power flow study in cases where many distribution circuits are being planned simultaneously, some distribution load buses are especially important or critical, or instances of potential difficulty in planning or operating a distribution system.

A distribution power flow study is essentially the same as the transmission voltage level counterpart. Certain peculiarities of distribution systems may be specifically considered for a production grade computer program designed for the analysis of distribution systems. These special considerations include the following:

1. *The predominantly radially configured network.* Most distribution circuits are strictly radially configured. This configuration can give difficulty in convergence since improper initialization will result in slow updating of V_{bus} in early iterations. One solution to this difficulty when a Newton–Raphson solution is used is to perform a few Gauss–Seidel iterations before executing the Newton–Raphson algorithm. It is also advisable to use an "educated guess" to initialize bus voltages and angle (such as initialization of S by assuming that power flow is approximately proportional to the difference between phase angles at adjacent buses).

2. *The distinctive form of load data.* Load data for distribution circuits are usually simply related to the capacity of the associated distribution transformer. All distribution buses to not peak at the same time, and peculiar results might be obtained if 100% of the distribution transformer capacity were used as the active demand. A reasonable load participation factor, usually in the range 60 to 80%, is applied to the transformer rating to obtain load data. Also, typical load power factors may be applied depending on the character of the load. Typical power factors are

Residential	0.95 lagging
Commercial	0.90 lagging
Light industrial (no induction motor loads)	0.87 lagging
Heavy industrial (with induction motor loads)	0.85 lagging

Obviously, more accurate data are used as available.

Table 5.3
Approximate Wire Sizes for Distribution Conductors

AWG Conductor Size	Effective Radius, r' (m)
No. 6	2.02×10^{-3}
No. 4	2.54×10^{-3}
No. 2	3.21×10^{-3}
1/0	3.61×10^{-3}
2/0	4.06×10^{-3}
3/0	4.55×10^{-3}
4/0	4.94×10^{-3}
250 kcmil	4.95×10^{-3}

3. *The distinctive form of line data.* Unlike transmission line data, which are often available in perunitized impedances, distribution data are usually in the form of wire sizes. Some distribution load flow programs accept line data directly in terms of wire size, circuit length, and spacing geometry. Table 5.3 may be of value in estimating conductor impedances in connection with the following approximate formulas for series line impedance in ohms per phase per foot, $\bar{z}_\ell$, and shunt line admittance in mhos per foot, $\bar{y}_\ell$, (both at 60 Hz, 3ϕ):

$$\bar{z}_\ell = 5.885 \times 10^{-2} \frac{\rho}{(r')^2} + j2.298 \times 10^{-5} \ln \frac{0.3048(GMD)}{r'} \tag{5.35}$$

$$\bar{y}_\ell = j \frac{6.388 \times 10^{-9}}{\ln[0.2374(GMD)/r']} \tag{5.36}$$

where r' is the effective conductor radius in meters, GMD is the geometric mean distance between conductors in feet, and ρ is given by

$$\rho = 2.83 \times 10^{-8} \qquad \textit{aluminum},\ 20\ ^\circ C$$

$$\rho = 3.17 \times 10^{-8} \qquad \textit{aluminum},\ 50\ ^\circ C$$

$$\rho = 1.77 \times 10^{-8} \qquad copper, \ 20\ ^{\circ}C$$

$$\rho = 1.97 \times 10^{-8} \qquad copper, \ 50\ ^{\circ}C$$

Note in Table 5.3 that conductor size is often given in American Wire Guage (AWG) sizes with the larger numbers (e.g., AWG No. 6) having a smaller radius than the lower numbers (e.g., AWG No. 1). The conductor AWG No. 00 is written 2/0 and is read "two ought." The conductor AWG No. 4/0 is approximately 250 kcmil in area (approximately 1/2 inch in diameter).

Interactive Power Flow Studies

Most power flow studies are batch jobs because input and output data are so voluminous as to preclude convenient interpretation by casual inspection or viewing a video display. There are instances in which an interactive study is desirable: this is a study in which condensed output is displayed after executing a power flow study using a given, base case data file as input. At this point, an interrupt is generated and control is given to the operator who may elect to change load or line data. This feature is particularly useful in distribution system planning. It is possible to resize conductors, assess the impact of load siting, and conveniently set transformer taps. For such applications, some special advantage may be made of the multiple run character of the power flow study. For example, initialization using the previously solved $[\delta\ V]^t$ is advisable and it may be possible to modify the existing Y_{bus} matrix (to reflect line data changes) rather than recalculate the matrix.

Bibliography

[1] B. Stott, "Decoupled Newton Load Flow," *IEEE Trans. Power Apparatus and Systems,* v. PAS-91, no. 5, September–October 1972, pp. 1955–1959.

[2] B. Stott and O. Alsac, "Fast Decoupled Load Flow," *IEEE Trans. Power Apparatus and Systems,* v. PAS-93, May–June 1974, pp. 859–869.

[3] B. Stott, "Review of Load Flow Calculation Methods," *Proc. IEEE,* v. 62, July 1974, pp. 916–929.

[4] F. Wu, "Theoretical Study of the Convergence of the Fast Decoupled Load Flow," *IEEE Trans. Power Apparatus and Systems,* v. PAS-96, no. 1, January–February, 1977, pp. 268–275.

[5] P. Rao, K. Rao, and J. Nanda, "An Empirical Criterion for the Convergence of the Fast Decoupled Load Flow Method," *IEEE Trans. Power Apparatus and Systems,* v. PAS-103, no. 5, May 1984, pp. 974–981.

[6] L. Kraft, "Harmonic Resonance in Electric Power Systems," Ph.D. thesis, Purdue University, West Lafayette, Ind., August 1984.

[7] E. Hines and H. Limmer, "Rapid Load Flow Program Using Superposition," *Proc., First Power Systems Computation Conference,* Queen Mary College, London, 1963.

[8] C. MacArthur, "Transmission Limitations Computed by Superposition," *AIEE Trans. Power Apparatus and Systems,* v. 80, 1961, pp. 827–831.

[9] H. Limmer, "Techniques and Applications of Security Calculations Applied to Dispatching Computers," *Proc., Power Systems Computation Conference,* Rome, 1969.

[10] A. El-Abiad and G. Stagg, "Automatic Evaluation of Power System Performance - Effects of Line and Transformer Outages," *AIEE Trans. Power Apparatus and Systems,* v. 81, 1963, pp. 712–716.

[11] D. Xia and G. T. Heydt, "Harmonic Power Flow Studies, Part I – Formulation and Solution," *IEEE Trans. Power Apparatus and Systems,* vol. PAS–101, no. 6, June 1982, pp. 1257–1265.

[12] D. Xia and G. T. Heydt, "Harmonic Power Flow Studies, Part II – Implementation and Practical Application," *IEEE Trans. Power Apparatus and Systems,* vol. PAS-101, no. 6, June 1982, pp. 1266-70.

[13] W. M. Grady and G. T. Heydt, "Determination of Harmonics in an AC Power System Caused by HVDC Converters," *J. Electric Machines and Power Systems, 1985,* v. 10, no. 1, pp.39-52.

[14] W. M. Grady, "Harmonic Power Flow Studies," Ph.D. thesis, Purdue University, West Lafayette, Ind., August 1983.

[15] A. Mahmoud and R. Shultz, "A Method for Analyzing Harmonic Distribution in AC Power Systems," *IEEE Trans. Power Apparatus and Systems,* v. PAS-101, no. 6, June 1982, pp. 1815-24.

[16] N. Miller, W. Price, M. Lebow, and A. Mahmoud, "A Computer Program for Multi-phase Harmonic Analysis," *1984 International Symposium on Power System Harmonics,* Worcester, Mass., October 1984.

[17] L. A. Kraft and G. T. Heydt, "Harmonic Voltage Resonance in Electric Power Systems," *1984 International Symposium on Power System Harmonics,* Worcester, Mass., October 1984.

[18] W. Shephard and P. Zand, *"Energy Flow and Power Factor in Nonsinusoidal Circuits,"* Cambridge University Press, Cambridge, 1979.

[19] G. Stagg and A. El-Abiad, *"Computer Methods in Power System Analysis,"* McGraw-Hill, New York, 1968.

[20] G. Heydt and W. Grady, "A *Z* Matrix Method for Fast Three Phase Load Flow Calculations," *Proc., Power Industry Computer Applications,* Cleveland, Ohio, 1973.

[21] D. Tarsi, "Analysis of Unbalanced Three Phase Load Flows," M.S.E.E. thesis, Purdue University, West Lafayette, Ind., January, 1967.

[22] B. Johnson, "Applications of the Pseudoinverse to the Load Flow Problem," M.S.E.E. thesis, Purdue University, West. Lafayette, Ind., May, 1985.

[23] J. Arillaga, C. Arnold, B. Harker, *Computer Modeling of Electrical Power Systems,* Wiley (Interscience Publication), Chichester, UK, 1983.

Exercises

Exercises that do not involve computer programming

5.1 A decoupled load flow study is to be done on a 1035 bus power system with 42 generators (one of which is the swing bus, the others are PV buses), 8 TCUL voltage regulated buses, and the remainder are load buses (all PQ). Describe the dimension of each section of the decoupled Jacobian matrix.

5.2 A certain triangular factorization and forward–backward solution technique for $Ax = b$ is used in a decoupled power flow study. The method required central processor time given by t_{cp},

$$t_{cp} = \frac{1}{2}n^3 + 2n^2 \quad \text{milliseconds}$$

for an n-dimensional A matrix. Calculate the advantage of using a decoupled load flow versus a conventional load flow. Assume that the system to be studied has $n+1$ buses, N iterations are required for a conventional load flow and $N+1$ iterations are required for a decoupled load flow. Graphically illustrate this advantage for $N =$ 2, 3, 4.

5.3 Consider a power system with a near unity, flat bus voltage profile. Find the conditions, if any, for which the off diagonal positions of the $\partial P/\partial V$ submatrix in J are greater than 10% of the off diagonal entries in the $\partial P/\partial \delta$ submatrix. This should be done for the purely reactive transmission system case and also for the case of a system with X/R equal to 10 throughout.

5.4 Derive the expression for $\rho_{\ell m,n}^{out\ ij}$ given in (5.15).

5.5 Equation (5.15) gives an expression for $\rho_{\ell m,n}$ if line ij is outaged. Find an expression for $\tau_{jk,\ell m}$ for the system if line ab is outaged (i.e., find $\tau_{jk,\ell m}^{out\ ab}$).

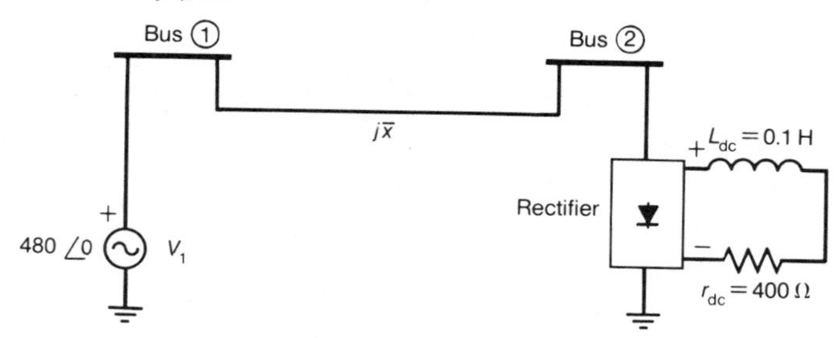

FIGURE P5.6 Power system with a full wave rectifier load.

5.6 Figure P5.6 shows a single phase ac power system serving a rectifier load. The source V_1, occurs at fundamental frequency (60 Hz) only. The line reactance shown is at this frequency. Actual voltages, currents, and reactances are shown in the figure.

a. For a full wave rectifier, calculate the bus current waveform and Fourier series at bus 2.

b. Calculate the line current waveform and Fourier series.

c. Calculate the total harmonic distortion (THD) of the line current and bus voltage

$$THD^2 = \frac{\sum_{i=2}^{\infty} a_i^2}{\sum_{i=1}^{\infty} a_i^2}.$$

Exercises that require programming

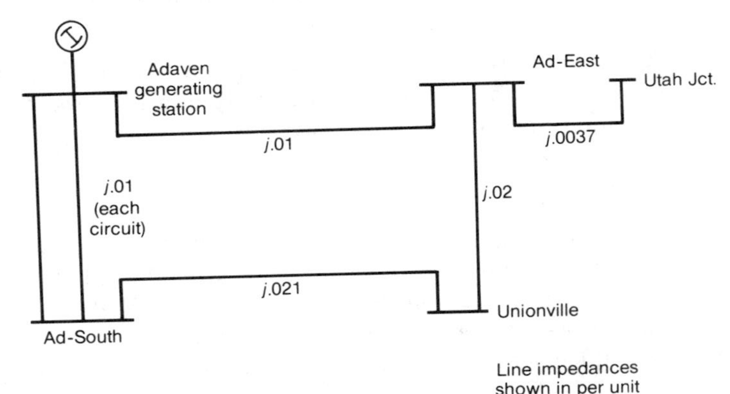

FIGURE P5.7 Adaven power system.

5.7 Figure P5.7 shows a five bus power system for which the swing bus is the Adaven generating station. In this exercise, the effectiveness of the $P-Q$ decoupling process will be investigated for the load schedule in the following table:

Bus	*Load* (MVA)
Ad-East	$100 + j20$
Ad-South	$20 + j12$
Utah Jct.	$22 - j15$
Unionville	$35 + j10$

In subsequent exercises, the effectiveness and validity of the linear load flow study and transportation method will be studied. In Figure P5.7, line data are perunitized on a 138 kV, 100 MVA base,

a. Using the Newton–Raphson method, solve the power flow problem finding V_{bus} and line flows for the given base case data. Use a 1% bus mismatch stopping criterion.

b. Repeat part (b), but this time zero the $\partial Q/\partial \delta$ and $\partial P/\partial V$ quadrants of the Jacobian matrix. This corresponds to a decoupled power flow solution. Compare bus voltages and line flows with the conventional Newton–Raphson study. Use 1% bus mismatch stopping criterion

c. Is the solution obtained in part (b) less accurate than the solution obtained in part (a)? Explain your answer.

5.8 This exercise illustrates the use of power transfer distribution factors using the five bus system shown in Figure P5.7 and described in Exercise 5.7. It will be necessary to perform a Newton–Raphson power flow solution using the base case data before proceeding with this exercise.

a. Consider the following changed case loading schedules:

Bus	Load (MVA)
Ad-East	$100 + j19$
Ad-South	$21 + j13$
Utah Jct.	$26 - j10$
Unionville	$34 + j10$

Run a full Newton–Raphson load flow study for the changed case using 1% bus mismatch stopping criterion. The solution obtained will be termed the "exact" solution for the changed case.

b. Write a computer program to form $Y_{bus}^{(swing)}$ and invert this matrix to obtain $Z_{bus}^{(swing)}$ for the given system. Using $Z_{bus}^{(swing)}$, calculate the ρ matrix of distribution factors.

c. Recognizing the changed case as a small change from the base case, calculate the change in line flows, $\Delta\overline{S}$, using the distribution factor method,

$$\Delta\overline{S} = \rho\Delta S_{bus} .$$

Compare the resulting line flows with the "exact" solution obtained in part (b).

d. Consider now an additional changed case which will be termed load schedule B. Load schedule B is the same as the changed case except that the active demand at the Ad-East bus is P_x. Using the appropriate element of the ρ matrix, find P_x such that the MVA loading of the line between the Adaven generating station and Ad-East goes above its rating. The rating of this line is 155 MVA.

e. In the overload application illustrated in part (d), note that the 155 MVA rating is approximate. Comment on the appropriateness of a distribution factor solution. How could this problem be solved "exactly"?

5.9 In this exercise, the five bus system of Figure P5.7 described in Exercise 5.7 will be solved one more time using the transportation method. It will be necessary to have solved the system using the base case data given in Problem 5.7 using the Newton–Raphson method. It will also be necessary to have access to an eigenvalue – eigenvector analysis subroutine to complete this problem.

a. Draw a diagram of this system showing the *active* power flow in each line.

b. Calculate the eigenvalues and eigenvectors of L^tL and LL^t.

c. Using the transportation method, calculate a feasible load flow solution and compare with part (a). Note: In a very small system such as this, the transportation method will not perform very well since line flows are strongly controlled by small voltage differences between buses. In a much larger system, transportation feasible solutions become more reasonable.

5.10 Figure P5.10 shows an elementary two bus power system. This simple system will be used to gain some insight into the two Newton–Raphson convergence criteria given in this chapter.

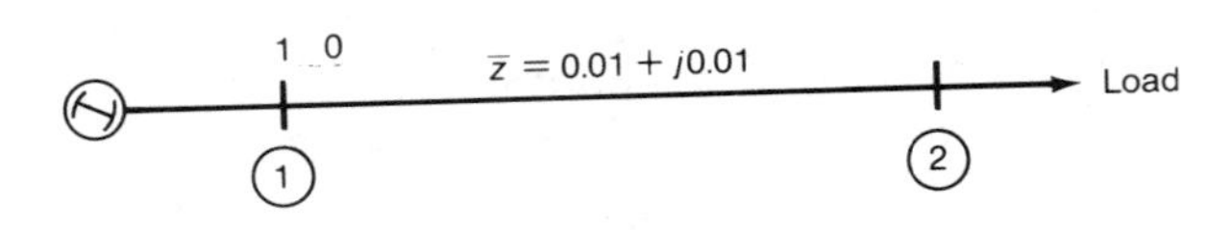

FIGURE P5.10

Although these criteria are not in general use for practical problems, they do give some qualitative assessment of the convergence process. For the system shown, form the 2 by 2 Jacobian matrix J and write a computer program to obtain the inverse Jacobian matrix, Ξ.

a. *Eigenvalue criterion.* Use Ξ in a computer program to obtain a load flow solution for the given system. Note that the maximum possible load at bus 2 occurs when the load impedance is $\bar{z}^*$. This maximum P_2 is −25 per unit. Run several power flow solutions in the range $0 \leq |P_2| \leq 25$ (e.g., $P_2 = 0, -2, -4, -8, -16, -25$). For each load flow study, print out the eigenvalues of Ξ at each iteration. Comment on the eigenvalue criterion that

$$|\lambda_i| < 1$$

guarantees convergence.

b. *Newton–Kantorovich criterion.* Repeat part (a), but this time, calculate the Newton–Kantorovich parameters a and c at each iteration. Also calculate b at the end of the first iteration. Print out the value of abc and comment on the criterion that

$$abc \leq 1/2$$

guarantees convergence.

c. *Effect of acceleration factors.* The effect of fractional acceleration factors is to enhance convergence in power flow studies that are slowly convergent or nonconvergent. Using a high-power load at bus 2 to illustrate this phenomenon, apply a fractional acceleration factor to enhance convergence. Very small (e.g. $\alpha = 0.01$) acceleration factors will cause a considerable increase in the number of iterations required for convergence. This is not a problem in this small illustrative system.

CHAPTER 6

Optimal Dispatch

6.1

OBJECTIVES OF OPTIMAL GENERATION DISPATCH

The terms optimal dispatch, optimal generation dispatch, economic optimal dispatch, optimal load flow study, and optimal power flow study are essentially synonyms which refer to a power flow type of calculation in which some quantity or quantities are optimized with respect to the generation schedule. The most usual formulation is that in which the fuel costs are minimized, but there are other formulations in which such factors as capital investment for shunt capacitors or the emission of pollutants is minimized. The term ***production costing*** is also used to refer to economic dispatch and calculation of fuel requirements under economic dispatch. The general term ***economic dispatch*** refers to formulations in which operating or investment costs are minimized.

The objectives of optimal dispatch are generally the minimization of one or more functions, C. These functions are broadly termed costs and may represent economic costs or parameters such as emission of pollutants.

The control variables, U, are those parameters that are adjusted to bring about the required optimization. Usually, the control variables are the generation levels at all generators except the swing machine. There are other possible choices for the control vector U.

In optimal dispatch problems, constraints are very important. As an example, in the absence of the constraint that the load must be met, it would be possible to obtain minimum fuel cost by shutting down the entire system. Meaningful results are obtained only by applying such constraints as the satisfaction of the load or the requirements of the Ohm's and Kirchhoff's laws. Equality constraints are generally of the form

$$G = 0,$$

where G is a vector. Sometimes inequality constraints apply

$$H \leq 0 .$$

An example of the latter is a requirement that at all PV buses, all Q limits must be met.

With the nomenclature set out above, a common formulation for an optimal power flow is to

$$\textit{minimize} \quad C(U, X, P) \tag{6.1}$$

with respect to the control vector U subject to

$$G(U, X, P) = 0 \tag{6.2}$$

$$H(U, X, P) \leq 0 \tag{6.3}$$

where X is the state vector of the system (usually the $[\delta \ \ V]^t$ vector in a Newton–Raphson formulation, but line currents and TCUL tap settings may also appear in X) and P is the given parameter vector which is usually the given load schedule. In (6.3), H is a vector, and the inequality refers to term-by-term inequalities.

The way in which the objective (6.1) is obtained depends on the specific application and the desired sophistication and accuracy of the solution. In this chapter, a very simple solution method is first presented followed by a modification that reflects inclusion of system losses. The former is the equal incremental cost rule and the latter is the method of loss or B-coefficients. These methods, although easy to understand and program, do not include the Kirchhoff laws in $G(U, X, P)$, and therefore the accuracy obtained may be inadequate. It is important to recognize the very high level of fuel costs: this high cost is motivation for attaining considerable accuracy in the economic optimal dispatch. Even a small percentage change in accuracy in an optimal dispatch may result in a considerable savings in fuel costs. For this reason, the principal attention of this chapter is on a gradient optimization technique known as the Dommel–Tinney method [1], which includes the Kirchhoff laws in $G(U, X, P)$. The treatment of optimal dispatch in this chapter is one of presentation of only the highlights of the subject; books have been devoted in their entirety on the subjects (representative texts are references [19–21]).

6.2 EQUAL INCREMENTAL COST RULE

The optimal dispatch problem is given by (6.1) through (6.3). Perhaps the simplest formulation is that in which the cost function, C, is simply the fuel cost

$$C(U, X, P) = C(U) . \tag{6.4}$$

Equation (6.4) is a greatly simplified formulation since the fuel cost is represented as a function only of generator settings. It is not possible to select *all* generator settings since the system losses are unknown and are, in fact, a function of the generator settings. In other words, one cannot specify the swing bus power since this power is dependent on the losses. If transmission losses are not considered, the solution of the optimal dispatch problem is greatly simplified. Also, (6.4) does not model generation voltage setting as a control variable. With these simplifications admitted, let the total cost C (scalar) be a function of vector U which has elements $u_1, u_2, \ldots, u_g$ which are g generator settings

$$C(U) = \sum_{i=1}^{g} f_{u_i}(u_i), \tag{6.5}$$

where $f_{u_i}(u_i)$ is the fuel cost of unit i operating at a power level of u_i. The constraints in this formulation, G, will simply be that the load demand is met. In the absence of losses,

$$G(U, X, P) = \sum_{i=1}^{g} u_i + \sum_{j=1}^{\ell} p_j, \tag{6.6}$$

where p_i are the ℓ load bus powers in injection notation (i.e., $p_i < 0$). Evidently,

$$G(U, X, P) = G(U, P)$$

$$= 0 . \tag{6.7}$$

The constrained problem is obtained by augmenting (6.7) into (6.5) using a single Lagrange multiplier, λ,

$$L(U, P) = C(U) - \lambda G(U, P) . \tag{6.8}$$

In (6.8), L, C, λ, and G are scalars, U is a g-vector, and P is an ℓ-vector. The constrained minimum occurs where

$$\nabla_u L(U, P) = 0.$$

In other words,

$$\nabla_u [\sum_{i=1}^{g} f_{u_i}(u_i) - \lambda \sum_{i=1}^{g} u_i - \lambda \sum_{j=1}^{\ell} p_j] = 0 . \tag{6.9}$$

Equation (6.9) is a vector expression with g rows. In row i,

$$\frac{\partial}{\partial u_i} f_{u_i}(u_i) = \lambda \tag{6.10}$$

for all i. Equation (6.10) states that at the optimum dispatch, each generator is operated such that its incremental fuel cost, $\partial f / \partial u$, is equal to the incremental fuel costs at each other machine. This result is known as the equal incremental cost rule, which is repeated here:

Equal incremental cost rule: A lossless system is optimally dispatched at a point where the incremental fuel cost of each generator, $\partial f / \partial u$, is equal to that at each other generator.

Expressions (6.9) and (6.10) do not include the effect of a machine being operated at its maximum rating. If u_k at machine k is at its upper limit, the derivative of the constant

$$\frac{\partial}{\partial u_k} f_{u_k}(u_k^{\max})$$

is zero, and this term is dropped out of (6.10). The same remark applies for a machine operated at minimum setting (or zero).

The total fuel cost of the optimally dispatched system is the sum of the individual fuel costs [Eq. (6.5)], but the incremental ***system*** fuel cost is the derivative of C with respect to the u's. All $\partial f/\partial u$ terms are equal to λ, and therefore the incremental system fuel cost is λ. The notation of λ for a Lagrange multiplier is widespread, and it is common to refer to the system incremental fuel cost for an optimally dispatched system as the "system λ."

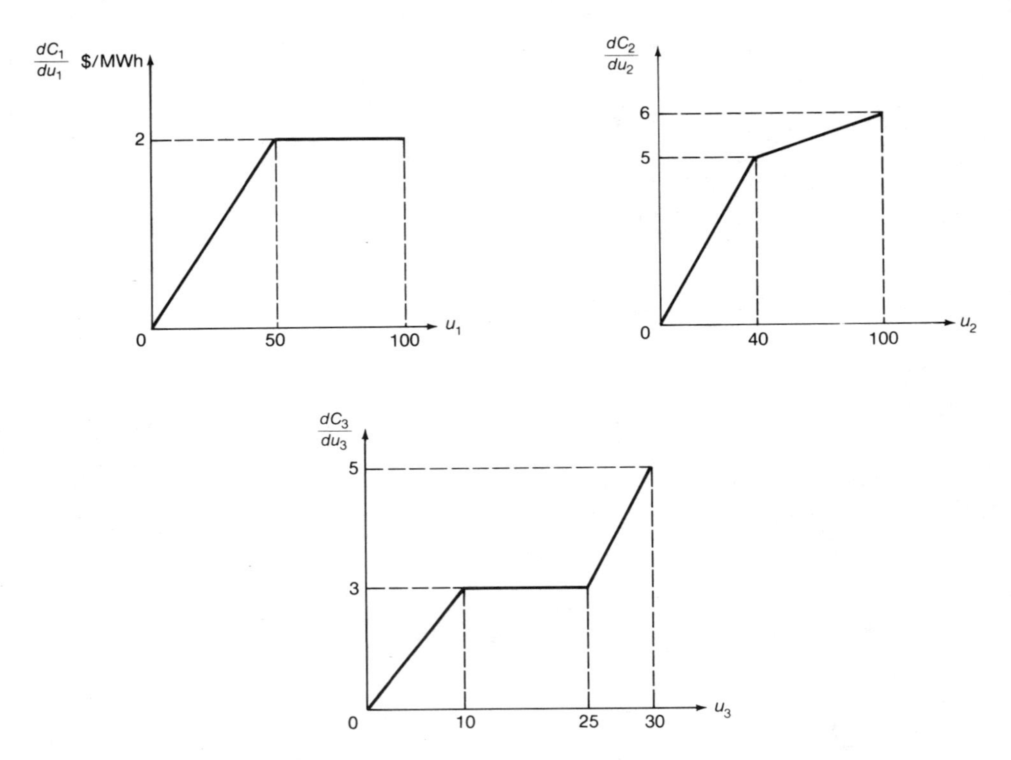

FIGURE 6.1 Fuel costs for a three-generator power system.

Example 6.1

Figure 6.1 shows the fuel costs of a three generator system. The system capacity is 100 + 100 + 30 = 230 MW. Losses and bus voltages are ignored. It is desired to find the optimal generation dispatch and system λ.

Solution

Let P_T denote the total power

$$P_T = u_1 + u_2 + u_2 \, .$$

When $P_T = 0$, all generators are set to zero. Now permit a small value of P_T (e.g., 1 MW). At this low level, it is clear that each machine, if it is operating, will be at a low level. Near $P_i = 0$,

$$\frac{\partial f_{u_1}}{\partial u_1} = \frac{2}{50} u_1$$

$$\frac{\partial f_{u_2}}{\partial u_2} = \frac{5}{40} u_2$$

$$\frac{\partial f_{u_3}}{\partial u_3} = \frac{3}{10} u_3 .$$

The equal incremental cost rule prescribes that these three incremental costs must be equal:

$$\frac{2}{50} u_1 = \frac{5}{40} u_2 = \frac{3}{10} u_3 = \lambda .$$

Thus, for $\lambda = 0.01$, for example,

$$u_1 = 0.25\ MW \quad u_2 = 0.08\ MW \quad u_3 = 0.033\ MW$$

$$P_T = 0.363\ MW .$$

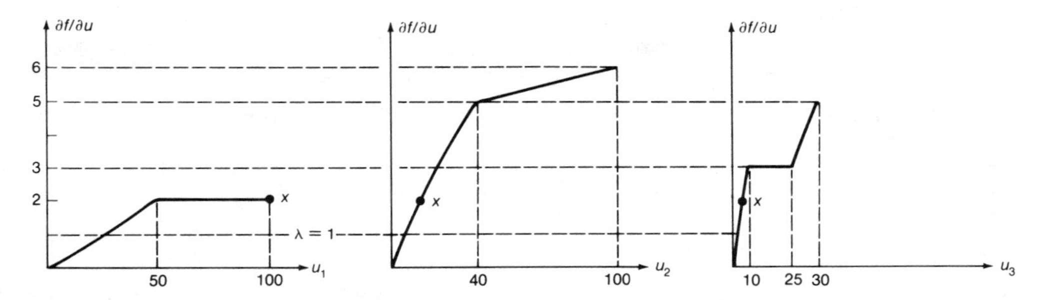

FIGURE 6.2 Application of the equal incremental cost rule.

It is clear that as λ rises, u_i will rise for each machine in a linear manner until one (or more) machines reach a point such that its $\partial f / \partial u$ expression is no longer as specified above. This is pictorially shown in Figure 6.2 with $\lambda = 1$ indicated. At $\lambda = 1$,

$$u_1 = 25 \qquad u_2 = 8 \qquad u_3 = 3.33$$

$$P_T = 36.33 \ .$$

It is clear from inspection of Figure 6.2 that as λ is permitted to rise to λ_a, the $\lambda = \lambda_a$ line will intersect each of the graphs in their initial region up to the point where $\lambda = \lambda_a = 2$. At this value of system λ, the fuel cost characteristic at machine 1 will change and $\partial f / \partial u$ at machine 1 will be 2 \$/MWh for $50 \le u_1 < 100$. The initial region ($0 < \lambda < 2^-$) is shown in Figure 6.3 labeled "I." For $2^- < \lambda < 2^+$, only machine 1 will change load. At $\lambda = 2^+$,

$$u_1 = 100 \quad u_2 = 16 \quad u_3 = 6.67$$

$$P_T = 122.67 \ .$$

This region is labeled "II" in Figure 6.3.

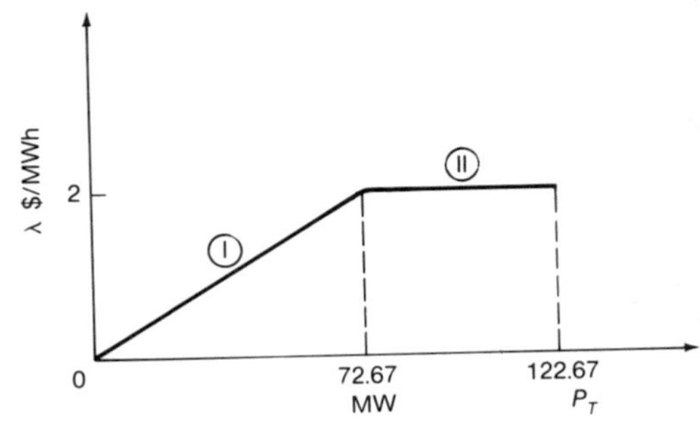

FIGURE 6.3 Solution regions in Example 6.1.

The term "cycle" is used for generating units to indicate a unit whose power output is changing. Over region I, units 1, 2, and 3 cycle. Over region II, only unit 1 cycles; units 2 and 3 are fixed. Note that at P_T = 122.67, the three units are operated at the point labeled X in Figure 6.2. At X, each incremental fuel cost is 2 \$/MWh.

When λ is permitted to exceed 2 \$/MWh, unit 1 remains at its rating (100 MW) and effectively drops out of consideration for further cycling. In the next region, III, units 2 and 3 cycle. The same linear behavior in $\partial f_{u_2}/\partial u_2$ and $\partial f_{u_3}/\partial u_3$ result in linear behavior of P_T versus λ. Region III ends at $\lambda = 3$, where the $\partial f/\partial u$ characteristic of unit 3 changes form. At $\lambda = 3^-$,

$$u_1 = 100 \ \ (fixed) \qquad u_2 = 24 \qquad u_3 = 10$$

$$P_T = 134 \ .$$

The next region is denoted as region IV and only unit 3 cycles in this region as λ traverses $3^- < \lambda < 3^+$, At $\lambda = 3^+$

$$u_1 = 100 \quad u_2 = 24 \quad u_3 = 25$$

$$P_T = 149.$$

In region V, unit 1 remains at rating, and units 2 and 3 cycle. For this region

$$\frac{\partial f_{u_2}}{\partial u_2} = \frac{5}{40}\, u_2 \qquad \frac{\partial f_{u_3}}{\partial u_3} = \frac{2}{5}\, u_3 - 7.$$

The expression for the incremental fuel cost at unit 3 is simply the equation of a straight line for that unit in the region $25 \leq u_3 \leq 30$. For region V

$$\frac{5}{40}\, u_2 = \frac{2}{5}\, u_3 - 7 = \lambda.$$

At $\lambda = 5$ this region terminates and the power levels are

$$u_1 = 100 \quad u_2 = 40 \quad u_3 = 30$$

$$P_T = 170.$$

At the end of region V, unit 3 has reached its rating, and the only remaining unit that can be cycled is unit 2. Therefore, in region VI, unit 2 cycles from 40 MW to 100 MW. At $\lambda = 6$,

$$u_1 = 100 \quad u_2 = 100 \quad u_3 = 30$$

$$P_T = 230 = system \;\; capacity.$$

All six regions are shown in Figure 6.4. Also shown in Figure 6.4 are the required settings of u_1, u_2, and u_3 for a given total demand, P_T. Notice that Figure 6.4 is a schedule of how u_1, u_2, u_3 should be set for any demand $0 \leq P_T \leq 230$.

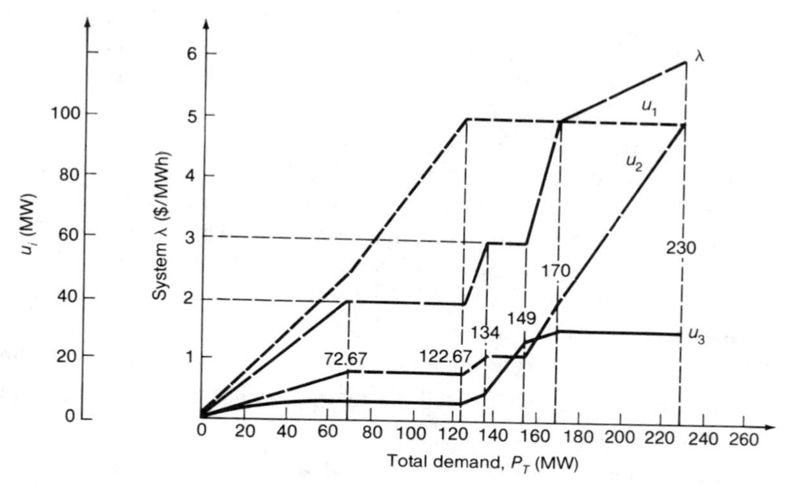

FIGURE 6.4 Solution for Example 6.1.

Nonzero Values of $\partial f/\partial u$ at $u = 0$

If the incremental cost of one or more units, $\partial f_{u_i}/\partial u_i$, is nonzero at $u_i = 0^+$, the equal incremental cost rule is still applied using the value of $\partial f_{u_i}/\partial u_i$ at $u_i = 0^+$ for the incremental fuel cost at $u_i = 0$.

Minimum Values of u_i, Base Load Units, and Restricted Operating Ranges

If a unit has a restricted operating range (e.g., a minimum value of u_i for which the unit may be operated), the equal incremental cost rule still applies. One convenient way to force $u_i \geq u_i^{\min}$ is to penalize $\partial f_{u_i}/\partial u_i$ to an arbitrarily high value for $u_i < u_i^{\min}$. If base load units are present, they are dispatched as either *on* or *off*, depending on whether the system λ is smaller than – or greater than the unit $\partial f/\partial u$, respectively. Base load units are characterized by low incremental generation costs (e.g., hydroelectric units).

Restricted operating ranges are similarly handled as arbitrarily high penalties on $\partial f/\partial u$ over the restricted range.

Computer Programming and Tabular Values of $\partial f/\partial u$

Usually, the incremental costs of individual units are not simple graphs as shown in Example 6.1. Rather, they are tabular in nature and they may even change with raw material costs, operating and ambient temperatures, and a variety of derating procedures. Tabular unit incremental costs are readily handled by digital computer by a series of interpolations and incrementing the system λ. The procedure is examined using the tabular incremental costs of each unit:

$$\begin{matrix} \text{DFDU}(1,1) & \text{DFDU}(2,1)\cdots\text{DFDU}(g,1) \\ \text{DFDU}(1,2) & \text{DFDU}(2,2)\cdots\text{DFDU}(g,2) \\ \cdots & \\ \text{DFDU}(1,r) & \text{DFDU}(2,r)\cdots\text{DFDU}(g,r) \end{matrix}$$

where DFDU(i, j) refers to the incremental fuel cost of unit i, $i = 1, 2, \ldots, g$, for each of r levels of power. There are r levels of quantization in power settings. If the input table for unit k is not the same length as that for unit ℓ, or if the tables are not synchronized because their resolutions are different (e.g., unit 1 is given in steps of 1 MW, but unit 2 is given in steps of 10 MW), steps must be taken to alleviate these problems. Synchronization of the tables is accomplished using interpolation. Resolution of discrepancies in the lengths of tables is accomplished by storing the length of each table, and before attempting to access data from a given table, reference to the table length is made. It is usually convenient to generate a set of data that "defines" the table: these data include u_i at the first entry of the table, Δu_i = the resolution of each row in the table in MW, and r_i = the row number of the last entry. These data are termed the reference vector or the reference data for the table.

The computerized procedure to implement the equal incremental cost rule is as follows:

1. Set the system lambda, ZLAMBDA, to zero or some arbitrarily low value.

2. Scan DFDU for unit 1 to find ZLAMBDA. The entry is found in E_1 and this represents $(\Delta u)E_1$ megawatts generated at unit 1.

3. Repeat step 2 at unit 2, 3, and so on, until all g units have been found. Store $(\Delta u)E_1$, $(\Delta u)E_2$, ..., $(\Delta u)E_g$. Note that the scanning of DFDU will require interpolation. Also, if ZLAMBDA is smaller than any entry in DFDU, the corresponding E value is zero. If ZLAMBDA is bigger than any entry in the DFDU table, return the maximum row number in the table as E. The maximum entries and resolution of the tables are stored in the reference vector for the table.

4. Having found the unit power settings for each unit for all given values of ZLAMBDA, it is possible to print out a table of ZLAMBDA, $(\Delta u)E_1 = u_1$; $(\Delta u)E_2 = u_2$;...; $(\Delta u)E_g = u_g$; and $\Sigma u = P_T = \Sigma\, i\; u_i$. Unfortunately, such a table would be evenly spaced in values of $\Delta\lambda$. Since this table is more likely to be used with P_T given, it is appropriate to regenerate the table with even spacing in ΔP_T. This is done by scanning the ZLAMBDA - $(\Delta u)E_1 - (\Delta u)E_2 - \cdots - (\Delta u)E_g - P_T$ tables with successively larger and evenly spaced values of P_T. The result of such interpolation is a table of $\lambda - u_1 - \cdots - u_g - P_t$ evenly spaced in P_T. This table is used by an operator to set generation levels given the system load. Alternatively, a computer may give each unit an automatic command for economic dispatch.

Production Cost

The system lambda is the incremental cost in dollars/(megawatthour) to produce the last megawatt for 1 hour. The total cost per hour is termed *production cost*, C_T,

$$C_T = \int_0^{P_T} \lambda(P)dP,$$

where C_T is in dollars per hour. The production cost is readily evaluated numerically using the trapezoidal rule when $\lambda(P)$ versus P is available in tabular form. When λ_i is the system lambda for total demand, $P_T = (P_T)_n$, by this method is

$$C_T \simeq \Delta P_T(1/2\lambda_n + \sum_{i=1}^{n-1} \lambda_i)$$

where ΔP_T is the resolution of the $\lambda - P_T$ table. Economic dispatch programs are commonly known as ***production costing*** programs, and sometimes commercially available programs will also have the capability to calculate fuel requirements, fuel types, and other information relating to unit fuel.

6.3 B-COEFFICIENTS

The equal incremental cost rule suffers from numerous flaws, which stem primarily from the fact that the Kirchhoff laws are not satisfied. Included in this observation is that losses are not considered. As a quick illustration, consider generating unit 1 at the end of a long, lossy transmission line and unit 2 centrally located near the load centers. The equal incremental cost rule does not penalize unit 1 even though a Watt generated at 1 is not as effective as a Watt generated at 2.

In the next section it will be evident that an attractive alternative approach is to augment the Kirchhoff laws into the optimization process. But this is not trivial, and a simpler approach based on penalty factors and the equal incremental cost rule may suffice. Prior to the advent of the Newton–Raphson method for power flow solutions, this method (known as the method of B–coefficients or loss coefficients) was in common use. The method is still widely used, although more accurate methods are slowly displacing this technique.

Reconsider (6.6) and (6.7) in such as to include total transmission system losses:

$$G(U, X, P) = \sum_{i=1}^{g} u_i + \sum_{j=1}^{\ell} p_j - \sum_{k=1}^{t} \bar{p}_k,$$

where $\bar{p}_k$ is the active power loss in transmission line k, and the sum is taken over t lines. The Lagrangian is

$$L(U, X, P) = C(U) - \lambda G(U, X, P).$$

To obtain an optimum,

$$\nabla_u L(U, X, P) = 0.$$

The result is

$$\frac{\partial}{\partial u_i} f_{u_i}(u_i) - \lambda + \lambda \frac{\partial}{\partial u_i} \sum_{k=1}^{t} \bar{p}_k = 0$$

for each $i = 1, 2, \ldots, g$. This equation may be rearranged as

$$\frac{\partial}{\partial u_i} f_{u_i}(u_i) = \lambda\left(1 - \sum_{k=1}^{t} \frac{\partial \bar{p}_k}{\partial u_i}\right)$$

so as to interpret the term $1 - \sum_{k=1}^{t} \frac{\partial \bar{p}_k}{\partial u_i}$ as a penalty factor on λ. The effect of this penalty factor is to cause $\partial f / \partial u$ to be effectively higher when there are losses. Perhaps a few further condensations in notation will make this result more evident. Let L_i be the penalty factor

$$L_i = 1 - \sum_{k=1}^{t} \frac{\partial \bar{p}_k}{\partial u_i}$$

and B_{ki} be the loss coefficient associated with transmission line k and generator i. Then

$$\frac{\partial}{\partial u_i} f_{u_i}(u_i) = \lambda L_i$$

$$L_i = 1 - \sum_{k=1}^{t} B_{ki} \,. \tag{6.11}$$

Then the modified equal incremental cost rule is restated as the following: *The method of B-coefficients: The optimal dispatch of a power system is obtained when the penalized incremental costs of each unit,* $1/L_i \; \partial f_i / \partial u_i$, *are equal to each other.*

Note that the $1/L_i$ multiplier is greater than 1 because the B–coefficients are positive [see Eq. (6.11)]. Hence it is not simply the unit incremental costs that are equated, but the penalized costs. The higher the loss coefficients, the greater the penalty.

The question that remains relates to how the B–coefficients must be calculated. In general, these coefficients are not constants. However, over a small range of generation swings, B_{ki} may be assumed nearly constant. There have been several proposed formulas for the B–coefficients. None are truly "standard," and two such formulas are derived and presented below.

A Formula for B–Coefficients

The loss coefficient B_{ij} is the incremental loss in line i for an increment in generation at bus j,

$$B_{ij} = \frac{\partial \bar{P}_{lost\ i}}{\partial P_j},$$

where P_j is the injected power at the generation bus j. The total incremental transmission system losses due to increment of generation at bus j is $\partial \bar{P}_L / \partial P_j$,

$$\frac{\partial \bar{P}_L}{\partial P_j} = \sum_{i=1}^{t} B_{ij},$$

where $\bar{P}_L$ is the total transmission loss and the sum is carried over all t lines. This $\partial \bar{P}_L / \partial P_i$ summation of incremental transmission losses is the summation term in (6.11). The total loss P_L is readily identified as being the total active power injected into the network

$$\overline{P}_L = Re\ (I_{bus}^t Z_{bus} I_{bus}^*)\ ,$$

which is a similar function used in connection with the discussion of the passivity of the network and the positive real character of Z_{bus}. Now note that $\overline{P}_L$ is due only to the resistive part of Z_{bus} namely R_{bus},

$$\overline{P}_L = [Re\ (I_{bus})]^t R_{bus}\ Re\ (I_{bus}) + [Im\ (I_{bus})]^t R_{bus}\ Im\ (I_{bus})\ .$$

The required incremental transmission loss is the derivative of this expression, which can be shown to be [18]

$$\frac{\partial \overline{P}_L}{\partial P_j} = \sum_{k=1}^{n_b} \sum_{\ell=1}^{n_b} \frac{\partial}{\partial P_j}[\alpha_{\ell k}(P_\ell P_k + Q_\ell Q_k) + \beta_{\ell k}(Q_\ell P_k - P_\ell Q_k)], \quad (6.12)$$

where n_b is the number of buses, P_ℓ and Q_ℓ denote the active and reactive power injection at bus ℓ, and $\alpha_{\ell k}$ and $\beta_{\ell k}$ are

$$\alpha_{\ell k} = \frac{\overline{R}_{\ell k}}{|V_\ell||V_k|} cos(\delta_\ell - \delta_k)$$

$$\beta_{\ell k} = \frac{\overline{R}_{\ell k}}{|V_\ell||V_k|} sin\ (\delta_\ell - \delta_k)\ .$$

Elgerd [18] has shown that (6.12) reduces to

$$\frac{\partial \overline{P}_L}{\partial P_j} = 2 \sum_{k=1}^{n_b} (P_k \alpha_{jk} - Q_k \beta_{jk})$$

$$+ \sum_{k=1}^{n_b} \sum_{\ell=1}^{n_b} [(P_\ell P_k + Q_\ell Q_k)\frac{\partial \alpha_{k\ell}}{\partial P_j} - (P_\ell Q_k - Q_\ell P_k)\frac{\partial \beta_{\ell k}}{\partial P_j}], \quad (6.13)$$

where the α and β terms are as defined earlier and the derivative terms are essentially Jacobian matrix entries

$$\left.\begin{aligned}
\frac{\partial \alpha_{k\ell}}{\partial P_j} &= \frac{\overline{R}_{k\ell}\ sin\ (\delta_k - \delta_\ell)}{|V_j||V_k||V_\ell|}\Big[\frac{1}{Y_{jk}|V_k|\ sin\ (\delta_k \delta_j + \theta_{jk})} \\
&\quad - \frac{1}{Y_{j\ell}|V_\ell|\ sin\ (\delta_\ell - \delta_j + \theta_{j\ell})}\Big] \\
\frac{\partial \beta_{\ell k}}{\partial P_j} &= \frac{\overline{R}_{\ell k}\ cos\ (\delta_\ell - \delta_k)}{|V_j||V_k||V_\ell|}\Big[\frac{1}{Y_{jk}|V_k|\ sin\ (\delta_k - \delta_j + \theta_{jk})} \\
&\quad - \frac{1}{Y_{j\ell}|V_\ell|\ sin\ (\delta_\ell - \delta_j + \theta_{j\ell})}\Big]
\end{aligned}\right\} \quad (6.14)$$

where Y and θ denote Y_{bus} entry magnitudes and angles.

When the derivatives of the α and β terms given in (6.14) are substituted into (6.13), the derivative $\partial \overline{P}_L / \partial P_j$ is readily calculated. Most required terms are available in the line list or Jacobian matrix. The double sum in (6.13) is usually small since the derivatives of α and β are small. If this double sum is discarded, the total incremental loss is simply

$$\frac{\partial \bar{P}_L}{\partial P_j} = 2 \sum_{k=1}^{n_b} (P_k \alpha_{jk} - Q_k \beta_{jk}) \,.$$

This approximate incremental transmission loss formula involves only line resistance, bus injection, and load flow data. Note also that

$$\alpha_{jk} = \sqrt{1 - \beta_{jk}^2} \,.$$

It was mentioned earlier that the B–coefficients are not constants. The usual procedure for their use is to assume constant values for the coefficients, perform an equal (penalized) incremental cost dispatch, and subsequently perform load flows for the dispatched cases. In this way, updated values of $|V|$ and δ are available for substitution into (6.14). The B–coefficients are then recalculated and the procedure is repeated if necessary.

A totally different approach is based on distributions factors. Although this approach is computationally more compact, it is based on the availability of power transfer distribution factors. These factors are not usually available, and they must be specially calculated.

Note that the B–coefficients are related to the distribution factors as follows:

$$B_{ij} = \frac{\partial \bar{P}_{lost\ i}}{\partial P_{generated\ at\ j}}$$

$$= \frac{1}{(pf)_j} \frac{\partial \bar{P}_{lost\ i}}{\partial |S_j|}$$

$$= \frac{2|\bar{I}_i|\bar{R}_i}{(pf)_j} \frac{\partial |\bar{I}_i|}{\partial |S_j|} ,$$

where $(pf)_j$ is the bus injection power factor at bus j. Thus

$$B_{ij} \simeq \frac{2|\bar{I}_i||\rho_{ij}|\bar{R}_i}{(pf)_j} \cdot \frac{1}{|V_{ti}|} , \tag{6.15}$$

where the voltage $|V_{ti}|$ is assumed to be the voltage at all points on the line i to ground. The two subscripts on the distribution factor in (6.15) denote line i and bus j. The assumptions in this development are those of the distribution factor method. Note that the line current magnitude $|\bar{I}_i|$ must be known from a power flow study. These line currents are available from a base case power flow study. As described above, the B–coefficients may be updated subsequent to a first pass obtained from a base case.

For the case of flat, unity bus voltage profile, (6.15) simplifies. The total incremental losses due to increment in generator j are

$$\frac{\partial \bar{P}_L}{\partial P_j} = \sum_{i=1}^{t} B_{ij} \simeq \frac{2}{(pf)_j} \sum_{i=1}^{t} |\bar{I}_i||\rho_{ij}|\bar{R}_i \,.$$

Implementing the B–Coefficient Method

There are several alternatives for implementing the B–coefficient method. Viewing the B–coefficient technique as a variation to the equal incremental cost method leads to the following algorithm:

1. Assume that all B-coefficients are zero (i.e., the no loss case).

2. Construct a conventional generation schedule based on the equal incremental cost rule.

3. For each row in the generation schedule table, calculate the B–coefficients associated with generator i and transmission line k (B_{ik}), and use (6.15) to calculate the penalty factor for unit i (L_i).

4. Modify the generation incremental cost, $\partial f/\partial u$, by multiplying by $1/L_i$. Thus the penalized cost of unit i is $1/L_i(\partial f_i/\partial U_i)$.

5. Repeat step 2 using the penalized generation incremental costs.

Steps 3 through 5 may be repeated to obtain better accuracy. Note that accuracy in optimal generation scheduling is especially important due to high production costs. A small error in generation scheduling could result in considerable financial expenditure.

Advantages and Disadvantages of the Method of B–Coefficients

The principal advantages of the method of B–coefficients relate to the simplicity of the method. Although the formulas for the B–coefficients may be involved, there are relatively few evaluations required. Memory requirements are low, and because the method is so similar to the equal incremental cost method, programming is not difficult.

The salient disadvantage of the method relates to accuracy. The Kirchhoff laws are not satisfied, and it is difficult to assess the degree to which they are not satisfied. The inaccuracy is high when the line loss $\bar{p}_k$ is high, and when $\bar{p}_k$ varies widely with generation schedule (i.e., B_{ik} is not constant). While the exact conditions of a specified accuracy are system and load dependent, it is typically the case that accuracy deteriorates at high line load, large reactive power flow, and high reactive power generation levels.

A further disadvantage of the method of B–coefficients (and the equal incremental cost rule) is that many types of constraints are inconvenient to consider. Because the reactive power flow is not modeled in these methods, the problems of bus voltage limits and generation var limits are difficult to consider.

6.4 GRADIENT DISPATCH METHODS

Many of the difficulties of the foregoing methods relate to the fact that the Kirchhoff laws are not modeled. In effect, the load flow equations must be modeled – this is a constraint that must appear in the Lagrangian. Let the power flow equations be represented as

$$G(U, X, P) = 0, \tag{6.16}$$

where G is the mismatch active and reactive power at all system buses except that the swing bus (mismatch Q at PV buses does not appear in G), U is the vector of elements that may be controlled to produce the required minimum cost, X is the system state variables, and P is the specified parameters. Typically, control vector U consists of the scheduled generation powers, the voltage set point of all voltage controlled buses (usually at all generator buses) and the phase angle set points of all controllable phase shifters. The state vector X usually consists of all other bus voltage magnitudes, all phase angles, and the tap setting on TCULs. The voltage magnitude at the swing bus and other voltage controlled buses does not appear in the state vector, X, since these magnitudes are control inputs and, as such, appear in U. The parameter vector, P, contains the specified active and reactive load levels. Using the notation

n_{pv} = number of PV buses

n_{pq} = number of PQ buses

n = total number of buses; that note $n = n_{pv} + n_{pq} + 1$

g = number of generators (assumed to be all PV buses)

τ = number of TCUL voltage controlled buses

the dimensions of the vectors in (6.16) are readily found (note that phase shifters are omitted from the present discussion):

G = mismatch vector of dimension $g + \tau + 2n_{pq}$

U = control vector of dimension $1+2g+\tau$

X = state vector of dimension $g+\tau + 2n_{pq}$

P = parameter vector of dimension $2n_{pq}$.

Each of these vectors is all real. The control vector is of the dimension indicated since it contains the swing bus voltage magnitude, the bus voltage magnitude at all other PV buses, and of course, the active power dispatched at all generating units.

The Lagrangian is

$$L(U, X, P) = C(U, X) + \lambda^t G(U,X, P), \tag{6.17}$$

where C is the cost function. In (6.17), the constraint term is written with the coefficient $+\lambda^t$. There is an ambiguity of sign for the Lagrange multiplier term; either $-\lambda^t$ or $+\lambda^t$ will work. Usually, $-\lambda^t$ is used in the equal incremental cost rule formulation in order to give $\lambda > 0$ at the solution. Thus "system lambda" will be a positive number. On the other hand, mathematicians often use the sign shown in (6.17), and this notation is in common use for the steepest descent method. Note that the majority of costs are incurred at generators remote from the swing bus. These costs reflect fuel costs for generation powers appearing in the contral vector U. However, the generation at the swing bus also results in a cost and the swing machine generation level does not occur in U; rather, the swing machine power must be calculated from the voltages at the terminals of the lines that have at least one terminal as the swing bus. Sometimes, the cost C is decomposed into two terms: C_1, the cost at the swing bus and

C_2, the cost of generation at other buses. The term C_1 is a function of both U and X; this is the case since the swing bus voltage is needed as well as some other bus voltages to calculate the swing machine power. The swing bus voltage appears in U and most other voltages in X. The cost C_2 is a function only of U. In (6.17), L and C are scalars, and λ^t is a $g + \tau + 2n_{pq}$ row vector.

The optimization problem is solved by setting several partial derivatives to zero

$$\frac{\partial L(U, X, P)}{\partial X} = 0 \tag{6.18}$$

$$\frac{\partial L(U, X, P)}{\partial U} = 0 \tag{6.19}$$

$$\frac{\partial L(U, X, P)}{\partial \lambda} = 0 \, . \tag{6.20}$$

Equation (6.18) is $g + \tau + 2n_{pq}$ partial derivatives of the scalar function L. This partial derivative is a $g + \tau + 2n_{pq}$ vector. Similarly, (6.19) represents $1 + 2g + \tau$ partial derivates equated to zero and (6.20) is $g + \tau + 2n_{pq}$ equations. In total, this simultaneous set is $1 + 4g + 3\tau + 4n_{pq}$ scalar equations. The unknowns are vectors X, U, and λ which are dimensionally consistent with the number of equations.

There are several techniques for solving these equations – one will be presented here and a second will be discussed later. Needless to say, as variations in the formulation are introduced, alternative advantages may be exploited by variations in numerical solutions. A few of these alternatives will also be discussed later.

Consider (6.18) using the formulation for the Lagrangian in (6.17),

$$\frac{\partial L(U, X, P)}{\partial X} = \frac{\partial C(U, X)}{\partial X} + (\frac{\partial G(U, X, P)}{\partial X})^t \lambda = 0. \tag{6.21}$$

[note that for constant, arbitrary vector B,

$$\frac{\partial}{\partial A} B^t D = (\frac{\partial D}{\partial A}) B;$$

this is used in (6.21)]. Solving for λ yields

$$\lambda = -[(\frac{\partial G(U, X, P)}{\partial X})^t]^{-1} \frac{\partial C(U, X)}{\partial X} \, . \tag{6.22}$$

Expand (6.19)

$$\frac{\partial L(U, X, P)}{\partial U} = \frac{\partial C(U, X)}{\partial U} + (\frac{\partial G(U, X, P)}{\partial U})^t \lambda$$

$$= \frac{\partial C(U, X)}{\partial U} - \left[\frac{\partial G(U, X, P)}{\partial U}\right]^t \cdot$$

$$((\frac{\partial G(U, X, P)}{\partial X})^t)^{-1} \frac{\partial C(U, X)}{\partial X} \, . \tag{6.23}$$

Equation (6.23) is recognized as the gradient of C and this gradient is used to obtain a numerical solution to the optimal dispatch problem using the method of steepest descent (i.e., the gradient method). Equation (6.23) is recognized as the required gradient by noting that the total differential of C is

$$dC(U, X) = (\frac{\partial C(U, X)}{\partial U})^t dU + (\frac{\partial C(U, X)}{\partial X})^t dX . \quad (6.24)$$

Noting that

$$G(U, X, P) = 0$$

gives

$$\frac{\partial G(U, X, P)}{\partial X} dX + \frac{\partial G(U, X, P)}{\partial U} dU = 0 \quad (6.25)$$

which reveals the following upon substitution of dX from (6.25) into (6.24),

$$dC(U, X) = \left[(\frac{\partial C(U, X)}{\partial U})^t - (\frac{\partial C(U, X)}{\partial X})^t (\frac{\partial G(U, X, P)}{\partial X})^{-1} \frac{\partial G}{\partial U} \right] dU . \quad (6.26)$$

Since

$$dC(U, X) = (\nabla C(U, X))^t dU ,$$

the long term in square brackets in (6.26) is recognized as the gradient of C, (∇C). This is exactly $\partial L / \partial U$ as given in (6.23).

The conclusion of this development is that the gradient of C is

$$\nabla C = \frac{\partial L}{\partial U} = \frac{\partial C}{\partial U} - (\frac{\partial G}{\partial U})^t ((\frac{\partial G}{\partial X})^t)^{-1} \frac{\partial C}{\partial X} . \quad (6.27)$$

This important result leads to the algorithm for the steepest descent solution of the optimal power flow problem. This is summarized below together with a few observations, alternative formulations, and alternative solution techniques.

Steepest Descent (Gradient) Solution to the Optimal Power Flow Problem

The essence of the steepest descent solution method for the optimal power flow problem is the reduction of C using (6.27) and simultaneously satisfying the load flow equations,

$$G(X, U, P) = 0 .$$

This is done as follows:

1. Guess a generation dispatch control vector, U.

2. Perform a conventional load flow study, forcing the mismatches to zero within a prescribed tolerance. The solution to this load flow study is X.

3. Calculate the gradient of the cost, C, using (6.27). Note that the $\partial G / \partial X$ terms are Jacobian entries; the $\partial G / \partial U$ terms are either Jacobian entries or unity. The latter is the case when the partial of active power mismatch is taken at bus i with respect to dispatched active power at bus i.

4. The gradient of C is a $1 + 2g + r$ vector each of whose entries correspond to a control, u_i. The vector $\nabla C / ||\nabla C||$ is a vector of unit length consisting of direction cosines of the curve $C(U, X)$ when plotted

in U-space. Hence the control u_i is modified using

$$u_i^{new} = u_i^{old} - k[row_i \frac{\nabla C}{||\nabla C||}]^{-1}, \tag{6.28}$$

where k determines the step size.

5. Having updated the control vector U, the process is repeated by performing another load flow study (step 2). The process stops when ∇C is sufficiently small (and the minimum C has been reached).

Selection of Step Size

The selection of parameter k in (6.28) should be made large when far from the solution and small when near to the solution. In approximate terms, k should be proportional to $||\nabla C||$.

Unfortunately, some elements of U are voltages, some are power settings. For this reason, it may not be reasonable to use the same k for all values of i in (6.28). Several near optimal techniques for selection of the step size exist.

The Calculation of ∇C

The success of many algorithms in power engineering depend on the appropriateness of sparsity programming and the avoidance of large matrix inversion. This is the case because most matrices commonly encountered in power engineering are of very high dimension. With these remarks in mind, consider the formula for ∇C, Eq. (6.27), and let

$$W = (\frac{\partial G}{\partial U})^t [(\frac{\partial G}{\partial X})^t]^{-1} . \tag{6.29}$$

Then

$$\frac{\partial G}{\partial X} W^t = \frac{\partial G}{\partial U} \tag{6.30}$$

and W^t can be found by triangularly factorizing $\partial G/\partial X$ and solving for column i of (W^t) using

$$[A_L][A_U]col_i(W^t) = col_i(\frac{\partial G}{\partial U}), \tag{6.31}$$

where A_L and A_U are the lower and upper triangular factors of $\partial G/\partial X$. Equation (6.31) must be repeated for $i = 1, 2, ..., 1 + 2g + r$. Having found all columns of W^t, ∇C is calculated using

$$\nabla C = \frac{\partial C}{\partial U} - W\frac{\partial C}{\partial X} . \tag{6.32}$$

Note that in (6.31), the triangular factors of $\partial G/\partial X$ are simply the triangular factors of the Jacobian matrix. Thus the computer code needed for the sparsity programming, triangular factorization, and Jacobian element calculations need not be repeated – these implementations are taken from the Newton–Raphson power flow study. However, it is necessary to perform $1 + 2g + r$ forward/backward substitutions to calculate W. It turns out that there is a better alternative, which involves the calculation of λ – this alternative is discussed later. The matrices $\partial G/\partial X$ and $\partial G/\partial U$ are sparse.

The vectors $\partial C/\partial U$ and $\partial C/\partial X$ are calculated from the input data for costs at each machine. If C is decomposed into the swing machine cost, C_1 and the other costs, C_2,

$$\frac{\partial C}{\partial U} = \frac{\partial C_1(U, X)}{\partial U} + \frac{\partial C_2(U)}{\partial U}$$

$$\frac{\partial C}{\partial X} = \frac{\partial C_1(U, X)}{\partial X} .$$

The vector $\partial C_2/\partial U$ consists of conventional incremental cost data, but the vectors $\partial C_1/\partial U$ and $\partial C_1/\partial X$ require both Y_{bus} data and incremental cost data.

An Alternative Solution Technique Using Calculation of λ

The selection of matrix W in (6.29) was made as the first two terms in the three term product in (6.27). If, instead, the last two terms are considered, namely $((\partial G/\partial X)^t)^{-1}\ \partial C/\partial X$, one recognizes this product as $-\lambda$ [see Eq. (6.22)]. Thus

$$\nabla C = \frac{\partial C}{\partial U} - (\frac{\partial G}{\partial U})^t [-\lambda] \tag{6.33}$$

$$(\frac{\partial G}{\partial X})^t [-\lambda] = \frac{\partial C}{\partial X} . \tag{6.34}$$

Now the solution method is to solve for $-\lambda$ in (6.34) using the triangular factors of $(\partial G/\partial X)^t$. Since $\partial C/\partial X$ is a vector, only one forward/backward substitution need be done, not $1 + 2g + r$ as in the previous method. The transposition of $\partial G/\partial X$ in (6.34) gives no trouble since the triangular factors required here are essentially the transposes of the triangular factors of the Jacobian matrix. This fact is readily verified by noting that when matrix A has triangular factors LU, the matrix A^t has factors $U^t L^t$. One concludes that the calculation of the vector $[-\lambda]$ is accomplished using available triangular factors, sparsity programming, and Jacobian entries.

Having found $[-\lambda]$, (6.33) is used to calculate ∇C. This gradient is used to minimize C.

Inclusion of Constraints

There are numerous advantages of the gradient formulation outlined above – and these will be summarized below – but perhaps one of the most important is that this formulation permits the inclusion of several types of constraints. These constraints are needed to obtain realistic solutions to the optimal dispatch problem. As a brief illustration, consider the voltage set points at generation buses. These set points are controls – but if allowed to be unconstrained, the optimal power flow study will not converge because the ever increasing economics of higher and higher voltage (and lower and lower $|I|^2\bar{r}$ losses) will drive voltage magnitude entries in the control vector to infinity. Similarly, certain power levels must be constrained to reflect maximum power limits of machines and transmission lines.

There are numerous ways to classify constraints on the optimal dispatch problem: ***voltage*** and ***current*** constraints; and ***equality*** and ***inequality*** constraints; constraints on ***power*** or ***voltage*** (or perhaps some other parameter in special applications); and constraints at ***buses*** or on *network* parameters. Not all of these are not discussed here, but a few general methods are shown for inequality constraints. Note that equality constraints are handled by augmenting those equalities into the Lagrangian.

The two most widely used techniques to handle inequality constraints are through the use of penalty functions and by application of the Kuhn–Tucker conditions. The former has the advantages of simplicity in programming but disadvantages of inexactness in modeling; the latter is more difficult to program, but offers accuracy commensurate with the other segments of the optimal dispatch algorithm. The choice of method and model depends on the application.

Penalty functions may be used to increase $C(U, X)$ as inequality constraints grow closer to violation. Consider the inequality constraint

$$M(U, X) \leq \overline{M}, \tag{6.35}$$

where M is a vector–valued function and $\overline{M}$ is a limit vector of same dimension. Inequality (6.35) holds when the inequality holds in each row; otherwise, it does not hold. Now consider f, a scalar–valued function of a vector–valued argument such that $f(M(U, X) - \overline{M})$ is very large whenever any row of $M(U, X) - \overline{M}$ is positive, is large whenever any row of $M(U,X) - \overline{M}$ is zero, and is small when $M(U, X) - \overline{M}$ is negative. Let f be a positive valued function illustrated by Figure 6.5. The function f is termed a ***penalty function,*** and it is used to penalize $C(U,X)$ to model inequality constraints,

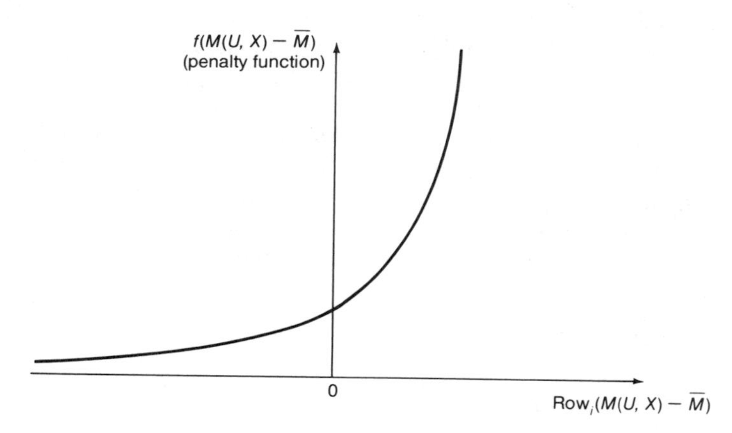

FIGURE 6.5 Penalty function.

$$L(U, X) = C(U, X) + f(M(U, X) - M) + \lambda^t G(U, X, P)\,. \tag{6.36}$$

Since most penalty functions, by virtue of the finite change in f as M exceeds $\overline{M}$, augment L with a finite penalty, the use of these functions does not usually preclude a violation of an inequality constraint. For this reason, an inequality constraint that is modeled using penalty functions is termed a *soft* constraint. A few typical penalty functions are

Exponential

$$f(m) = e^{km} \qquad k > 0$$

Power Times Exponential

$$f(m) = m^{2n} e^{km} \qquad n > 0, \quad k > 0$$

Composite

$$f(m) = am^{2n} e^{km} + be^{km} \qquad n > 0, \quad k > 0, \quad a > 0, \quad b > 0$$

The selection of penalty function is heuristic and sometimes involves a trial-and-error process. The function should be differentiable over the range of argument that may be encountered in $\partial L / \partial U$ and $\partial L / \partial X$.

Because the use of a penalty function does not guarantee that inequality constraints are satisfied, there is considerable motivation to seek an alternative to this technique. The most commonly used method relies on the Kuhn–Tucker conditions which result in the satisfaction of the imposed limits; for the latter reason, the limits imposed are termed ***hard*** limits. Consider the scalar cost function $C(U, X)$ which is to be minimized. Equality constraints of the form

$$G(U, X, P) = 0$$

are imposed, and inequality constraints of the form

$$H(U, X, P) \leq 0$$

are also applied. This is a vector inequality which holds when the scalar inequalities of each row hold. The Lagrangian in this case is

$$L(U, X, P) = C(U, X) + \lambda^t G(U, X, P) + \mu^t H(U,X,P) , \tag{6.37}$$

where λ^t is a row vector of dimension commensurate with G and consists of the Lagrange multipliers and μ^t is a row vector of dimension equal to the dimension of vector H. Vector μ consists of elements termed ***dual variables***. There is an ambiguity of sign for both the λ^t and μ^t terms – either a positive or a negative sign will suffice. The Kuhn–Tucker conditions are

$$\frac{\partial L(U,X,P)}{\partial U} = 0 \tag{6.38}$$

$$\frac{\partial L(U,X,P)}{\partial X} = 0 \tag{6.39}$$

$$\frac{\partial L(U, X, P)}{\partial \lambda} = 0 \tag{6.40}$$

augmented by an additional k_μ conditions where k_μ is the dimension of vectors μ and H. These additional conditions are termed the ***exclusion conditions***,

$$\mu^t H(U, X, P) = 0 . \tag{6.41}$$

The practical significance of (6.41) is that ***either***

$$\{\mu_i = 0\} \tag{6.42a}$$

or

$$\{\frac{\partial L}{\partial \mu_i} = H_i = 0\}, \tag{6.42b}$$

here $i = 1, 2, ..., k_\mu$. The usual solution algorithm is to solve (6.38) through (6.40) simultaneously with $\mu = 0$ as in the case without inequality constraints. Then H_i is checked for a violation of the constraint. In cases of violation, (6.42b) is used and one solves for μ. This process must be done for each permutation of violations of inequality constraints. For each case, the minimum C is stored and the global solution is deduced as the case of minimum C over the finite number of solved cases.

When H_i does not violate its inequality constraint, μ_i is zero. When H_i does violate its inequality constraint, μ_i becomes a Lagrange multiplier. A brief example will illustrate the procedure before examining the more complex power engineering application.

Example 6.2

It is desired to find the minimum value of $C(x, y)$,

$$C(x, y) = x^2 + y^2$$

subject to equality constraint g,

$$g(x, y) = x^2 - y + 1 = 0,$$

and inequality constraint h,

$$h(x, y) = 6x + y + 2 \leq 0.$$

Solution

In the absence of the inequality constraint, the Lagrangian is L,

$$L(x, y) = C(x, y) + \lambda g(x, y) .$$

Using the given functions f and g,

$$\frac{\partial L}{\partial x} = 2x + 2x\lambda = 0$$

$$\frac{\partial L}{\partial y} = 2y - \lambda = 0 .$$

The simultaneous solution yields only one real-valued solution,

$$x = 0 \quad y = 1 \quad \lambda = 2 \quad C(0, 1) = 1.$$

When the inequality constraint is added,

$$L(x, y) = x^2 + y^2 + \lambda(x^2 - y + 1) + \mu(6x + y + 2),$$

where μ is a dual variable. For $\mu = 0$ one finds the aforementioned trial solution. At this trial solution, h is checked for a violation

$$h(0, 1) = 3.$$

This value of h violates the stated inequality condition. Therefore, the condition that $\mu = 0$ is removed and in its place one adds

$$\frac{\partial L}{\partial \mu} = 0.$$

Hence

$$\frac{\partial L}{\partial x} = 0 = 2x + 2x\lambda + 6\mu$$

$$\frac{\partial L}{\partial y} = 0 = 2y - \lambda + \mu$$

$$\frac{\partial L}{\partial \lambda} = 0 = x^2 - y + 1$$

$$\frac{\partial L}{\partial \mu} = 0 = 6x + y + 2.$$

The simultaneous solution is

$$x = -3 \pm \sqrt{6} \qquad \lambda = 37 \pm \frac{33\sqrt{6}}{2}$$

$$y = 16 \mp 6\sqrt{6} \qquad \mu = 69 \pm \frac{57\sqrt{6}}{2}$$

where all the upper signs correspond to one solution, and all the lower signs correspond to another. The upper sign is found to yield the minimum C,

$$C = 487 - 198\sqrt{6}$$

$$\simeq 2.001.$$

As expected, the imposition of the inequality constraint causes the minimum value of C to be greater than that without the inequality constraint.

It is interesting to compare this result with a penalty function approach. This is done in Table 6.1 for two different penalty functions.

Table 6.1

Comparison of Solutions to Example 6.2 Using the Kuhn–Tucker and Penalty Function Methods

Method	Penalty Function	x	y	λ	C
Kuhn–Tucker	None	-0.551	1.303	3.417	2.001
Penalty function[a]	$e^{10(6x+y+2)}$	-0.599	1.359	3.696	2.206
Penalty function[a]	$(6x + y + 2)^2 e^{10(6x+y+2)}$	-0.540	1.330	3.390	2.061
No inequality constraint	—	0.000	1.000	2.000	1.000

[a] Solution for x, y, λ obtained numerically. Values of x, y, λ, C shown at minimum C condition.

Practical Application of the Kuhn–Tucker Method to the Steepest Gradient Solution of the Optimal Power Flow Problem

The following are additional guidelines for the practical application of the Kuhn–Tucker method for the steepest gradient solution of the optimal power flow problem:

1. *Types of inequality constraints:* The most usual types of inequality constraints are upper bus voltage limits at generation and load buses, lower bus voltage limits at load buses, var limits at generation buses, maximum active power limits corresponding to generating unit ratings, minimum active power limits corresponding to lower limits at some generators, and maximum line loading limits. Note that generation bus voltage limits usually must be implemented when U contains generation bus voltages, to avoid unbounded increase in these voltage magnitudes. Also, limits on tap settings of TCULs and phase shifter settings should be applied.

2. *Gradient step size in the presence of TCULs:* The gradient step size as discussed above is designed for per-unitized parameters which are generally near unity. When TCULs are present, there may be anomalous conditions which occur since the step size is too big or two small. In such cases, rather than working with the tap setting t_i, a factor α_{ti} is applied and $\alpha_{ti} t_i$ is used in calculations. A similar condition occurs in phase shifting transformers.

3. *Uncertainty in fuel availability and cost:* In some cases, optimal power flow studies are run using load forecast data for future conditions. This may be done for cases as much as 20 years in the future. This is done to obtain production cost projections and to evaluate alternative generation configurations. This may also be done to assess the dependency on neighbors for generation interchange and to determine whether adequate generation reserve is maintained in all parts of the system under economic dispatch conditions. Whenever studies are run for future cases, the uncertainty in transmission, generation, fuel, and economic parameters should be considered. If load forecast data are used for bus loads, it may be prudent to repeat the study for a 20% increase in load and a 20% decrease in load. Similarly, fuel availability and cost should be considered for cases in the future. These parameters could be modeled as stochastic processes and the statistics of the optimal dispatch could be calculated under assumption of stationary statistics, but the method is complex and not in common use [2,3].

4. *Frequency of performing optimal power flow studies:* Optimal power flow studies are used in some system planning procedures. These studies are often done using load forecast data for future cases. A much more common application of optimal dispatch algorithms is in *system operation.* The motivation for the use of economic dispatch methods in system operation is quite different from system planning procedures: in system planning studies, feasibility and projected costs are considered. In system operation, minimum operating cost is the primary consideration (considering reliability, reserve, and other operating requirements). The frequency of performing optimal power flow studies (i.e., how often must they be run?) is dictated by the latter motivation.

Usually, an economic optimal dispatch study will be run once or twice a day in order to obtain the approximate generating schedule for that day. As the day progresses, various on-line procedures are used to update the generation schedule. This update could occur on a minute-to-minute basis. The update algorithm must be a high speed method, and the equal incremental cost rule (or a variant) is often used.

A common dispatch procedure is to run an economic optimal dispatch using the gradient method for a 12–hour period which includes the peak load condition. Fuel is scheduled using the results of this study. As the 12–hour period progresses, a minute-by-minute update is made and the generating units are adjusted automatically and on-line. Data are presented to the system operator minute by minute so that a malfunction may be detected. Also, automatic alarms are implemented to annunciate problems (e.g., an alarm is used to annunciate the case of system λ out of range – typically, $\pm 20\%$ of the value calculated using the gradient solution). In the case of malfunction of the automatic generation raise and lower controls, the operator will manually dispatch the system, probably using the gradient solution. Under the automatic operating mode, the minute-by-minute update is made using the equal incremental cost rule. Numerous instances of system specific operating conditions occur. Among these are special treatment of base load units, manual dispatch of some older units, and special consideration of hydro requirements. Near the end of the 12–hour period, another full gradient solution is run using the most recent forecast load data. The process of minute-by-minute automatic updates is repeated.

There is no universal, industry-wide technique or schedule for economic optimal dispatch. The conditions for a small system of two machines may be very different from those for a large system of 50 or more generating units. Complicating the picture may be contractual requirements with neighbors and power pools. Also, the economic dispatch of a company may be done in consort with one or more other companies in order to derive additional economic advantages. Such power pools usually have agreements among its members relating to system reliability, power plant ownership, operating economics, and operating policies.

Accuracy of the Gradient Method

The gradient method of solution of the optimal dispatch problem bears numerous similarities to the Newton–Raphson method of power flow solution with regard to solution accuracy. The optimal dispatch contains one load flow study per step in the direction of the negative gradient. The bus power mismatch tolerated in each solved power flow study is typically 0.01 per unit, and this tolerance is probably less than that warranted by the accuracy of line and load data. Beyond this error, there is inaccuracy associated with the values of $||\nabla C||$ which is permitted at the stopping of the gradient steps. The tolerance in $||\nabla C||$ depends on the units of the cost function and the nature of the U vector. A typical gradient step stopping criterion is to terminate the procedure when $||\nabla C||$ is less than $0.01c_f$ dollars per hour per unit of power, where c_f is a typical fuel cost per hour.

The salient advantages of the gradient formulation of the optimal power flow study are modeling and constraint capability, and solution accuracy.

OPTIMAL REACTIVE POWER AND OTHER DISPATCH METHODS

The most common dispatch procedures are economic optimal dispatch methods of the type outlined in the preceding sections. There are a few other methods intended for specialized applications. Two of these specialized procedures are discussed here. Also, optimal reactive power dispatch is discussed. The latter bears numerous dissimilarities to the economic optimal dispatch despite the commonality of optimization. Many power engineers do not consider var dispatch methods as authentic optimal dispatch techniques; the reader should judge this question of terminology on its own merits.

Minimum Pollution Dispatch

The environmental requirements of some areas may dictate that pollution and other environmental impact factors be considered in generation dispatch. One approach is to augment the fuel cost terms in the Lagrangian with ***costs*** that reflect pollutant emission. The term ***cost*** is used in the broad sense here since it is often very difficult to associate a firm cost with a given pollutant. The new cost, C', is given by

$$C'(U, X) = C(U, X) + \psi(U, X),$$

where ψ reflects the pollutant costs associated with the total generation production.

Environmental requirements may also be modeled as inequality constraints. Typically the total generation at coal fired stations in a given region should be less than a specified limit. This type of constraint is readily handled in H [see Eq. (6.37)].

Unfortunately, one of the most insideous factors of pollutant emissions relates to the temporal character rather than the instantaneous level of the emission. This observation leads to the conclusion that the ***cost*** of pollutants is more a function of the definite integral of U and X over a specified period, T_p,

$$C' = C(U, X) + \psi(\int_{t-T_p}^{t} U dt, \int_{t-T_p}^{t} X \, dt) .$$

Perhaps the easiest way to include the temporal properties of emissions is to consider function ψ as a scalar such that $\psi_i(U_i, X_i)$ is the total emissions in a specified region at hour i. Using ψ_0 to denote the present total emissions, ψ_{-1} the total emissions 1 hour ago, and so on, the sum

$$\sum_{i=-T_p}^{-1} \psi_i(U_i, X_i)$$

accounts for the total emission over the past T_p hours. The subscripts on U and X refer to optimal dispatch solutions at the time i. Then the emission at time $i = 0$ (i.e., at the present) is constrained by

$$\sum_{i=-T_p}^{-1} \psi_i(U_i, X_i) + \psi_0(U_0, X_0) \leq \overline{\psi}, \tag{6.43}$$

where $\overline{\psi}$ is the upper limit of total emissions in the past $T_p + 1$ hours.

The subscript "0" may be dropped and the superscript k added to denote several different areas in which emissions are to be limited. Thus

$$L(U, X, P) = C(U, X) + \lambda^t G(U, X, P) + \sum_k \mu_k [\sum_{i=-T_p}^{-1} \psi_i^k(U_i, X_i)$$

$$+ \psi^k(U, X) - \overline{\psi}^k] + \nu^t H(U, X) . \tag{6.44}$$

In (6.44), μ_k is the scalar dual variable associated with inequality (6.43) in region k, and ν is a vector of dual variables associated with inequalities H. Equation (6.44) is solved as (6.37) using the exclusion conditions

$$\mu_k [\sum_{i=-T_p}^{-1} \psi_i^k(U_i, X_i) + \psi^k(U, X) - \overline{\psi}^k] = 0 \quad (\text{all } k)$$

$$\nu^t H(U, X) = 0 .$$

This approach to optimal pollutant emission dispatch is a quantization approach for ψ^k. References [4–6] document other approaches and illustrations of applications. Perhaps due to the heuristic nature of the quantification of pollution costs, this type of optimal dispatch is not in common use.

Alert and Emergency State Dispatches

To this point, optimal dispatch has been viewed as economic in some sense. It is reasonable to ask whether there are conditions in which economic operation is not desired. The answer is that there are conditions such as emergencies and alert states which require a dispatch to restore or help restore normal operation. Also, it is possible that even under normal operating conditions, considerations may dictate that generating units be dispatched in a way as to comply with reliability requirements. These requirements may be set by the utility company or a "power pool" of which the utility company is a member. Although such dispatch strategies have not been widely used in the past, increased interest and concern for reliable operation of large power systems has lead to the development of algorithms for this purpose. A very brief outline of the objectives of these strategies is given here.

The *normal operating state* of an electric power system is the operating condition in which all important bus voltages are within specified operating range, and all important line loads are within their long term loading limits. In some instances, a specified generation reserve is added to these requirements. The *alert operating state* is that state in which some bus voltages requirements or long term line load limits are violated. Also, there may be a violation of a generation reserve requirement. The *emergency operating state* refers to the case of loss of load (i.e., outage) at some system buses, or deviation in system frequency, or cascade tripping of lines, or negative generation margin (load $>$ generation). The terminology of a fourth state, the *restorative operating state,* may be useful to refer to an operating state in which corrective measures are being made to return either an alert or emergency status to the normal operating state. It is in this restorative mode in which alert and emergency dispatches are used to:

1. Redistribute the power flow through circuits which are not used to capacity.
2. Minimize the bus voltage angular difference across each heavily loaded transmission circuit.
3. Introduce damping to system states which are in oscillation.
4. Adjust interchange power to restore the load/generation balance.
5. Shed load where appropriate.

Traditionally, these objectives have been accomplished manually. Automatic action, often classified as ***control*** rather than ***dispatch,*** may be very short in duration or may extend over an hour or more. The shorter of these actions is best termed "emergency state control," and this is discussed in detail in [7–9]; the longer algorithms more closely resemble dispatch strategies, and these are discussed in [10–12]. These techniques are too lengthy to discuss in detail here; however, emergency control strategies generally involve the correction of energy imbalance through rather "heroic" measures (e.g., venting steam at the main values at a generating station; damping generation swings by switching a "breaking resistor" as a load [13]). Emergency dispatch methods are less drastic, and correspondingly less effective in the short term. The usual strategy of an emergency dispatch is to redistribute line loads, decrease bus voltage phase angle differences at the terminals of overloaded transmission circuits, and perform multiple power flow studies to assess alternative dispatches.

Reactive Power Dispatch Strategies

Usually, the term ***optimal dispatch*** refers to active power dispatch. However, in some cases, the reactive power dispatch is also of economic importance, and the dispatch of reactive power is always of considerable operating importance. The reason that reactive power dispatch assumes a somewhat secondary role compared to the active power analog is that since the inception of ac power systems, it has been operating and design policies to "make up reactive power" locally. This policy refers to one or more of the following procedures: use of shunt capacitors at load buses to locally correct power factor; use of shunt capacitors at distribution and transmission/distribution substations to correct power factor of local circuits; the use of synchronous condensers and other means to support transmission buses at a given voltage magnitude; and the generation of reactive power at central generating stations to hold the station bus at a fixed voltage. Each of these policies results in relatively low reactive power flows and interchanges. Although these procedures are often satisfactory and they are certainly simple, it is natural to ask whether they are optimal in that the required bus voltage are supported (i.e., held within range) using a minimum number of vars.

At the generation and transmission level, the Dommel–Tinney gradient solution is useful for the calculation of reactive power dispatch. This reactive power is generated entirely by synchronous generators. The models for bus voltage regulation appear in the power flow equations, $G(U, X, P)$. The reactive power limits appear either in the inequality constraints, $H(U, X)$, or in the power flow solution which is solved at each iteration in the steepest gradient descent. Shunt capacitors, however, are not readily modeled in the Lagrangian because the objective of optimal shunt capacitor placement may not be represented as a fuel cost. There are several alternative approaches discussed in detail in [14–16]. Reference [17] contains a summary of several methods. Since placement of shunt capacitors is especially important in distribution and subtransmission circuits, a method particularly applicable to low voltage circuits is chosen for description here.

A Linearized Method for Optimal Var Dispatch Using Linear Programming

Consider a subtransmission or distribution system in which certain bus voltages of interest form vector $|V|$. Assume that the increase in bus voltage magnitude is linearly proportional to injected reactive power at several buses; the vector ΔQ denotes the latter injection. The dimensions of vectors $|V|$ and ΔQ may not be the same since the capacitor placement may occur at a different number of buses compared to the buses at which $|V|$ is to be supported. Then

$$\Delta|V| = B\,\Delta Q \tag{6.45}$$

where B is recognized as elements the inverse of the $\partial Q/\partial|V|$ portion of the Jacobian (under the assumptions of superposition and decoupled power flow). To ensure high enough bus voltage,

$$|V| + \Delta|V| \geq |V_{\text{min}}|,$$

where $|V_{\text{min}}|$ is a vector of minimum bus voltage magnitudes and $|V|$ is the "base case" (i.e., no capacitive compensation) bus voltage profile. The $\Delta|V|$ term comes from capacitive compensation. Hence

$$|\Delta V| = B\,\Delta Q \geq |V_{\text{min}}| - |V|. \tag{6.46}$$

Again, the concept of a vector inequality in (6.46) is said to hold when each scalar row holds. Further, it is desired to minimize c_q,

$$c_q = c^t\,\Delta Q, \tag{6.47}$$

where c^t is a row vector of 1's of commensurate dimension with ΔQ. The cost function, c_q, is a scalar. The minimization of c_q subject to (6.46) is accomplished by linear programming (see Appendix A). In most linear programming formulations, the inequality constraints are written with the solution vector appearing on the "smaller than" side [i.e., opposite to inequality (6.46)], and the index that is extremized is maximized rather than minimized. Both problems are avoided by working with $\Delta Q'$ rather than ΔQ, where

$$\Delta Q' = K - \Delta Q\,.$$

In this discussion, ΔQ entries are assumed to be positive for shunt capacitive compensation.

It is possible also to introduce capacitor costs which depend on the size of the unit. This is done by allowing other than unity weighting in c^t. Upper limits may also be introduced, but these are usually not needed.

Example 6.3

Consider a transmission system in which three buses experience low voltage. The bus voltage magnitudes at these three buses must be incremented by Δv_1, Δv_2, and Δv_3 where

$$\Delta v_1 \geq 0.04 \quad \Delta v_2 \geq 0.05 \quad \Delta v_3 \geq 0.06 .$$

The Δv's are to be obtained by shunt capacitor placement at buses 2 and 17. The sensitivities of Δv_1, Δv_2, and Δv_3 with respect to Δq_2, and Δq_{17} are identified as coefficients of the equations

$$\Delta v_1 = 0.01\ \Delta q_2 + 0.03\ \Delta q_{17} \geq 0.04$$

$$\Delta v_2 = 0.08\ \Delta q_2 + 0.05\ \Delta q_{17} \geq 0.05$$

$$\Delta v_3 = 0.02\ \Delta q_2 + 0.01\ \Delta q_{17} \geq 0.06 .$$

In these expressions, Δq_2 and Δq_{17} are "capacitive" (i.e., they are positive for the capacitive case). These six sensitivities are from the inverse Jacobian matrix. Find the values of Δq_2, Δq_{17} such that $\Delta q_2 + \Delta q_{17}$ is minimum and the required Δv's are attained.

Solution

The problem as stated is to minimize c_q,

$$c_q = (1 \quad 1)\ \Delta Q,$$

where $(\Delta Q)^t = (\Delta q_2 \ \Delta q_{17})$. This minimization is subject to the constraint

$$B\ \Delta Q \geq \underline{V}$$

$$B = \begin{bmatrix} 0.01 & 0.03 \\ 0.08 & 0.05 \\ 0.02 & 0.01 \end{bmatrix}$$

$$\underline{V} = \begin{bmatrix} 0.04 \\ 0.05 \\ 0.06 \end{bmatrix} .$$

This is not quite the same as the linear programming formulation given in Appendix A since the inequality constraint is of the reverse sign and the "cost function," c_q, in this formulation is to be minimized. Both problems are obviated by allowing

$$\Delta Q' = K - \Delta Q$$

such that

$$B\ \Delta Q' \leq BK - \underline{V} .$$

In this formulation, the quantity c_q',

$$c_q' = (1 \quad 1)\ \Delta Q'$$

is maximized. Vector K is chosen such that all entries of $BK - \underline{V}$ are positive. For example, for $K = (50\ 50)^t$, this problem is equivalent to maximizing c_q' subject to

$$B\ \Delta Q' \leq \begin{bmatrix} 1.96 \\ 6.95 \\ 1.44 \end{bmatrix} = BK - \underline{V}.$$

Figure 6.6 shows the initial simplex tableau with Δq_x, Δq_y, and Δq_z being slack variables. At the initial trial vertex,

$$\Delta q_x = 1.96 \quad \Delta q_y = 6.95 \quad \Delta q_z = 1.44$$

$$\Delta q_1' = 0 \qquad \Delta q_2' = 0 .$$

			1	1	0	0	0
Δq_x	0	1.96	0.01	0.03	1	0	0
Δq_y	0	6.45	0.08	0.05	0	1	0
Δq_z	0	1.44	0.02	0.01	0	0	1

FIGURE 6.6 Initial simplex tableau for Example 6.3.

Since

$$\Delta Q = K - \Delta Q'$$

$$= \begin{bmatrix} 50 \\ 50 \end{bmatrix} - \Delta Q' ,$$

one finds that

$$\Delta q_2 = 50 \quad \Delta q_{17} = 50 \quad c_q = 100.$$

At this point, the F row is calculated and the $C^T - F$ row is also calculated. Figure 6.7 shows the results with the θ column and key axes marked.

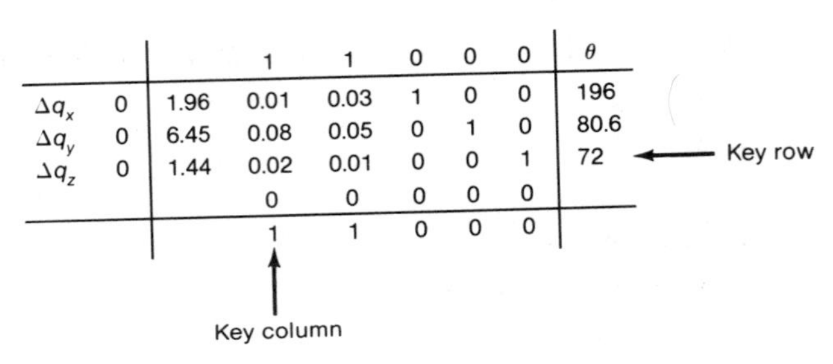

			1	1	0	0	0	θ
Δq_x	0	1.96	0.01	0.03	1	0	0	196
Δq_y	0	6.45	0.08	0.05	0	1	0	80.6
Δq_z	0	1.44	0.02	0.01	0	0	1	72 ← Key row
			0	0	0	0	0	
			1 ↑ Key column	1	0	0	0	

FIGURE 6.7 Key row and key column selection.

The key column is column 1, and therefore $\Delta q'_1$ replaces Δq_z in the trial vertex. The body of the tableau is Kron reduced around the 1–z pivot, and the result is shown in Figure 6.8. This figure also shows the determination of the new key axes. The cost at this stage is calculated as

$$\Delta q_x = 1.24 \quad \Delta q_y = 0.69 \quad \Delta q_z = 0$$

$$\Delta q'_1 = 72 \quad \Delta q'_2 = 0 \quad \Delta q_2 = -22 \quad \Delta q_{17} = 50$$

$$c_q = 28.$$

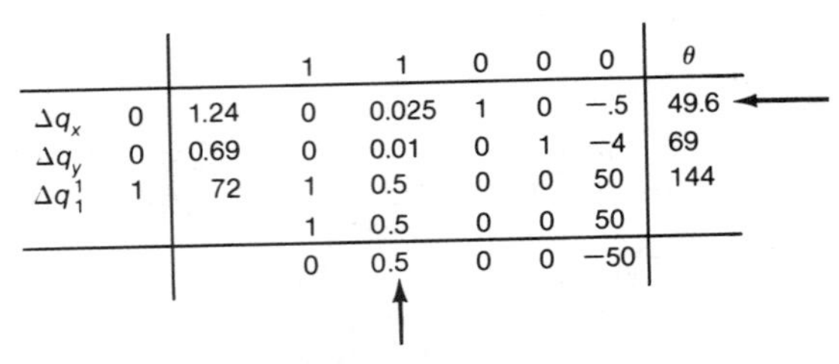

			1	1	0	0	0	θ
Δq_x	0	1.24	0	0.025	1	0	−.5	49.6 ←
Δq_y	0	0.69	0	0.01	0	1	−4	69
Δq_1^1	1	72	1	0.5	0	0	50	144
			1	0.5	0	0	50	
			0	0.5 ↑	0	0	−50	

FIGURE 6.8 Intermediate simplex tableau.

The second column –first row form the key axis in Figure 6.8. The details of the last iteration are omitted, but Figure 6.9 shows the result. The cost is calculated as

			1	1	0	0	0
$\Delta q_2'$	1	49.6	0	1	40	0	−20
Δq_y	0	0.194	0	0	−0.4	1	−3.8
Δq_1	1	47.2	1	0	−20	0	60
			1	1	20	0	40
			0	0	−20	0	−40

FIGURE 6.9 Final simplex tableau.

$$\Delta q_x = 0 \qquad \Delta q_y = 0.194 \qquad \Delta q_z = 0$$

$$\Delta q_1' = 47.2 \qquad \Delta q_2' = 49.6 \qquad \Delta q_2 = 2.8 \qquad \Delta q_{17} = 0.4$$

$$c_q = 3.2.$$

Thus the reactive voltamperes 2.8 and 0.4 (capacitive) should be sited at buses 2 and 17, respectively. Note that the corresponding values of Δv are

$$\Delta v_1 = 0.04 \quad \Delta v_2 = 0.244 \quad \Delta v_3 = 0.06.$$

In this case, it is important to check $|v_2|$ for possible overvoltage due to the placement of shunt capacitance at bus 2. The $C^t - F$ row in Figure 6.9 indicates that this is the final step.

Var Dispatch Using Distribution Factors

It is possible to use the advantages of linearization in the distribution factor formulation for reactive power dispatch. Unfortunately, bus voltage magnitudes are not readily calculable using distribution factors, and therefore this approach cannot be used to dispatch var to support bus voltages. The distribution factor approach is useful for minimizing line loss, $\overline{P}_L$,

$$\overline{P}_L = Re\ (\overline{I}^t \overline{Z} \overline{I}^*), \tag{6.48}$$

where $\overline{I}$ is a vector of line currents and $\overline{Z}$ is a diagonal matrix of line impedances. Since it is rarely required to dispatch Q_{bus} to minimize $\overline{P}_L$, this approach is *not* in common use. It is possible to augment (6.48) to include voltage support objectives. The formulation in which $\overline{P}_L$ is minimized is quite straightforward and entails calculation of Q_{bus} based on

$$\nabla_{Q_{bus}} [\frac{1}{2}(\overline{I}^t \overline{Z} \overline{I}^* + (\overline{I}^t \overline{Z} \overline{I}^*)^*)] = 0\ . \tag{6.49}$$

Allow

$$\overline{S} \simeq \overline{I};$$

thus

$$\nabla_{Q_{bus}} [(\overline{S})^H \overline{Z} \overline{S}] + \nabla_{Q_{bus}} [(\overline{S})^H \overline{Z}^* (\overline{S})^H] = 0\ .$$

Note that S_{bus} should be replaced by S_{bus} in "schedule B" (i.e., the schedule in which optimal Q_{bus} is dispatched). Also note that

$$\frac{\partial \overline{S}_i}{\partial_j Q_k} = \rho_{i,k} \qquad \overline{S}^B = \overline{S}^A + \rho(jQ_{bus})\ .$$

This gives

$$\rho^H \bar{Z}\bar{S}^B - \rho^t \bar{Z}(\bar{S}^B)^* - \rho^t \bar{Z}^*(\bar{S}^B)^* + \rho^H \bar{Z}^* \bar{S}^B = 0 .$$

Solve for Q_{bus} noting that the primitive line resistance matrix is $\bar{R}$,

$$\bar{R} = Re\ (\bar{Z}) = \frac{1}{2}(\bar{Z} + \bar{Z}^*) ,$$

$$Q_{bus} = -[Re\ (\rho^H \bar{R} \rho)]^{-1}\ Im\ (\rho^H \bar{R}\bar{S}^A) . \tag{6.50}$$

Equation (6.50) gives the optimal var dispatch to minimize line losses. No voltage support is guaranteed. If some buses are unavailable for var dispatch, (6.50) could be revised to accommodate this difficulty; this is done by deleting the columns of ρ which correspond to unavailable buses. In [16], a method to include voltage support is described.

A Gradient Method of Optimal Reactive Power Dispatch

The preceding linearized algorithms are easily programmed, and simple to use, but accuracy is not easily calculated. In recent years, commercially available reactive power dispatch analysis and operating software has moved in the direction of a unified philosophy for the treatment of both active and reactive power. By such a unified approach, consistent modeling and solution accuracy results. When the gradient approach is modified to include voltage support constraints and capacitor costs, the accuracy of this procedure is consistent with the Newton–Raphson power flow study and optimal generation dispatch. The salient required modification of the optimal generation dispatch algorithm is the inclusion of a $c_c^t Q_c$ term in the cost function (c_c^t is a row vector of incremental costs, $/per unit var, for the required shunt capacitors and synchronous condensers, and Q_c is a vector of the var injections of these devices). The cost function, C, includes this term and the Lagrangian is formulated as

$$L(U, X, P) = C(U, X) + \lambda^t G(U, X, P) + \mu^t H(U, X) . \tag{6.51}$$

The vector Q_c consists of two parts, one corresponding to injected var at generation buses, and the other at load buses. The former is included in X and the latter in U. Thus, $C(U, X)$ is a function of U for several reasons:

1. Due to fuel costs at buses other than the swing bus (i.e., active power dispatch).
2. Due to capacitor siting at buses other than the swing bus (i.e., reactive power dispatch).

Also $C(U, X)$ is a function of X:

1. Due to fuel costs at the swing bus (i.e., active power dispatch).
2. Due to capacitor siting a load buses (i.e., reactive power dispatch).

The second required modification is that inequality constraints H must include bus voltage support terms. Since these bus voltage magnitudes occur in the state vector X, this inclusion presents no difficulties. Equation (6.51) is solved as described earlier, the difference being in the formulation of C and H.

AUTOMATIC GENERATION CONTROL

Generation is obtained from generators within the power system and from interties with neighbors. The methods discussed in this chapter are used to dispatch system generators. When the method of B coefficients or gradient dispatch is used, the system demand minus the power derived via interties is generated. Let P_d denote the demand power, $\bar{P}_L$ the total system losses, and P_S the total intertie power ($P_S > 0$ denotes net injection of active power). Then one would expect that P_{gen} will be the total, optimally dispatched generation,

$$P_{gen} = P_d + \bar{P}_L - P_S \ .$$

There are usually constraints imposed by legal contracts and agreements on the level of P_S; these agreements consider economics, technical constraints imposed by ratings of the interties, and reliability considerations. It is possible that the quantity ACE,

$$\mathrm{ACE} = P_{gen} - P_d - \bar{P}_L + P_S \ , \tag{6.52}$$

termed the *area control error,* may be nonzero for short periods of time. If the ACE is positive, there is excessive generation and machines will accelerate, while negative ACE results in machine deceleration. The ACE, expressed in megawatts, is a very important system parameter: ACE is usually displayed to operators together with system lambda. The ACE is calculated using (6.52), where the generation, intertie, and some demand terms are obtained from telemetered field data. These data are collected using sensors known as *remote terminal units* and the data collection system is termed the *supervisory control and data acquisition* system. Usually, some components of (6.52) must be obtained from the optimal dispatch, and some must be estimated. This topic is considered in Chapter 9 in connection with a discussion of data acquisition systems and state estimators.

There is a relationship between the ACE and system frequency, f,

$$\frac{df}{d(\mathrm{ACE})} \simeq K \ .$$

Usually, Δf is measurable by measuring bus voltage angles and noting the integral relationship between frequency and phase angle.

The desired value of area control error in the long term is zero – but positive or negative ACE may be used in the short term to correct the short term demand/generation history. The control of ACE is known as *area interchange control* and the methodology in common use is to augment the load flow equations with an additional equation setting the ACE to the desired value. This method is used in power system operation. Britton [22] gives further details and practical experience.

Operators may wish to know the maximum possible value of P_S in order to anticipate unit outages. A series of power flow studies may be run off-line which, with system experience, will be usable to estimate the maximum interchange power available. Limits that must be considered in estimating this maximum include tie line ratings, generation limits of neighboring companies, and transmission ratings within the power system.

A useful linear formulation to find the interchange power maximum is described in Appendix A; the method is based on distribution factors and linear programming.

Bibliography

[1] H. W. Dommel, and W. F. Tinney, "Optimal Power Flow Solutions," *IEEE Trans. Power Apparatus and Systems,* v. PAS–87, no. 10, October 1968, pp. 1866-1876.

[2] G. L. Viviani, "Stochastic Optimal Power Flow Studies," Ph.D. thesis, Purdue University, West Lafayette, Ind., 1981.

[3] G. L. Viviani and G. T. Heydt, "Stochastic Optimal Energy Dispatch," IEEE Trans. Power Apparatus and Systems, v. PAS-100, July 1981, pp. 3221–3228.

[4] J. Lamont and M. Gent, "Environmentally Oriented Dispatching Techniques," *Proc., Eighth IEEE Power Industry Computer Application Conference,* Minneapolis, Minn., June, 1973.

[5] J. Delson, "Controlled Emission Dispatch," *IEEE Trans. Power Apparatus and Systems,* v. PAS-93, September–October 1974, pp. 1359–1366.

[6] A. Tsuji, "Optimal Fuel Mix and Load Dispatching Under Environmental Constraints," *IEEE Trans. Power Apparatus and Systems,* v. PS-100, May 1981, pp. 2357–2364.

[7] L. Fink and K. Carlsen, "Operating Under Stress and Strain," *IEEE Spectrum,* March 1978, pp. 48–53.

[8] J. Zaborzky and A. Subramanian, "Monitoring Evaluation, and Control of Power System Emergencies," *Conference on Systems Engineering for Power,* Henniker, N.H., August 1977.

[9] J. Zaborzky, A. Subramanian, T. Tarn, and K. Lu, "A New State Space for Emergency Control in the Interconnected Power System," *IEEE Trans. Automatic Control,* v. AC-22, August 1977, pp. 507–517

[10] B. Stott and E. Hobson, "Power System Security Control Calculations Using Linear Programming," *IEEE Trans. Power Apparatus and Systems,* v. PAS-97, October 1978, pp. 1712–1731.

[11] R. Minghetti, "Corrective Generation/Load Rescheduling by Multi-stage Nonlinear Programming," M.S.E.E. thesis, University of Illinois, Urbana, Ill., May 1983.

[12] J. Blaschak, G. Heydt, and J. Bright, "A Generation Dispatch Strategy for Power Systems Operating Under Alert Status," *IEEE Summer Power Meeting,* Vancouver, B.C., July 1979.

[13] D. Lubkeman and G. Heydt, "Transient Stability Enhancement in Multimachine Power Systems Using Braking Resistors," *J. Electric Machines and Power Systems,* v. 9, no. 1, January 1984.

[14] J. Carpentier, "Contributions à l'étude du dispatching écionomique," *Bull. Société Francaise des Electriciens,* Sé. B, v. B3, 1962, pp. 431–447.

[15] H. Nicholson and M. Sterling, "Optimum Dispatch of Active and Reactive Generation by Quadratic Programming," *IEEE Trans. Power Apparatus and Systems,* v. PAS-82, 1963, pp. 644–654.

[16] G. T. Heydt and W. M. Grady, "A Matrix Method for Optimal VAr Siting," *IEEE Trans. Power Apparatus and Systems,* v. PAS-94, July–August 1975, pp. 1214–1222.

[17] H. Happ, "Optimal Power Dispatch – A Comprehensive Survey," *IEEE Trans. Power Apparatus and Systems,* v. PAS-96, May–June 1977, pp. 841–854.

[18] O. Elgerd, *Electric Energy Systems Theory, An Introduction,* McGraw–Hill, New York, 1971.

[19] L. Kirchmayer, *Economic Operation of Power Systems,* Wiley, New York, 1958.

[20] L. Kirchmayer, *Economic Control of Interconnected Systems,* Wiley, New York, 1959.

[21] M. El-Hawary and G. Christensen, *Optimal Economic Operation of Electric Power Systems,* Academic Press, New York, 1979.

[22] J. Britton, "Improved Area Interchange Control for Newton's Method Load Flows," *IEEE Trans. Power Apparatus and Systems,* v. PAS-88, October 1969, pp. 1577–1581.

Exercises

Exercises that do not require computer programming

6.1 Using the equal incremental cost rule, find the optimal dispatch strategy for the two machine system whose incremental costs are shown in Figure P6.1. The machine capacities are 200 and 300 MW, respectively. Your answer should include P_1 and P_2 plotted versus P_{demand} and the system λ plotted versus P_{demand}.

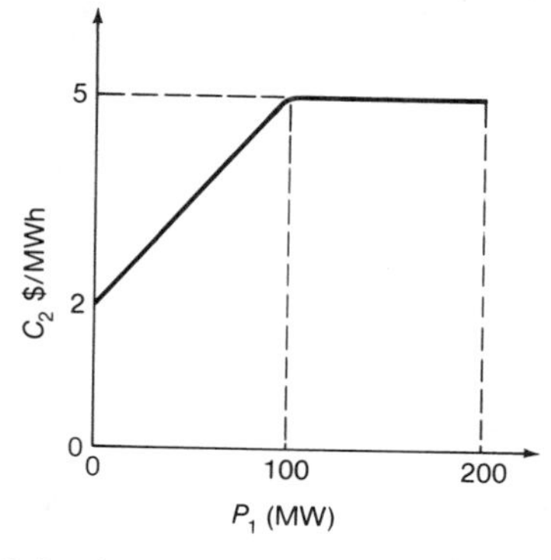

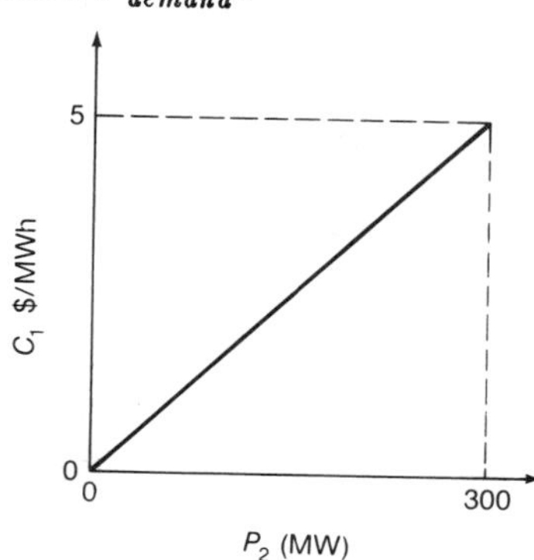

FIGURE P6.1 Incremental generation costs at two units.

6.2 In Exercise 6.1, transmission losses are not considered. However, through the use of the method of B–coefficients, the effects of losses may be estimated.

a. Let the penalties applied to c_1 and c_2 be L_1 and L_2,

$$L_1 = \frac{1}{(1 - P_1/500)} \qquad L_2 = \frac{1}{(1 - P_2/800)}$$

Repeat the economic dispatch using the penalized costs, $L_i C_i$, rather than simply the costs.

b. In part (a), the penalty factors are conveniently a function only of the generation level of the machine being penalized. This is not usually the case. Repeat the economic dispatch using

$$L_1 = \frac{1}{1 - 0.0015P_1 - 0.0008P_2}$$

$$L_2 = \frac{1}{1 - 0.005P_1 - 0.000425P_2} .$$

6.3 Consider the power system defined by

Line	Line impedance (pu)
1–2	0.01 + j0.10
1–3	0.02 + j0.10
2–3	0.00 + j0.05
3–4	0.03 + j0.09

Consider bus 1 as the swing bus and consider bus 4 as a generation bus. For a given load schedule, the bus voltage profile is essentially flat and at rating. The generators may be modeled as Thevenin equivalent reactances of 12% (each). These are ground ties at buses 1 and 4. The solution bus voltage phase angles at buses 2, 3, and 4 are $-1\,^\circ$, $-1.2\,^\circ$, and $-0.1\,^\circ$, respectively. Calculate all the B coefficients for this system.

6.4 Find the minimum value of the function c,

$$c(x, y) = 2x^2 + 3y^2$$

a. subject to the equality constraint

$$g(x, y) = 0 = x - y^2 - 1$$

b. subject to the two equality constraints

$$g_1(x, y) = 0 = x^2 - y^2 - 1$$

$$g_2(x,y) = 0 = x - y - 1$$

c. subject to one equality constraint

$$g(x, y) = 0 = x^2 - y^2 - 1$$

and one inequality constraint,

$$h(x, y) = 2 - x - y \leq 0 .$$

In each case, calculate $c_{\min}$, the Lagrange multipliers, and the dual variables.

6.5 Find the minimum value of the function c,

$$c(x, y) = x^3 + 2y^3$$

subject to two inequality constraints:

$$h_1(x, y) = 1 - x - y \leq 0$$

$$h_2(x, y) = 2 - 3x - y \leq 0.$$

a. After completing your study, consider the solution

$$x = 10^9 + 1$$

$$y = -10^9$$

for this problem. Explain the value of c evaluated using dual variables, and explain the results of calculating $c(x, y)$ at the stated values of x and y.

b. In part (a) you should discover that the fact that x and y can be negative creates a difficulty. The cubic behavior of c in terms of its arguments is not really the difficulty; in power engineering problems, parameters such as x and y will usually be constrained by

$$h_3(x, y) = -x \leq 0 \qquad h_4(x, y) = -y \leq 0 .$$

When c contains only even powers of its arguments, these additional constraints are unnecessary.

Using all four inequality constraints, find the minimum value of c. (Note: with four inequality constraints, 16 cases must be checked. Four of those cases are ruled out immediately since they result in $x = y = 0$, which violates the h_1 and h_2 conditions. Several other cases are similarly ruled out.)

6.6 In this problem, the effects of various penalty functions on the minimization of

$$C(x, y) = (x - 1)^2 + (y - 2)^2$$

subject to

$$h(x, y) = -x + 3 \leq 0$$

are examined. The inequality constraint is to be considered by penalizing C.

a. Consider the candidate penalty function

$$f(m) = \begin{cases} 10^9 & m > 0 \\ 0 & m \leq 0. \end{cases}$$

Can this function be used as a penalty function? If not, find the requirements on f to allow its use as a penalty function.

b. Use the function

$$f(m) = e^{10m}$$

as a penalty function with

$$m = -x + 3.$$

Compare results with the exact solution as obtained using dual variables and the Kuhn–Tucker conditions.

c. Repeat part (b) using

$$f(m) = m^2 e^{10m}.$$

d. Repeat (b) using

$$f(m) = e^{10m}(1 + m^2).$$

6.7 Consider a power system in which per unit bus voltages at 1, 2, 3, and 4 are 0.915, 0.905, 0.905, and 0.890, respectively. It is required to raise these bus voltages to at least 95% of rating by adding shunt capacitors. This capacitive compensation will be done at buses 2, 5, and 12. The inverse Jacobian gives

$$\Delta v_1 = 0.010\Delta q_2 + 0.028\Delta q_5 + 0.001\Delta q_{12}$$

$$\Delta v_2 = 0.077\Delta q_2 + 0.020\Delta q_5 + 0.001\Delta q_{12}$$

$$\Delta v_3 = 0.020\Delta q_2 + 0.005\Delta q_5 + 0.002\Delta q_{12}$$

$$\Delta v_4 = 0.010\Delta q_2 + 0.001\Delta q_5 + 0.010\Delta q_{12}$$

Find the minimum dispatch of Δq to obtain the required voltages. [Note: Δq in the Δv expression is assumed to be capacitive, that is, $\Delta q_2 = 1$ implies that $S_2 = 0 + j1$ in injected notation, and bus 2 is capacitive (−j1 in load notation).]

Exercises that involve computer programming

6.8 The following table contains the incremental costs of generating units 1, 2, and 3.

P_{out} (MW)	C_1 (\$/MWH)	C_2 (\$/MWH)	C_3 (\$/MWH)
0	1.00	1.00	1.00
10	3.00	4.00	2.90
20	3.05	4.00	3.10
30	3.10	4.01	3.30
40	3.15	4.01	3.50
50	3.21	4.03	3.80
60	3.26	4.04	4.15
70	3.33	4.05	4.55
80	3.39	4.07	—
90	3.48	4.08	—
100	3.60	4.10	—

Units 1 and 2 are rated at 100 MW each, and unit 3 is rated at 70 MW. Write a program to read these unit heat rate data and calculate the economic optimal dispatch using the equal incremental cost rule. Also print out the system lambda. (Note: It will probably be worthwhile to write a short subroutine which can interpolate a table:

Parameter A	Parameter B
xxx	aaa
iii	?
yyy	bbb.

Be careful to consider the case in which aaa and bbb are the same.)

6.9 In Exercise 6.8, the tabulated heat rates were conveniently evenly spaced. Consider now the four machine system. The incremental costs of the Amoli and Rawhide generating stations are:

Power Output (MW)	Cost at Amoli ($/MWH)	Cost at Rawhide ($/MWH)
0	5.04	3.00
50	6.10	3.00
100	6.31	3.00
150	6.53	3.19
200	6.87	3.50
250	7.17	4.00
350	7.88	5.01
450	—	6.10
550	—	7.31

The incremental costs of the Smoke Desert and Sparks central generating stations are:

Power output (MW)	Cost at Smoke Desert ($/MWH)	Cost at Sparks C.G.S. ($/MWH)
0	5.07	3.00
13	6.09	3.00
41	8.10	3.00
72	8.99	3.05
80	9.08	—
100	9.12	—

The unit ratings are

Unit	Rating (MW)
Amoli	376
Rawhide	550
Smoke Desert	100
Sparks CGS	75

a. Using the equal incremental cost rule, obtain the optimal system dispatch and λ versus P_{demand}.

b. Calculate the daily cost of servicing the load shown in Figure P6.9.

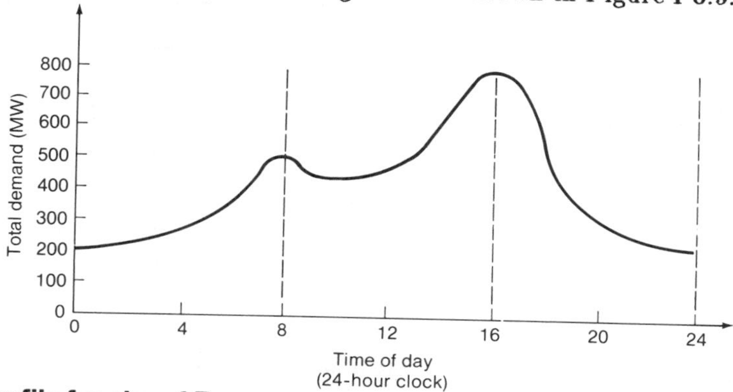

FIGURE P6.9 Load profile for city of Reno.

c. The system capacity as stated is 1101 MW. Consider this as the long term machine ratings. Suppose that 12% additional capacity can be obtained from each unit. Repeat the economic system dispatch to a total demand of 1233.1 MW.

d. Consider now a forced outage of the Amoli generating station. Consider serving the city of Reno with load profile shown in Figure P6.9. Recalculate the daily cost of serving this load. Explain your result. In your considerations for the explanation of results, you should note that the outage of low cost units result in higher production costs since more costly units are now brought into the generation schedule.

In Exercise 6.10, the steepest gradient method is illustrated for optimal power dispatch. To avoid complexity, inequality constraints are not considered. Therefore, the control vector U can not contain bus voltage set points.

6.10 Figure P6.10 shows a six bus, two machine, 800 MVA (capacity) power system. The bus at the Schmitzville generating station is the swing bus, which is held at rated voltage on the high voltage side. Similarly, the Johnson City generator is operated at rated bus voltage. The peak and off-peak (lean period) loads are:

Bus	Peak Load (MVA)	Off-Peak Load (MVA)
Electricville	$300+j100$	$100+j0$
Bodie	$90+j0$	$30+j0$
Big John	$100-j40$	$30+j0$
Verdi	$20+j0$	$5+j2$

a. Perform a conventional power flow study for the system at both the peak and off-peak periods. (Assume a reasonable generating schedule).

b. Consider the unit generating costs given as

$$\textit{Schmitzville}: \quad c(p) = 10 - 6.25 \times 10^{-5}(p - 400)^2.$$

$$\textit{Johnson City}: \quad c(p) = 0.03p.$$

where c_i are expressed in \$/MWh and the p_i are in megawatts. Calculate the economic optimal dispatch starting with the U and X vectors used in part (a).

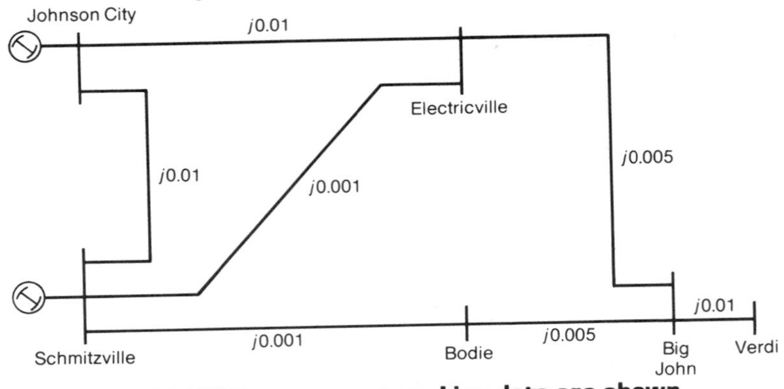

FIGURE P6.10 Six-bus 800-MVA power system. Line data are shown in per unit on 100-MVA, 345-kV base.

CHAPTER 7

Fault Studies

7.1

THE NATURE OF POWER SYSTEM FAULTS

It is appropriate to begin this chapter with a remark that a good deal of power system planning and design is centered around events which are either unlikely, or short in duration, or both. For example, transmission circuits must be capable of handling the peak though even load this condition may be short in duration. A similar design philosphy occurs in generation planning. The reader is no doubt aware that a considerable effort is expended in minimizing the occurrence of power system faults. Protective relaying is designed to minimize the effects of faults when they do occur. The design of these protective relay circuits requires a knowledge of the expected fault current levels. The selection of interruption devices (circuit breakers, fuses) also critically depends on the fault current interruption requirements. Further, even in systems with well designed protective relays, power system transmission circuits and distribution circuits must be planned with the levels of fault currents in view. In each of these areas, fault studies are performed by relay engineers, system planners, and others for design purposes and to identify potentially harmful conditions.

A fault is a short circuit, and the term *bolted fault* refers generally to the case of zero fault impedance. A three phase fault usually implies that the three phases and the neutral are shorted. When fault impedance is present, it is expressed in ohms per phase in the equivalent wye, or simply in per unit. High impedance faults refer to conditions in which the fault current levels are not much different from load current magnitudes. For 69 to 250 kV transmission circuits, high impedance faults generally refer to fault impedances of about 40 Ω/phase (this is *not* a standard).

Fault studies are power system studies in which the fault current levels, fault capacity (prefault voltage times fault current), and postfault system voltages are calculated. Other quantities that may be calculated include line currents and voltage phase angles during the fault. Like most power system studies, fault studies are done with output variable precision of at least three decimal places; the student should question this accuracy because faults, like other unplanned events, rarely occur under precise conditions. The fault impedance frequently fluctuates widely during the fault, and may include both nonlinear and stochastic phenomena. Also, the behavior of the transmission system during a fault is difficult to model accurately since components may be operating outside their normal operating range. Nonetheless, line data at 50°C temperature rise are commonly employed, and transformer nameplate reactances are used (saturation effects cause higher equivalent positive sequence reactances in many cases, thus making typical fault calculations conservative). Finally, the behavior of loads during fault conditions is difficult to model because of unknown aspects of the volt-ampere characteristics of those loads under anomalous bus voltage conditions. Usually, loads are modeled simply as constant S or constant Z (or some combination). These are the same assumptions made for power flow studies. Additionally, fault studies, like load flow studies, rely on sinusoidal steady state assumptions.

With all these observations, it is probably optimistic to expect accuracy within 10% in fault studies. The commonality of the accuracies of calculation methods, however, make possible the comparison of bus fault capacities, fault currents, and other quantities. Fault studies give a best estimate for the calculation of relay settings, breaker requirements, fuse selection, series reactor placement, stability assessment, and other parameters relating to circuit interruption.

7.2 THEVENIN EQUIVALENT CIRCUIT OF A FAULTED POWER SYSTEM

Figure 7.1 is a pictorial of a simplified, faulted power system. The most common approach to the analysis of faulted power systems is to resolve the network and generators into a Thevenin equivalent circuit as viewed from the faulted bus [8]. With this objective, each of the components of the pictorial are examined.

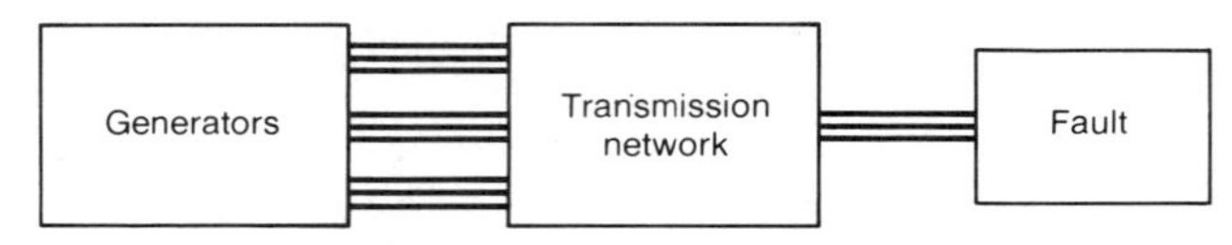

FIGURE 7.1 Simplified, faulted power system.

Generators

A synchronous generator experiencing a three phase bolted fault at its terminals has the familiar phase current waveform shown in Figure 7.2. If the machine has salient poles with direct axis reactance x_d, transient direct axis reactance x_d', and subtransient direct axis reactance x_d'', each of the phase currents will have the form [1]

$$i_a(t) = \sqrt{2}\frac{E_{afo}}{x_d} \cos(\omega t + \theta_0) + \sqrt{2}E_{afo}\left(\frac{1}{x'_d} - \frac{1}{x_d}\right)\epsilon^{-t/T_d'} \cos(\omega t + \theta_0)$$

$$+\sqrt{2}E_{afo}\left(\frac{1}{x_d''} - \frac{1}{x_d'}\right)\epsilon^{-t/T_d''} \cos(\omega t + \theta_0) . \tag{7.1}$$

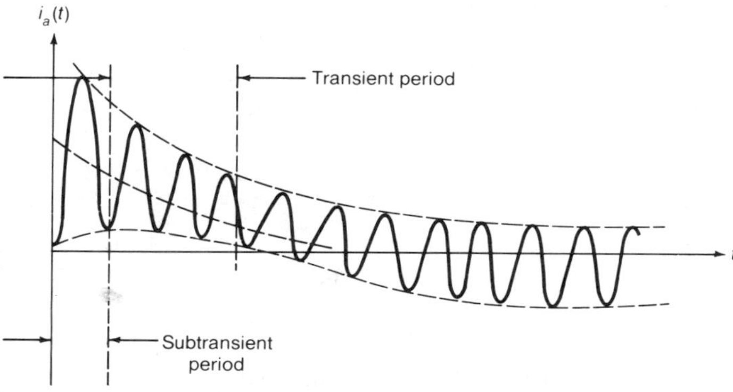

FIGURE 7.2 Phase current for a faulted synchronous machine.

The ϕA current is chosen as an illustration. In (7.1), E_{afo} is the prefault internal armature voltage (rms, magnitude only),

$$E_{afo} = \frac{1}{\sqrt{2}}\omega L_{af} I_{fo},$$

where ω is the power frequency, θ_0 the phase angle at the instant of the fault, T'_d and T_d'' the direct axis transient and subtransient time constants, L_{af} is the mutual inductance (direct axis) between ϕA and the field winding, and I_{fo} the prefault field current. The dc offset shown in Figure 7.2 does not appear in (7.1); only the symmetrical (ac) term is indicated because the dc components of the transient are effectively blocked in the network by transformers. The phase angle θ_0 is chosen as zero to give the worst case (largest) current. If the fault study is to focus on the very short period directly following the fault, the $\epsilon^{-t/T_d''}$ term in (7.1) becomes the dominant factor. The primary use of fault studies relates to circuit protection and interruption. Circuit breakers are usually activated by protective relays, which are devices that often act late in the subtransient period. Although fuses generally operate much later, subtransient data are nonetheless used because conservative results are obtained; in the subtransient period, high machine currents occur, and if these levels are used in designs, the lower transient and steady state periods will be accommodated. The transient period immediately follows the subtransient period and is typically 500 ms long (i.e., 8 to 10 cycles on a 60 Hz base). Occasionally, data from this period are used. The transient machine currents are much lower than those in the subtransient period. Still smaller are the steady state currents.

In the subtransient period,

$$\epsilon^{-t/T'_d} \simeq 0$$

and (7.1) becomes

$$i_a(t) = \sqrt{2}\frac{E_{afo}}{x_d''} \cos \omega t. \tag{7.2}$$

Thus the Thevenin equivalent impedance is simply jx_d''.

The generator voltage to be used for fault studies is E_{afo}; fortunately, however, it is not necessary to know this quantity since a more readily available voltage in the network is used to estimate the Thevenin equivalent voltage. Remarks below further relate to this point.

The electromechanical dynamics of generators are usually much longer in duration than the electrical transients, and these phenomena are not considered in fault study. Fault studies are usually sinusoidal steady state studies modeled in the subtransient period. This will not be the case in stability studies – a topic considered in Chapter 8.

Network

Figure 7.3 is a pictorial of the network at which bus k will be faulted. The generation buses are shown as ground ties of jx_d''. It may be necessary to convert the per unit impedance base of these ground ties to be consistent with the transmission system base (often 100 MVA). Thus a 600 MVA machine with 25% subtransient reactance will appear nearly as a short circuit (i.e., j0.041 per unit) on the system base. In the absence of other information, machines are occasionally considered as short circuits for the purpose of calculating the Thevenin equivalent impedance. If only the transient reactance is known, use half this value as x_d''. If only the synchronous reactance is known, estimate the value of x_d'' as 10% of the synchronous reactance.

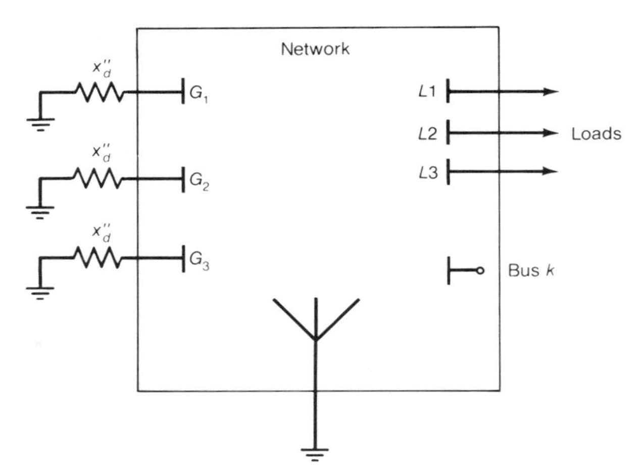

FIGURE 7.3 Network model for a faulted power system.

The network as viewed from bus k is simply $(Z_{bus})_{kk}$ when the independent voltage sources are shorted. If the impedances of the loads are known, they may be used in the formation of Z_{bus}, but this is usually unnecessary since these impedances are usually quite high compared to the transmission circuit impedances. It is usual to omit loads in calculating Z_{bus}.

Attention now turns to the Thevenin voltage. When bus k is open circuited, the prefault voltage, $V_k^{prefault}$, appears here. This is available from a load flow study (or it may be estimated using $| V_k^{prefault}| \simeq 1.0$). Note that although the loads are often omitted in calculation of $(Z_{bus})_{kk}$, their effect on $V_k^{prefault}$ is included since this prefault voltage is obtained from a load flow study under prefault conditions.

Fault

A 3ϕ bolted fault is modeled as a short circuit. If fault impedance is used in the balanced case, z_f occurs as a ground tie at the faulted bus. For the imbalanced case considered later, we will have much more to say about fault models.

The Thevenin Equivalent

Figure 7.4 shows the Thevenin equivalent circuit of a faulted power system under the simplifications cited in this article. Bus k is the faulted bus, z_f is the fault impedance, and the superscript "pre" on V_k refers to the prefault voltage. Evidently, using injection notation

$$I_k^{post} = \frac{-V_k^{pre}}{(Z_{bus})_{kk} + z_f} \tag{7.3}$$

and the superscript "post" denotes postfault.

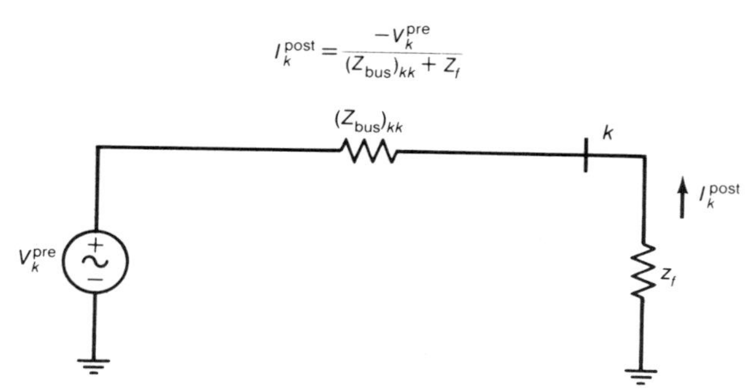

FIGURE 7.4 Thevenin equivalent circuit of a faulted power system.

7.3 POST FAULT VOLTAGES

The fault current I_k^{post} causes changes in system voltage according to

$$\Delta V_{bus} = Z_{bus} \begin{bmatrix} 0 \\ 0 \\ I_k^{post} \\ 0 \end{bmatrix} . \tag{7.4}$$

In (7.4), it is assumed that during the fault, the only anomalous current is the fault current (i.e., load currents are fixed). To calculate ΔV_{bus}, triangular factors of Y_{bus} are used:

$$LU = Y_{bus} \tag{7.5}$$

$$LU\Delta V_{bus} = \begin{bmatrix} 0 \\ 0 \\ I_k^{post} \\ 0 \end{bmatrix}. \tag{7.6}$$

The change in bus voltages is calculated by forward and backward substitution. Subsequently,

$$V_{bus}^{post} = V_{bus}^{pre} + \Delta V_{bus} . \tag{7.7}$$

If the fault has zero impedance, v_k^{post} will be zero.

The *fault capacity* at bus k is defined as the product of the magnitude of the fault current at bus k, $|I_k^{post}|$, and the prefault voltage magnitude at bus k, $|V_k^{pre}|$

$$F_k = |I_k^{post}||V_k^{pre}| . \tag{7.8}$$

The fault capacity at a bus is a common measure of the intensity of a fault because like most power system components, circuit interruption equipment cost is more closely proportional to the volt ampere product of voltage and current ratings rather than either one of these parameters. Fault capacity, however, is only one of many ratings of a circuit breaker. Other specifications include voltage rating, long term and short term nominal load current, interruption current, and several parameters describing the speed of interruption.

Example 7.1

Figure 7.5 shows a power system for which Z_{bus} referenced to ground is

$$\begin{bmatrix} j0.2500 & j0.2500 & j0.2500 & j0.2500 & j0.2500 \\ & j0.2575 & j0.2525 & j0.2550 & j0.2550 \\ & & j0.2575 & j0.2550 & j0.2550 \\ & \text{Symmetric} & & j0.2600 & j0.2600 \\ & & & & j0.2644 \end{bmatrix}$$

per unit (100 MVA base) calculated with a 25% reactive ground tie at bus G1, which is the swing bus. Bus G1 is assumed to be an infinite bus. Generator G2 has 27% subtransient reactance, is rated at 185 MVA, and

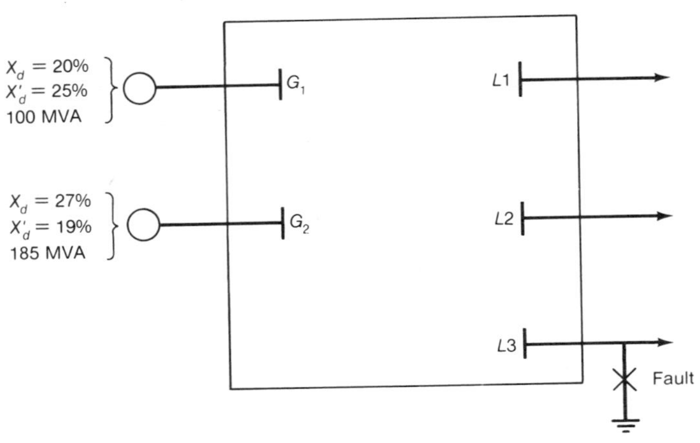

FIGURE 7.5 Power system used in Example 7.1.

is not yet represented in this matrix. It will be necessary to include the effects of generator G2 by accounting for its subtransient reactance. The axis ordering is G1, G2, L1, L2, L3. Find the fault current and capacity for a bolted fault and 40 Ω/phase fault at bus L3. The transmission voltage is 138 kV. Also, repeat this problem with zero impedance ties to ground at generation buses (rather than jx_d'' ties). Table 7.1 is a prefault power flow solution.

Table 7.1
Power Flow Solution for Example 7.1

Bus	Prefault Voltage (kV)
G1	138.2
G2	138.0
L1	137.1
L2	137.1
L3	135.0

Solution

This problem is first solved for a bolted fault at each of the buses. The matrix given already has a subtransient reactance tie at G1. If jx_d'' is tied to ground at G2, the stated 27% reactance must be converted to a 100 MVA base. Note that it is usually necessary to convert the generator subtransient reactance data to the per unit base used for the network. Sometimes conversion of this reactance is required due to both S_{base} and V_{base}. In this case,

$$x_d'' = (0.27)\frac{100}{185} .$$

This ground tie is added to Z_{bus} and the resulting matrix is converted to reference bus G1 since that bus is an infinite bus. The diagonal entries of $Z_{bus}^{(G1)}$ are

$$z_{G2G2} = j0.0074 \qquad z_{L1L1} = j0.0074 \qquad z_{L2L2} = j0.01007$$

$$z_{L3L3} = j0.01437 .$$

The prefault bus voltage data are reduced to a 138 kV base and the fault capacity at each bus (except that at G1) is calculated

$$F_{G2} = \frac{138}{138}\left(\frac{1}{0.0074}\right)$$

$$= 135.14 \text{ per unit.}$$

The other fault capacities are

L1	133.38 per unit
L2	98.01 per unit
L3	66.60 per unit.

Conversion of these capacities to MVA is done using a 100 MVA base.

If both G1 and G2 were tied with a zero ohm impedance to ground, the impedance matrix referenced to G1 becomes

$$\begin{bmatrix} j0.00667 & j0.00333 & j0.00333 \\ & j0.00667 & j0.00667 \\ \text{symmetric} & & j0.01185 \end{bmatrix}$$

(the axes are L1, L2, L3). The diagonal entries of this matrix are used to find the fault capacities,

L1	148.95 per unit
L2	148.95 per unit
L3	84.33 per unit.

The additional fault capacity occurs since the system circuit impedances are lower due to the zero ohm ground ties.

The problem is now repeated for a 40 Ω/phase (resistive) fault impedance,

$$z_f = 40 \frac{S_{base}}{V_{base}^2} = 0.21 \text{ per unit.}$$

When G1 and G2 experience ground ties which are generator subtransient reactances, the fault capacities are

Bus	Fault Capacity (pu)
G2	4.76
L1	4.73
L2	4.73
L3	4.65

If zero ohm ties at G1 and G2 are used, the fault capacities are

Bus	Fault Capacity (pu)
L1	4.73
L2	4.73
L3	4.65

7.4 DIGITAL CALCULATION OF FAULTS

For a fault at bus k, the calculation of the fault current requires only the diagonal element of Z_{bus} in the k axis. Of course, a prefault load flow study is also required. Let L and U be the triangular factors of Y_{bus}. Lower left triangular L and upper right triangular U are as sparse as Y_{bus}. Then

$$LUZ_{bus} = I,$$

where I is the identity matrix. Let

$$W = UZ_{bus} . \tag{7.9}$$

Then the equation

$$L \; col_i(W) = col_i(I) \tag{7.10}$$

is solved by forward substitution in column i. The entire i column of W must be calculated. Having found this column, (7.10) is used:

$$col_i(W) = U \; col_i(Z_{bus}), \tag{7.11}$$

but in this instance, backward substitution is used only up to the point where the diagonal element, $(Z_{bus})_{ii}$, is found. The process is repeated for

$i = 1, 2, 3, ..., n$, or for whatever buses at which a fault study is to be performed.

If postfault voltages are to be calculated, some of the calculations of (7.9) and (7.11) are usable, and the lower and upper triangular factors are directly usable. Matrices L and U are sparsity programmed. There are two broad reasons for raising the points of triangular factorization and sparsity programming in this section. Triangular factorization affords both time and memory advantages. Time advantages occur since (7.10) needs to be done only for part of matrix W, and (7.11) is solved only for desired entries of column i of Z_{bus}. Triangular factorization in connection with sparsity programming offers very large savings in memory since the full Z_{bus} matrix is not calculated. Only sparse matrices are needed. Sparsity programming will also speed calculations since the use of a "next pointer" to solve (7.10) and (7.11) will avoid multiplications with zero multipliers. For example, a next pointer for L in (7.10) will indicate to which column to skip to find the next nonzero entry of L.

7.5 IMBALANCED FAULTS—SYMMETRICAL COMPONENT TRANSFORMATION

Only positive sequence voltages and currents occur in balanced three phase circuits with positive sequence excitation. For this reason, balanced three phase faults on balanced three phase systems are studied in the simple way described in the preceding sections. The familiar one-line diagrams used for this purpose are the same as those used for load flow studies. These analyses are positive sequence studies only. If the transmission system, the operating condition, or the fault is imbalanced, the simple positive sequence analysis fails since other sequences may occur. In this section the case of either imbalanced prefault operating conditions and/or imbalanced faults is considered. The transmission network is considered to be balanced and its model entails the same network for both positive and negative sequence signals. To the conclusion of this paragraph, however, briefly consider the much more complicated case of an imbalanced transmission network. Some EHV and UHV transmission circuits are designed as untransposed systems and therefore the assumption of a balanced transmission network becomes invalid. Under such conditions, a variety of approaches are applied which are tradeoffs of various degrees between accuracy and computational complexity. These include: (1) analog simulation, (2) assumption of a network model which is a "hybrid" between the positive and negative sequence models, and (3) full three phase representation and solution without the use of symmetrical components.

Consider a balanced three phase network with admittance and impedance matrices Y_{bus} and Z_{bus}. Let these matrices have n axes. If each element of these matrices, Y_{ij} for example, is examined in detail, bus i may be recognized as three circuit nodes corresponding to the three phases. Similarly, bus j may be regarded as three circuit nodes. Thus Y_{ij}, a scalar entry in Y_{bus}, may be regarded as a 3 by 3 submatrix, $Y_{ij}^{3\phi}$, of the $3n$ by $3n$ matrix, $Y_{bus}^{3\phi}$. The submatrix $Y_{ij}^{3\phi}$ is formed using the

Y_{bus} building algorithm (i.e., the diagonal entry $Y_{i(a)i(a)}$ is the sum of the admittances connected to ϕA of bus i, $Y_{i(a)j(a)}$ is the negative of the admittance joining buses i and j in ϕA, etc.). The inverse of the three phase bus admittance matrix is the three phase bus impedance matrix, $Z_{bus}^{3\phi}$. For a balanced transmission network, each submatrix of $Y_{bus}^{3\phi}$ and $Z_{bus}^{3\phi}$ is of the form D,

$$D = \begin{bmatrix} s & m & m \\ m & s & m \\ m & m & s \end{bmatrix} .$$

This is a consequence of the fact that the same series impedance occurs in each phase of the circuit, and the same mutual admittance occurs for $\phi A - \phi B$, $\phi B - \phi C$, $\phi C - \phi A$. The eigenvalues of D are readily found from

$$det \ (D - \lambda I) = 0 . \tag{7.12}$$

Equation (7.12) is termed the characteristic equation of D and is found to be the cubic expression,

$$- \lambda^3 + s\lambda^2 + \lambda(- 3s^2 + 2m^2) + s^3 - 2m^2 s + 2m^3 = 0 .$$

The roots of this cubic are

$$\lambda = s - m$$
$$\lambda = s - m$$
$$\lambda = s + 2m .$$

These roots are the eigenvalues of D and are labeled the positive sequence, negative sequence, and zero sequence eigenvalues, respectively. The corresponding eigenvectors of D are found from

$$(D - \lambda_+ I)e_+ = 0 \qquad ||e_+|| \neq 0 \tag{7.13}$$

$$(D - \lambda_- I)e_- = 0 \qquad ||e_-|| \neq 0 \tag{7.14}$$

$$(D - \lambda_0 I)e_0 = 0 \qquad ||e_0|| \neq 0. \tag{7.15}$$

The complex, orthonormal set of eigenvectors is not unique and depends on the selection of certain elements of the eigenvectors. One common choice is

$$e_+ = \frac{1}{\sqrt{3}} \begin{pmatrix} 1 \\ a^2 \\ a \end{pmatrix} \quad e_- = \frac{1}{\sqrt{3}} \begin{pmatrix} 1 \\ a \\ a^2 \end{pmatrix} \quad e_0 = \frac{1}{\sqrt{3}} \begin{pmatrix} 1 \\ 1 \\ 1 \end{pmatrix} ,$$

where

$$a = 1 \underline{/120^\circ} .$$

We will have more to say about this choice of eigenvectors and others in the next section. The reason for multiple solutions to (7.13) through (7.15) is that these equations are each of the form

$$Ax = 0,$$

where $det(A) = 0$. Such a set of equations has an infinite number of solution vectors, x. The motivation for the choice of eigenvalues shown is based on physical observation. There is no salient mathematical reason for this selection.

When the eigenvectors e_+, e_-, and e_0 are arranged in a matrix,

$$M = (e_+ | e_- | e_0) \,,$$

M is termed the modal matrix of D and this matrix will diagonalize D in a similarity transformation,

$$M^{-1}DM = \Lambda = diag\ (\lambda_+, \lambda_-, \lambda_0).$$

A consequence of the orthonormality of the eigenvectors is

$$M^H = M^{-1} \,.$$

If D is the 3 by 3 submatrix, $Y_{ij}^{3\phi}$, the equation

$$I = Y_{ij}^{3\phi} V$$

is decoupled by defining

$$V = MV' \tag{7.16}$$

$$I = MI' \tag{7.17}$$

since

$$MI' = Y_{ij}^{3\phi} MV'$$

results in

$$I' = [M^{-1} Y_{ij}^{3\phi} M]\, V'$$

$$= \Lambda_{ij}\, V' \tag{7.18}$$

where

$$\Lambda_{ij} = diag\ (\lambda_{+ij}, \lambda_{-ij}, \lambda_{oij}) \,.$$

The λ's are the three eigenvalues of $Y_{ij}^{3\phi}$.

Equation (7.18) is decoupled since I' in row 1 depends only on V' in row 1. If I' is composed of I_+, I_-, and I_o, and the vector V' is similarly composed, (7.18) is simply

$$I_+ = \lambda_{+ij}\, V_+ \tag{7.19}$$

$$I_- = \lambda_{-ij}\, V_- \tag{7.20}$$

$$I_0 = \lambda_{oij}\, V_o \,. \tag{7.21}$$

Equations (7.19) through (7.21) are the Ohm's law expressions for three sequence circuits, each responding only to its own sequence current, and each circuit is uncoupled to the other two. The λ's in (7.19) through (7.21) are the sequence admittances of the circuit.

The transformation suggested here is the *symmetrical component transformation.* It is not the only selection of M. Other transformations will be discussed in the subsequent sections. All transformations that decouple the three phase circuit must obey (7.13) through (7.15). The reason for the popularity of the symmetrical component transformation is based in part on history and in part on a physical interpretation of I' and V'. The topic of symmetrical components used to occupy a significant place in undergraduate power engineering curricula. When viewed as a transformation, however, it is possible to employ numerous useful properties of matrix algebra and thereby avoid lengthy discussions of the symmetrical component approach to problems. References [5–6] are classics in this area.

The philosophy of (7.16) and (7.17) is now applied to the entire V_{bus} and I_{bus} vectors. Using the symmetrical component transformation, the equation

$$I_{bus}^{3\phi} = Y_{bus}^{3\phi} V_{bus}^{3\phi}$$

becomes

$$I_{bus}^{+-o} = T^t Y_{bus}^{3\phi} T V_{bus}^{+-o}, \tag{7.22}$$

where

$$T = \begin{bmatrix} M & 0 & 0 & \cdots & 0 \\ 0 & M & 0 & \cdots & 0 \\ 0 & 0 & M & \cdots & 0 \\ \cdots & & & & \\ 0 & 0 & 0 & \cdots & M \end{bmatrix}$$

$$MI_{bus}^{+-o} = I_{bus}^{3\phi}$$

$$MV_{bus}^{+-o} = V_{bus}^{3\phi} .$$

In other words, each triplet of $I_{bus}^{3\phi}$ and $V_{bus}^{3\phi}$ is transformed using the symmetrical component transformation. The resulting coefficient of V_{bus}^{+-o} in (7.22) is

$$Y_{bus}^{+-o} = T^t Y_{bus}^{3\phi} T$$

$$= \begin{bmatrix} diag\ (\lambda_{+11}, \lambda_{-11}, \lambda_{o11}) & diag\ (\lambda_{+12}, \lambda_{-12}, \lambda_{o12}) & \cdots \\ diag(\lambda_{+21}, \lambda_{-21}, \lambda_{o21}) & diag\ (\lambda_{+22}, \lambda_{-22}, \lambda_{o22}) & \\ & \cdots & \end{bmatrix} .$$

Each 3 by 3 block of $Y_{bus}^{3\phi}$ has been diagonalized. Thus row 1 of (7.22) is coupled only to rows 4, 7, 10, and so on. Row 2 is coupled only to rows 5. 8, 11, and so on, Similarly, the zero sequence rows are coupled. If rows 1, 4, 7, 11, and so on, are extracted, the positive sequence relation is obtained,

$$I_{bus}^{+} = Y_{bus}^{+} V_{bus}^{+} . \tag{7.23}$$

Similarly,

$$I_{bus}^{-} = Y_{bus}^{-} V_{bus}^{-} \tag{7.24}$$

$$I_{bus}^{o} = Y_{bus}^{o} V_{bus}^{o} . \tag{7.25}$$

Equations (7.23) through (7.25) represent three decoupled sequence networks. Examination of these networks indicates:

1. When only balanced conditions are considered throughout, $Y_{bus}^{+} = Y_{bus}^{-}$, and this is the familiar bus admittance matrix used in single-line analysis.

2. The $Z_{bus}^{3\phi}$ matrix is block-by-block decoupled by the same symmetrical component transformation and the analogs of (7.23) through (7.25) are

$$V_{bus}^{+} = Z_{bus}^{+} I_{bus}^{+} \tag{7.26}$$

$$V_{bus}^- = Z_{bus}^- I_{bus}^- \tag{7.27}$$

$$V_{bus}^o = Z_{bus}^o I_{bus}^o \,. \tag{7.28}$$

3. If no rotating elements are modeled in Z_{bus} or Y_{bus},

$$Z_{bus}^+ = Z_{bus}^-$$

$$Y_{bus}^+ = Y_{bus}^- \,.$$

4. The symmetrical component transformation is not the only transformation that may be used to obtain the sequence matrices Z_{bus}^+, Z_{bus}^-, Z_{bus}^o, and the admittance counterparts. No matter what transformation is used to decouple the three phase circuit, one obtains the same decoupled circuits.

5. The symmetrical component transformation, M, as given above is

$$M = \frac{1}{\sqrt{3}} \begin{bmatrix} 1 & 1 & 1 \\ a^2 & a & 1 \\ a & a^2 & 1 \end{bmatrix} .$$

Since the ordering of the sequences as "+, -, 0" is not mandatory, one may obtain an alternative transformation by reordering the sequences. Thus the ordering "-, +, o" results in

$$M = \frac{1}{\sqrt{3}} \begin{bmatrix} 1 & 1 & 1 \\ a & a^2 & 1 \\ a^2 & a & 1 \end{bmatrix} .$$

6. The ambiguity in characteristic 5 is further compounded by noting that M need not be orthonormal. That is, the eigenvectors of D need not be normalized. Thus cM will diagonalize D if M diagonalizes D (where c is any nonzero, complex constant). This observation explains why the symmetrical component transformations taken from the literature may differ from that given here. This ambiguity is actually further extended by noting that the selection of the phase ordering "ϕA, ϕB, ϕC" is not necessary. This point is considered further in the exercises at the end of the chapter.

The case of balanced faults was considered earlier by considering the fault current at bus k, I_k^{post}, as being sufficient to force V_k to zero (for a bolted fault). This approach was applied to the one line diagram, which is, in fact, the positive sequence circuit. The same approach is applied for imbalanced faults: the phase fault currents I_{ak}^{post}, I_{bk}^{post} are calculated to force the appropriate imbalanced postfault voltage condition. As an illustration, consider a ϕA to neutral fault. In this case, the phase currents must be such as to produce zero postfault bus voltage on ϕA. In this case, the ϕB and ϕC fault currents are zero. The three conditions

$$V_{ak}^{post} = 0 \qquad I_{bk}^{post} = 0 \qquad I_{ck}^{post} = 0$$

yield sufficient information to solve the fault (i.e., to find the ϕA fault current and postfault voltages on the unfaulted phases as well as the postfault conditions elsewhere and the sequence voltages and currents). Using symmetrical components, these calculations are readily accomplished by

1. Calculating the sequence postfault voltages and currents at the faulted bus. These will be three unknown voltages, and three unknown currents, and the fault conditions will yield three relations. Hence one would expect the net number of unknowns as three. In the $\phi A-N$ fault indicated above, the unknowns are I_{ak}^{post}, V_{bk}^{post}, V_{ck}^{post}.

2. The sequence bus impedance matrices Z_{bus}^{+}, Z_{bus}^{-}, Z_{bus}^{o} must be known or calculated from $Z_{bus}^{3\phi}$. In the absence of rotating elements, $Z_{bus}^{+} = Z_{bus}^{-}$. Since the sequence networks are decoupled, the following postfault conditions are applied

$$V_k^{post\ +} = V_k^{pre\ +} + Z_{kk}^{+} I_k^{post\ +} \tag{7.29}$$

$$V_k^{post\ -} = V_k^{pre\ -} + Z_{kk}^{-} I_k^{post\ -} \tag{7.30}$$

$$V_k^{post\ o} = V_k^{pre\ o} + Z_{kk}^{o} I_k^{post\ o}. \tag{7.31}$$

3. The three relations in (7.29) - (7.31) contain three unknowns. Simultaneous solution yields all the sequence parameters at the faulted bus and the inverse symmetrical component transformation yields the postfault phase variables at bus k.

4. Postfault voltages at buses other than the fauled bus are readily calculated using

$$V_{bus}^{post\ +} = V_{bus}^{pre\ +} + Z_{bus}^{+} I_{bus}^{post\ +} \tag{7.32}$$

$$V_{bus}^{post\ -} = V_{bus}^{pre\ -} + Z_{bus}^{-} I_{bus}^{post\ +} \tag{7.33}$$

$$V_{bus}^{post\ o} = V_{bus}^{pre\ o} + Z_{bus}^{o} I_{bus}^{post\ o}. \tag{7.34}$$

These sequences voltages when inverse transformed yield the required phase voltages.

5. Fault capacity at the faulted bus is F_{fault},

$$F_{fault} = || V_{ak}^{pre} (I_{ak}^{post})^* + V_{bk}^{pre} (I_{bk}^{post})^* + V_{ck}^{pre} (I_{bk}^{post})^* || . \tag{7.35}$$

If an orthonormal symmetrical component transformation is used, (7.35) may be more conveniently written as

$$F_{fault} = || V_k^{pre\ +} (I_k^{post\ +})^* + V_k^{pre\ -} (I_k^{post\ -})^* + V_k^{pre\ o} (I_k^{post\ o})^* || . \tag{7.36}$$

Example 7.2

For a certain power system with

$$Z_{bus}^{+} = Z_{bus}^{-}$$

a phase–to–ground fault occurs at bus 25. The faulted phase is ϕB. Before the fault, the bus is at rated voltage, zero phase angle. The sequence impedances from Z_{bus} are

$$z_{25\ 25}^{+} = j0.10 \qquad z_{25\ 25}^{o} = j0.30$$

(per unit). Find the fault current and capacity.

Solution

The prefault voltage at bus 25 is V^{pre}

$$V_{abc}^{pre} = \begin{bmatrix} 1 \\ a^2 \\ a \end{bmatrix} \qquad V_{+-o}^{pre} = \frac{1}{\sqrt{3}} \begin{bmatrix} 1 & a & a^2 \\ 1 & a^2 & a \\ 1 & 1 & 1 \end{bmatrix} \begin{bmatrix} 1 \\ a^2 \\ a \end{bmatrix} = \begin{bmatrix} \sqrt{3} \\ 0 \\ 0 \end{bmatrix}.$$

The postfault voltage at bus 25 is

$$V_{abc}^{post} = \begin{bmatrix} x \\ 0 \\ y \end{bmatrix} \qquad V_{+-o}^{post} = \frac{1}{\sqrt{3}} \begin{bmatrix} x + a^2 y \\ x + ay \\ x + y \end{bmatrix}.$$

The postfault current at bus 25 is

$$I_{abc}^{post} = \begin{bmatrix} 0 \\ I_f \\ 0 \end{bmatrix} \qquad I_{+-o}^{post} = \frac{1}{\sqrt{3}} \begin{bmatrix} aI_f \\ a^2 I_f \\ I_f \end{bmatrix}.$$

The three unknowns, x, y, I_f, are found by simultaneous solution of

$$\frac{1}{\sqrt{3}} \begin{bmatrix} x + a^2 y \\ x + ay \\ x + y \end{bmatrix} = \begin{bmatrix} j0.10 & 0 & 0 \\ 0 & j0.10 & 0 \\ 0 & 0 & j0.30 \end{bmatrix} \frac{1}{\sqrt{3}} \begin{bmatrix} aI_f \\ a^2 I_f \\ I_f \end{bmatrix} + \begin{bmatrix} \sqrt{3} \\ 0 \\ 0 \end{bmatrix}.$$

In condensed form,

$$\begin{bmatrix} 1 & a^2 & -j0.1a \\ 1 & a & -j0.1a^2 \\ 1 & 1 & -j0.3 \end{bmatrix} \begin{bmatrix} x \\ y \\ I_f \end{bmatrix} = \begin{bmatrix} 3 \\ 0 \\ 0 \end{bmatrix}.$$

The solution is

$$\begin{bmatrix} x \\ y \\ I_f \end{bmatrix} = \begin{bmatrix} 0.88\underline{/16^\circ} \\ 0.88\underline{/104^\circ} \\ 4.24\underline{/-30^\circ} \end{bmatrix} \quad V_{abc}^{post} = \begin{bmatrix} 0.88\underline{/16^\circ} \\ 0 \\ 0.88\underline{/104^\circ} \end{bmatrix} \quad I_{abc}^{post} = \begin{bmatrix} 0 \\ 4.24\underline{/-30^\circ} \\ 0 \end{bmatrix}.$$

The fault capacity, F_{fault}, is readily calculated to be 4.24 per unit.

Example 7.3

Repeat Example 7.2 for a phase–phase fault between phases A and B.

Solution

The prefault conditions are as stated above, but the postfault voltages and currents are

$$V_{abc}^{post} = \begin{bmatrix} x \\ x \\ y \end{bmatrix} \qquad V_{+-o}^{post} = \frac{1}{\sqrt{3}} \begin{bmatrix} (1 + a)x + a^2 y \\ (1 + a^2)x + ay \\ 2x + y \end{bmatrix}$$

$$I_{abc}^{post} = \begin{bmatrix} -I_f \\ I_f \\ 0 \end{bmatrix} \qquad I_{+-o}^{post} = \frac{1}{\sqrt{3}} \begin{bmatrix} (-1 + a)I_f \\ (-1 + a^2)I_f \\ 0 \end{bmatrix}.$$

The three unknowns, x, y, and I_f, are found by simultaneous solution of

$$\begin{bmatrix} 1+a & a^2 & -j0.1(-1+a) \\ 1+a^2 & a & -j0.1(-1+a^2) \\ 2 & 1 & 0 \end{bmatrix} \begin{bmatrix} x \\ y \\ I_f \end{bmatrix} = \begin{bmatrix} 3 \\ 0 \\ 0 \end{bmatrix}.$$

The result is

$$V_{abc}^{post} = \begin{bmatrix} 2.59 \angle{-30^\circ} \\ 2.59 \angle{-30^\circ} \\ 3 \angle{-60^\circ} \end{bmatrix} \qquad I_{abc}^{post} = \begin{bmatrix} 8.66 \angle{120^\circ} \\ 8.66 \angle{-60^\circ} \\ 0 \end{bmatrix}$$

and the fault capacity is 15.0 per unit.

Note that in this example as well as Example 7.2, the coefficient of the unknown vector $[x \;\; y \;\; I_f]^t$ is readily identified as certain coefficients in V_{+-o}^{post} and sequence impedances times I_{+-o}^{post}.

Example 7.4

Consider the system described in Example 7.2, now with a phase–phase–ground fault in phases A and B. Calculate the fault parameters and comment on the effects of fault impedance when a 138 kV, 100 MVA base is used.

Solution

The prefault conditions, again, are as described in Example 7.2 and the postfault conditions are (for $z_f = 0$)

$$V_{abc}^{post} = \begin{bmatrix} 0 \\ 0 \\ x \end{bmatrix} \qquad V_{+-o}^{post} = \frac{1}{\sqrt{3}} \begin{bmatrix} a^2 x \\ ax \\ x \end{bmatrix}$$

$$I_{abc}^{post} = \begin{bmatrix} I_{fa} \\ I_{fb} \\ 0 \end{bmatrix} \qquad I_{+-o}^{post} = \frac{1}{\sqrt{3}} \begin{bmatrix} I_{fa} + aI_{fb} \\ I_{fa} + a^2 I_{fb} \\ I_{fa} + I_{fb} \end{bmatrix}.$$

The sequence equations are

$$\begin{bmatrix} a^2 & -j0.1 & -j.1a \\ a & -j0.1 & -j.1a^2 \\ 1 & j0.3 & -j.3 \end{bmatrix} \begin{bmatrix} x \\ I_{fa} \\ fb \end{bmatrix} = \begin{bmatrix} 3 \\ 0 \\ 0 \end{bmatrix}$$

and the solution is

$$V_{abc}^{post} = \begin{bmatrix} 0 \\ 0 \\ 1.28 \angle{120^\circ} \end{bmatrix} \qquad I_{abc}^{post} = \begin{bmatrix} 8.92 \angle{106.1^\circ} \\ 8.92 \angle{313.9^\circ} \\ 0 \end{bmatrix}.$$

The fault capacity is 17.14 per unit (1714 MVA).

When a nonzero fault impedance is considered, the postfault voltage on the faulted phases (see Figure 7.6) is

$$V_{an} = V_{bn} = -(I_{fa} + I_{fb})z_f \; .$$

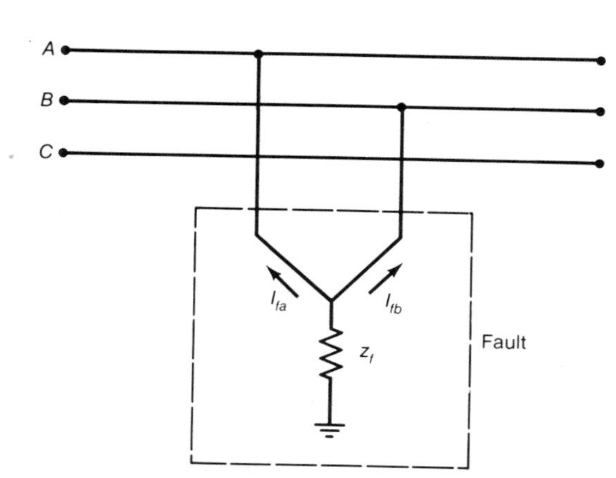

FIGURE 7.6 Phase-to-phase-to-ground fault (Example 7.4).

Denoting the phase–to–neutral voltage on the unfaulted phase as x results in

$$V_{+-o}^{post} = \frac{1}{\sqrt{3}} \begin{bmatrix} -(1+a)(I_{fa}+I_{fb})z_f + a^2x \\ -(1+a^2)(I_{fa}+I_{fb})z_f + ax \\ -2(I_{fa}+I_{fb})z_f + x \end{bmatrix}.$$

The fault current is denoted in the same way as in the $z_f = 0$ case. The sequence equations are

$$\begin{bmatrix} a^2 & -z_f(1+a)-j0.1 & -z_f(1+a)-j0.1a \\ a & -z_f(1+a^2)-j0.1 & -z_f(1+a^2)-j0.1a^2 \\ 1 & -2z_f - j0.3 & -2z_f - j0.3 \end{bmatrix} \begin{bmatrix} x \\ I_{fa} \\ I_{fb} \end{bmatrix} = \begin{bmatrix} 3 \\ 0 \\ 0 \end{bmatrix}.$$

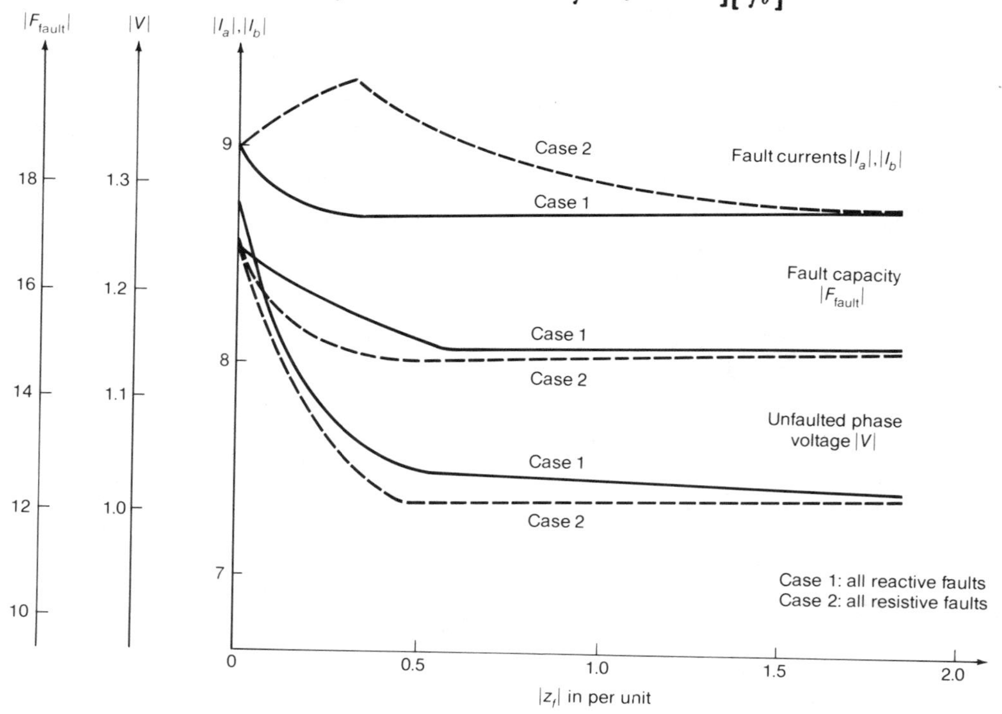

FIGURE 7.7 Results of a phase-phase-ground fault study.

The solution is readily found by a simple computer program. Because z_f is a complex quantity, it is difficult to depict the results. Figure 7.7 shows representative results for the cases of all reactive fault impedance (case 1) and all resistive fault impedance (case 2). At $z_f = 0$, the same results as found earlier are obtained. Since

$$I_a^{post} = 8.92 \underline{/106.1^\circ} \qquad I_b^{post} = 8.92 \underline{/313.9^\circ}$$

the total ground current is $4.29 \underline{/30^\circ}$. As $|z_f|$ increases in either case 1 or case 2, the fault capacity decreases and eventually reaches the phase to phase value (15 per unit). Note that the maximum fault current for the purely resistive case does not occur at $|z_f| = 0$. All parameters studied approach the $\phi A - \phi B$ fault case for large $|z_f|$ and these values agree with those from Example 7.3. As $|z_f|$ becomes very large,

$$I_a^{post} \simeq -I_b^{post} .$$

7.6

IMBALANCED FAULTS—OTHER TRANSFORMATIONS

In the preceding section, the orthonormal symmetrical component transformation was used. When the ordering of the sequences is "+, -, 0," and the ordering of the phases is "ϕA, ϕB, ϕC," this transformation (illustrated on a triplet of phase currents) is

$$I^{+-o} = \frac{1}{\sqrt{3}} \begin{bmatrix} 1 & a & a^2 \\ 1 & a^2 & a \\ 1 & 1 & 1 \end{bmatrix} I^{abc} .$$

It has already been pointed out that there is an ambiguity in the symmetrical component transformation in that reordering of sequences and/or phases will produce many other perfectly usable transformations. Also, if orthonormality of M is not demanded, the transformation matrix may be multiplied by an arbitrary, nonzero, complex constant. Beyond these observations on producing alternative symmetrical component transformations, it is interesting to investigate other similarity transformations which diagonalize D,

$$D = \begin{bmatrix} s & m & m \\ m & s & m \\ m & m & s \end{bmatrix} .$$

This investigation is done here for the general case of n_ϕ phases.

Consider now the general problem of diagonalizing n_ϕ by n_ϕ matrix D

$$(D)_{ij} = \begin{cases} s & i = j \\ m & i \neq j \end{cases} .$$

The eigenvalues of D are found from the solution of

$$det\ (D - \lambda I) = 0$$

to be

$$\lambda_\ell = \begin{cases} s - m & \ell = 1, 2, \ldots, n_\phi - 1 \\ s + (n_\phi - 1)m & \ell = n_\phi . \end{cases}$$

The eigenvector problem in this case is (for $\ell = 1, 2, \ldots, n_\phi - 1$)

$$\begin{bmatrix} m & m & \cdots & m \\ m & m & & \\ & \cdots & & \end{bmatrix} e_\ell = 0 \tag{7.37}$$

$$\| e_\ell \| \neq 0$$

and (for $\ell = n_\phi$)

$$\begin{bmatrix} -(n_\phi - 1)m & m & \cdots & m \\ m & -(n_\phi - 1)m & \cdots & m \\ \cdots & & & \end{bmatrix} e_\ell = 0 \tag{7.38}$$

$$\| e_\ell \| \neq 0 .$$

Since the coefficient of e_ℓ in (7.37) and (7.38) is singular, there is no unique solution for e_ℓ. Two types of solutions are very well known: the symmetrical component transformation and the Clarke's component transformation [6]. The symmetrical component transformation in the three phase case was considered earlier and is considered further in the polyphase case in an exercise at the end of this chapter. The Clarke component transformation is

$$e_\ell = \begin{bmatrix} 0 \\ 0 \\ 0 \\ \frac{n_\phi - \ell}{\sqrt{q}} \\ \frac{-1}{\sqrt{q}} \\ \frac{-1}{\sqrt{q}} \\ \vdots \\ \frac{-1}{\sqrt{q}} \end{bmatrix} \begin{matrix} \text{row } 1 \\ \\ \text{row } \ell - 1 \\ \text{row } \ell \\ \text{row } \ell + 1 \\ \text{row } \ell + 2 \\ \\ \text{row } n_\phi \end{matrix} \qquad \ell = 1, 2, \ldots, n_\phi - 1$$

$$e_\ell = \begin{bmatrix} \frac{1}{\sqrt{n_\phi}} \\ \frac{1}{\sqrt{n_\phi}} \\ \vdots \\ \frac{1}{\sqrt{n_\phi}} \end{bmatrix} \qquad \ell = n_\phi, \tag{7.39,}$$

where

$$q = (n_\phi - \ell)^2 + n_\phi - \ell . \tag{7.41}$$

Thus the three phase, six phase, and eight phase Clarke transformations are

$$\begin{bmatrix} \frac{2}{\sqrt{6}} & 0 & \frac{1}{\sqrt{3}} \\ \frac{-1}{\sqrt{6}} & \frac{1}{\sqrt{2}} & \frac{1}{\sqrt{3}} \\ \frac{-1}{\sqrt{6}} & \frac{-1}{\sqrt{2}} & \frac{1}{\sqrt{3}} \end{bmatrix} \tag{7.42}$$

$$\begin{bmatrix} \frac{5}{\sqrt{30}} & 0 & 0 & 0 & 0 & \frac{1}{\sqrt{6}} \\ \frac{-1}{\sqrt{30}} & \frac{4}{\sqrt{20}} & 0 & 0 & 0 & \frac{1}{\sqrt{6}} \\ \frac{-1}{\sqrt{30}} & \frac{-1}{\sqrt{20}} & \frac{3}{\sqrt{12}} & 0 & 0 & \frac{1}{\sqrt{6}} \\ \frac{-1}{\sqrt{30}} & \frac{-1}{\sqrt{20}} & \frac{-1}{\sqrt{12}} & \frac{2}{\sqrt{6}} & 0 & \frac{1}{\sqrt{6}} \\ \frac{-1}{\sqrt{30}} & \frac{-1}{\sqrt{20}} & \frac{-1}{\sqrt{12}} & \frac{-1}{\sqrt{6}} & \frac{1}{\sqrt{2}} & \frac{1}{\sqrt{6}} \\ \frac{-1}{\sqrt{30}} & \frac{-1}{\sqrt{20}} & \frac{-1}{\sqrt{12}} & \frac{-1}{\sqrt{6}} & \frac{-1}{\sqrt{2}} & \frac{1}{\sqrt{6}} \end{bmatrix}$$

$$\begin{bmatrix} \frac{7}{\sqrt{56}} & 0 & 0 & 0 & 0 & 0 & 0 & \frac{1}{\sqrt{8}} \\ \frac{-1}{\sqrt{56}} & \frac{6}{\sqrt{42}} & 0 & 0 & 0 & 0 & 0 & \frac{1}{\sqrt{8}} \\ \frac{-1}{\sqrt{56}} & \frac{-1}{\sqrt{42}} & \frac{5}{\sqrt{30}} & 0 & 0 & 0 & 0 & \frac{1}{\sqrt{8}} \\ \frac{-1}{\sqrt{56}} & \frac{-1}{\sqrt{42}} & \frac{-1}{\sqrt{30}} & \frac{4}{\sqrt{20}} & 0 & 0 & 0 & \frac{1}{\sqrt{8}} \\ \frac{-1}{\sqrt{56}} & \frac{-1}{\sqrt{42}} & \frac{-1}{\sqrt{30}} & \frac{-1}{\sqrt{20}} & \frac{3}{\sqrt{12}} & 0 & 0 & \frac{1}{\sqrt{8}} \\ \frac{-1}{\sqrt{56}} & \frac{-1}{\sqrt{42}} & \frac{-1}{\sqrt{30}} & \frac{-1}{\sqrt{20}} & \frac{-1}{\sqrt{12}} & \frac{2}{\sqrt{6}} & 0 & \frac{1}{\sqrt{8}} \\ \frac{-1}{\sqrt{56}} & \frac{-1}{\sqrt{42}} & \frac{-1}{\sqrt{30}} & \frac{-1}{\sqrt{20}} & \frac{-1}{\sqrt{12}} & \frac{-1}{\sqrt{6}} & \frac{1}{\sqrt{2}} & \frac{1}{\sqrt{8}} \\ \frac{-1}{\sqrt{56}} & \frac{-1}{\sqrt{42}} & \frac{-1}{\sqrt{30}} & \frac{-1}{\sqrt{20}} & \frac{-1}{\sqrt{12}} & \frac{-1}{\sqrt{6}} & \frac{-1}{\sqrt{2}} & \frac{1}{\sqrt{8}} \end{bmatrix}.$$

Equations (7.39) and (7.40) are readily verified by substitution into (7.37) and (7.38), respectively. Only the three phase Clarke transformation above has practical significance, although polyphase power circuits are occasionally encountered and proposals for six phase transmission circuits may yield significant new applications [2–4].

The salient difference between the symmetrical component and the Clarke component transformations is that the Clarke transformation is entirely real. Thus, even nonsinusoidal phase voltages and currents may be transformed using Clarke's components, whereas the usual sinusoidal steady state assumptions are made in the phasor, complex arithmetic of symmetrical components. Reference [7] contains a detailed discussion of the exploitation of this real transformation in such applications as calculation of traveling wave characteristics in three phase transmission lines.

It is important to note that Clarke's component transformation also has the ambiguities of form as the symmetrical component transformation. Also note that since the eigenvalues of D do not depend on the forms of its own modal matrix; the sequence impedances and admittances used in the method of symmetrical components are ***identical*** to those used in the method of Clarke's components. In the literature of 3ϕ Clarke's components, the terms alpha, beta, and zero sequence are used in place of positive, negative, and zero sequence (which are reserved for symmetrical components). For the Clarke's transformations shown above,

$$M^{-1} = M^t$$

as a consequence of the normalization of the eigenvectors [Eq. (7.39)].

Example 7.5

In this example, the illustration in Example 7.2 is reworked using Clarke's transformation.

Solution

The prefault voltage is

$$V_{abc}^{pre} = \begin{bmatrix} 1 \\ a^2 \\ a \end{bmatrix} \qquad V_{\alpha\beta 0}^{pre} = \begin{bmatrix} \frac{2}{\sqrt{6}} & \frac{-1}{\sqrt{6}} & \frac{-1}{\sqrt{6}} \\ 0 & \frac{1}{\sqrt{2}} & \frac{-1}{\sqrt{2}} \\ \frac{1}{\sqrt{3}} & \frac{1}{\sqrt{3}} & \frac{1}{\sqrt{3}} \end{bmatrix} \begin{bmatrix} 1 \\ a^2 \\ a \end{bmatrix} \qquad V_{\alpha\beta 0}^{pre} = \begin{bmatrix} \frac{\sqrt{3}}{\sqrt{2}} \\ -j\frac{\sqrt{3}}{\sqrt{2}} \\ 0 \end{bmatrix}.$$

Note that $V_{\alpha\beta 0}$ is found by premultiplying V_{abc} by the transpose of the matrix in (7.42). This transposed matrix is its inverse. The postfault voltage is

$$V_{abc}^{post} = \begin{bmatrix} x \\ o \\ y \end{bmatrix} \qquad V_{\alpha\beta o}^{post} = \begin{bmatrix} \frac{2x - y}{\sqrt{6}} \\ \frac{-y}{\sqrt{2}} \\ \frac{(x + y)}{\sqrt{3}} \end{bmatrix}.$$

The postfault currents are

$$I_{abc}^{post} = \begin{bmatrix} 0 \\ I_f \\ 0 \end{bmatrix} \qquad I_{\alpha\beta o}^{post} = \begin{bmatrix} \frac{-I_f}{\sqrt{6}} \\ \frac{I_f}{\sqrt{2}} \\ \frac{I_f}{\sqrt{3}} \end{bmatrix}.$$

The component equations are found from

$$V_{\alpha\beta o}^{post} = Z_{\alpha\beta o} I_{\alpha\beta o}^{post} + V_{\alpha\beta o}^{pre} \tag{7.43}$$

and $Z_{\alpha\beta o}$ is recognized as being identical to Z_{+-o}. This is a consequence of the fact that eigenvalues of a matrix are unique. Equation (7.43) becomes

$$\begin{bmatrix} \frac{2}{\sqrt{6}} & \frac{-1}{\sqrt{6}} & \frac{j0.1}{\sqrt{6}} \\ 0 & \frac{-1}{\sqrt{2}} & \frac{-j0.1}{\sqrt{2}} \\ \frac{1}{\sqrt{3}} & \frac{1}{\sqrt{3}} & \frac{-j0.3}{\sqrt{3}} \end{bmatrix} \begin{bmatrix} x \\ y \\ I_f \end{bmatrix} = \frac{\sqrt{3}}{2} \begin{bmatrix} 1 \\ -j1 \\ 0 \end{bmatrix}$$

and the solution is

$$\begin{bmatrix} x \\ y \\ I_f \end{bmatrix} = \begin{bmatrix} 0.88 \angle 16^\circ \\ 0.88 \angle 104^\circ \\ 4.24 \angle -30^\circ \end{bmatrix}.$$

This solution is recognized as the same as that found in Example 7.2. The fault capacity may be calculated using phase variables or Clarke's components. The latter is illustrated using

$$V_{\alpha\beta o}^{pre} = \begin{bmatrix} \frac{\sqrt{3}}{\sqrt{2}} \\ -j\frac{\sqrt{3}}{\sqrt{2}} \\ 0 \end{bmatrix} \qquad I_{\alpha\beta o}^{post} = \begin{bmatrix} 1.73 \angle 150^\circ \\ 3.00 \angle -30^\circ \\ 2.45 \angle -30^\circ \end{bmatrix}$$

$$\begin{aligned} F_{fault} &= \| V_\alpha^{pre} (I_\alpha^{post})^* + V_\beta^{pre} (I_\beta^{post})^* + V_o^{pre} (I_o^{post})^* \| \\ &= \left\| \frac{3}{\sqrt{2}} \angle -150^\circ + \frac{3\sqrt{3}}{\sqrt{2}} \angle -60^\circ \right\| \\ &= 4.24 \text{ per unit,} \end{aligned}$$

which agrees with the previous determination.

Bibliography

[1] A. Fitzgerald, C. Kingsley, and S. Umans, *Electric Machinery,* McGraw Hill, New York, 1983.

[2] N. Bhatt, S. Venkata, W. Guyker, and W. Booth, "Six-Phase (Multi-phase) Power Transmission Systems: Fault Analysis," *IEEE Trans. Power Apparatus and Systems,* v. PAS-96, no. 3, May–June 1977, pp. 758–767.

[3] S. S. Venkata, N. Bhatt, and W. Guyker, "Six-Phase (Multi-phase) Power Transmission Systems: Concept and Reliability Aspects," *IEEE Summer Power Meeting,* July 1976, Portland, Oreg.

[4] L. O. Barthold, and H. Barnes, *High Phase Order Transmission,* CIGRE Study Committee Report no. 31, London, 1972.

[5] C. L. Fortescue, "Method of Symmetrical Coordinates Applied to the Solution of Polyphase Networks," *Trans. AIEE,* v. 37, 1918, p. 1027.

[6] E. Clarke, *Circuit Analysis of AC Power Systems, Symmetrical and Related Components,* Wiley, New York, 1943.

[7] E. Clarke, *Circuit Analysis of AC Power Systems,* General Electric Co., Schenectady, N.Y., 1950.

[8] P. M. Anderson, *Analyses of Faulted Power Systems,* Iowa State University Press, Ames, Iowa, 1976.

Exercises

Exercises that do not involve computer programming

7.1 Let M be the orthonormal symmetrical component transformation. Note that

$$det\ (M) = 1\ .$$

Show that if F_{fault} is the fault capacity at bus k defined as

$$F_{fault} = \sum_{\ell = a,b,c} || V_{k\ell}^{pre} (I_{k\ell}^{post})^* ||,$$

then F_{fault} may also be calculated using

$$F_{fault} = \sum_{\ell = +,-,o} || V_{k\ell}^{pre} (I_{k\ell}^{post})^* ||\ .$$

That is, show that the symmetrical components may be used to calculate the fault capacity directly.

7.2 Demonstrate that the following matrix transformations are variations of the symmetrical component transformation. In each case state the ordering of the sequences and phases.

a. $M = \begin{bmatrix} a & a & a \\ 1 & a^2 & a \\ a^2 & 1 & a \end{bmatrix}$ b. $M = \begin{bmatrix} a & a & a \\ a & a^2 & 1 \\ a & 1 & a^2 \end{bmatrix}$

c. $M = \begin{bmatrix} 1 & 1 & 1 \\ a & 1 & a^2 \\ a^2 & 1 & a \end{bmatrix}$

7.3 In this exercise, the polyphase symmetrical component transformation will be developed. For this purpose, define

$$b = 1\ \underline{/(2\pi/n_\phi) rad}$$

as the analog of a used in three phase circuits. The number of phases is denoted as n_ϕ.

a. Find the eigenvalues of D,

$$(D)_{ij} = \begin{cases} s & i = j \\ m & i \neq j \end{cases}$$

for $i, j = 1, 2, \ldots, n_\phi$.

b. Using the ordering of the eigenvalues given in Section 7.6, show that the eigenvetors of D must satisfy

$$m \sum_{k=1}^{n_\phi} (e_\ell)_k = 0$$

for $\ell \neq n_\phi$, and

$$m \sum_{\substack{k=1 \\ \neq i}}^{n_\phi} (e_\ell)_k - (n_\phi - 1)m(e_\ell)_i = 0$$

for $\ell = n_\phi$. The latter expression must be shown to hold for all i, $1 \leq i \leq n_\phi$.

c. There is no unique solution to the simultaneous equations obtained in part (b). This is the case since

$$det\ (D - \lambda I) = 0\ .$$

Hence one must select some elements of the eigenvectors of D. In addition to the equations obtained in part (b), the eigenvectors must satisfy

$$e_i^H e_j = \begin{cases} 1 & i = j \\ 0 & i \neq j\ . \end{cases}$$

Let e_1 be $k_1[b^{n_\phi}\ b^{n_\phi - 1}\ \ldots\ b_1]^t$, and let e_{n_ϕ} be $k_\ell[1\ 1\ \ldots\ 1]^t$.

Find k_1, and k_ℓ. Then find e_2 using k_1 in the $(e_2)_1$ position; this entails finding a relationship among the remaining $n_\phi - 1$ elements of e_2. Since the root $\lambda_2 = s - m$ occurs with multiplicity $n_\phi - 1$ and e_2 must be (1) of unit length and (2) orthogonal to e_1 and e_{n_ϕ}, one would expect that having selected $(\ell_2)_1$, there remain $n_\phi - 2$ elements left to be selected. Examining the first two and last columns of M, we obtain

$$M = k \begin{bmatrix} 1 & 1 & \cdots & 1 \\ b^{n_\phi - 1} & x & & 1 \\ b^{n_\phi - 2} & x & & 1 \\ \cdots & & & \cdots \\ b & x & & 1 \end{bmatrix}.$$

Let e_2 be of the form

$$e_2 = k[1\ c^q\ c^{2q}\ \ldots\ c^{(n_\phi - 1)q}]$$

(where c is some suitably chosen complex number such as b or b^{-1}) and apply the orthogonality condition between e_2 and e_1 and also between e_2 and e_{n_ϕ}. You will find several valid values of q for which these conditions hold. Subsequently, form e_3 with a different value of q and repeat the process until M is obtained.

d. Verify that

$$M = \frac{1}{2}\begin{bmatrix} 1 & 1 & 1 & 1 \\ b^3 & b^2 & b & 1 \\ b^2 & 1 & b^2 & 1 \\ b & b^2 & b^3 & 1 \end{bmatrix}$$

is a valid orthonormal four phase symmetrical component transformation ($b = 0 + j1$).

e. Find a valid orthonormal six phase symmetrical component transformation.

7.4 Find Z^+, Z^-, and Z^o for the following two bus, three phase systems:

$$\text{a.} \quad j\begin{bmatrix} 0.05 & 0.01 & 0.01 & 0.03 & 0 & 0 \\ & 0.05 & 0.01 & 0 & 0.03 & 0 \\ & & 0.05 & 0 & 0 & 0.03 \\ & & & 0.07 & 0.01 & 0.01 \\ & & & & 0.07 & 0.01 \\ & \text{symmetric} & & & & 0.07 \end{bmatrix}$$

$$\text{b.} \quad j\begin{bmatrix} 0.09 & 0.08 & 0.08 & 0.02 & 0.01 & 0.01 \\ & 0.09 & 0.08 & 0.01 & 0.02 & 0.01 \\ & & 0.09 & 0.01 & 0.01 & 0.02 \\ & & & 0.10 & 0.07 & 0.07 \\ & & & & 0.10 & 0.07 \\ & \text{symmetric} & & & & 0.10 \end{bmatrix}$$

7.5 Some rotating elements have impedance and admittance matrices of the form

$$D = \begin{bmatrix} s & m_1 & m_2 \\ m_2 & s & m_1 \\ m_1 & m_2 & s \end{bmatrix},$$

where $m_1 \neq m_2$. In this exercise, the effectiveness of Clarke's transformation and the symmetrical component transformation are examined on this type of matrix.

a. Using the three phase Clarke's transformation,

$$M = \frac{1}{\sqrt{3}}\begin{bmatrix} \sqrt{2} & 0 & 1 \\ -\sqrt{1/2} & \sqrt{3/2} & 1 \\ -\sqrt{1/2} & -\sqrt{3/2} & 1 \end{bmatrix}$$

(note that this is identical to the Clarke's transformation given in the text, but written in a slightly different form), attempt to diagonalize D in a similarity transformation,

$$D' = M^{-1}DM \ .$$

Note that $M^{-1} = M^t$. The Clarke's components are known as alpha, beta, and zero sequences. For the M given, the ordering of the transformed variables is α, β, 0.

b. Repeat part (a) using the symmetrical component transformation.

c. Comment on the advantages of the symmetrical component transformation over Clarke's transformation. Use results of parts (a) and (b) to formulate this comment.

7.6 For some bulk power transmission applications, there may be advantages of using six phase circuits rather than three phase. One form of the symmetrical component transformation is,

$$M = \frac{1}{\sqrt{6}} \begin{bmatrix} 1 & 1 & 1 & 1 & 1 & 1 \\ b^5 & b^4 & b^3 & b^2 & b & 1 \\ b^4 & b^2 & 1 & b^4 & b^2 & 1 \\ b^3 & 1 & b^3 & 1 & b^3 & 1 \\ b^2 & b^4 & 1 & b^2 & b^4 & 1 \\ b & b^2 & b^3 & b^4 & b^5 & 1 \end{bmatrix} .$$

This form is unitary,

$$M^{-1} = M^H .$$

The ordering of the transformed variables is "one sequence", "two sequence", etc.

a. For a six phase transmission system such that

$$[(Z^{6\phi}_{\text{bus}})_{\text{bus } i \text{ bus } j}]_{\text{phase } k \text{ phase } \ell} = \begin{cases} s_{ij} & k = \ell \\ m_{ij} & k \neq \ell \end{cases}$$

find the one sequence Z_{bus} matrix, the two sequence Z_{bus} matrix, and so on. Note that for the M given, the order of the sequences is 1, 2, 3, 4, 5, 0.

b. Consider a line–to–neutral fault on phase A of the six phase system at bus k. Phases B, C, D, E, and F remain unfaulted. If the prefault voltage at bus k is $1.0 \underline{/0}$ per unit consisting of only one sequence, find the form of the postfault voltage both in phase variables and in symmetrical components.

c. Find the form of the postfault current at bus k in symmetrical components.

d. Verify that there are, in effect, six unknowns in the problem stated in part (b). Write the form of the six equations which are needed to solve for these unknowns.

e. Repeat parts (b) through (d) for the case of a ϕA to ϕB fault.

7.7 When nonlinear loads appear at power system buses, harmonics occur in the load current due to the distortion of the sinusoidal wave. This distortion nonetheless results in a periodic signal.

a. Show that all harmonics of order 4, 7, 10, 13, 16, 19, ... are positive sequence. Assume that the supply voltage is strictly balanced and the fundamental component of the supply voltage is positive sequence.

b. Under the same assumption in part (a), show that harmonics of order 2, 5, 8, 11, 14, 17, 20, ... are negative sequence.

c. Repeat part (a) to show that harmonics of order $3n$ are zero sequence.

d. Consider now an n_ϕ phase power supply which is balanced and designed to supply the fundamental power frequency at positive sequence. Let this source energize a full wave rectifier as shown in Figure P7.7. Suppose that the individual diodes are ideal and there are no even harmonics in the supply side (ac circuit) currents. Further, assume that the n_ϕ phase power is created from a three phase balanced supply and n_ϕ is an integer multiple of 6 (n_ϕ = 6, 12, 18, ...). This type of rectifier is known as an $2n_\phi$ pulse rectifier.

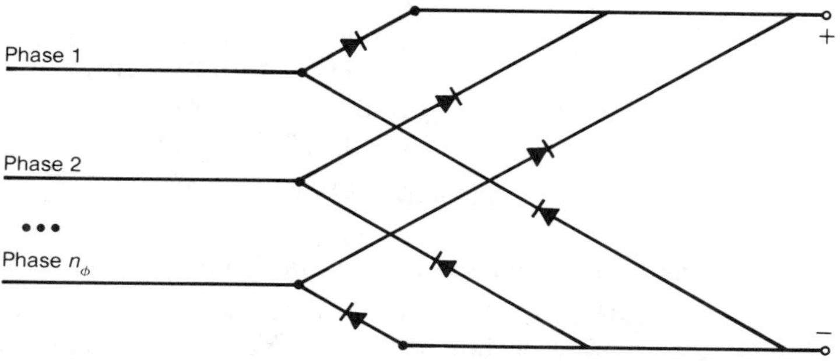

FIGURE 7.7 $2n_\phi$ pulse rectifier.

The polyphase supply for this rectifier is created by transformers whose primaries are wye connected. For $n_\phi = 3$ (i.e., a six pulse rectifier), the transformer bank is simply wye connected resulting in three individual phase conductors energizing the rectifier. For $n_\phi = 6$, a 12 pulse rectifier results and the 6 phase supply is created by one wye delta transformer bank and one wye wye bank. In this case, the 6 phase supply voltages to neutral are equal in amplitude and separated by $\pi/3$ radians in phase. For a 12 phase supply (i.e., 24 pulse rectifier), the transformer banks are wye wye connected and wye delta connected as before, but also an additional bank is wye-zig zag connected to give the required 30 ° phase shift to positive sequence voltages. For various values of n_ϕ, show that the harmonic currents present in the ac three phase supply are only those indicated by an "x" in the following table:

	Number of Rectifier Phases											
AC Harmonics	6	12	18	24	30	36	42	48	54	60	66	72
5	x											
7	x											
11	x	x										
13	x	x										
17	x	•	x									
19	x	•	x									
23	x	x	•	x								
25	x	x	•	x								
29	x	•	•	•	x							
31	x	•	•	•	x							
35	x	x	x	•	•	x						
37	x	x	x	•	•	x						
41	x	•	•	•	•	•	x					
43	x	•	•	•	•	•	x					
47	x	x	•	x	•	•	•	x				

Exercises that require computer programming

7.8 Figure P7.8 shows a seven bus power system. The prefault bus voltage profile is

Coal City	$0.99+j0.00$
Pearson	$0.97-j0.01$
Gary	$0.92-j0.09$
Hope	$0.92-j0.09$
Thomas	$0.91-j0.10$
Boiler	$0.91-j0.10$
Route 26	$0.92-j0.09$

(all in per unit on a 138 kV base). The line impedances are shown on a 100 MVA base. The transient reactance of the Coal City generator is very small on a 100 MVA base.

a. Calculate the fault current at each bus (except Coal City). Assume a bolted fault. You may assume that the machine terminals at Coal City are an infinite bus.

b. Calculate the fault capacity at each bus. Express your answer in GVA.

c. Repeat parts (a) and (b) using a fault impedance of 40 Ω per phase.

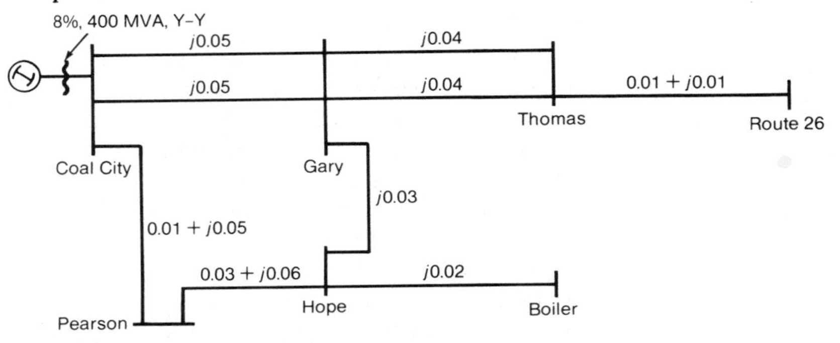

FIGURE P7.8 Coal City Power System

7.9 Figure P7.9 shows a 138 kV power system with line impedances on a 100 MVA base. Transformer reactances are shown on their own base. The prefault bus voltages are:

Aurora	$1.06+j0.00$
Kahoka	$1.06+j0.01$
Rawhide	$1.03+j0.00$
Roger	$1.00-j0.04$
Gabbo	$0.99-j0.03$
Sparky	$0.99-j0.07$
Big Paul	$0.97-j0.10$
Rancho	$0.98-j0.08$

and the machine reactances and ratings are as follows:

	x_d	x_d'	x_d''	rating
Aurora 1	80%	15%	10%	800 MVA 37 kV
Aurora 2	80%	15%	10%	800 MVA 37 kV
Kahoka	70%	27%	10%	800 MVA 37 kV
Gabbo	62%	19%	10%	1000 MVA 37.5 kV

The transformers at Aurora are rated 800 MVA each and are connected delta wye, 37 kV/138 kV. The transformer at Gabbo is connected delta wye, 37.5 kV/138 kV. The transformer at Kahoka is connected delta wye, 37 kV/138 kV.

a. Find the bolted fault current and capacity at each system bus.

b. The fault capacity at Rawhide is to be reduced to or below 160 GVA. Investigate the placement of series reactors in the system to produce this result. The reactors must be of minimum reactance to yield a minimum cost solution.

c. Find the postfault system voltages for a bolted fault at the Rancho bus.

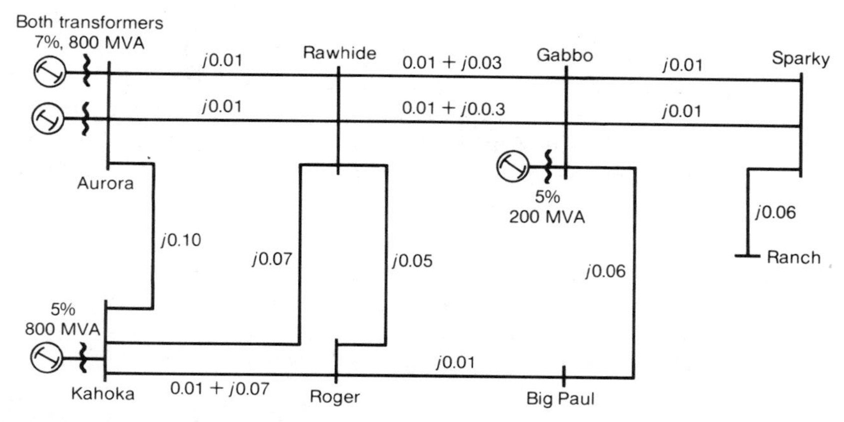

FIGURE P7.9 Aurora Power System

7.10 Often, transmission expansion results in higher fault capacities at system buses. Since circuit interruption equipment costs rise approximately as the GVA interruption capability, fault capacity is often used as a measure of expected interruption equipment costs at buses.

In this problem, the system shown in Figure P7.9, and described in Problem 7.9, is to be upgraded by the addition of a 250 kV circuit between the Aurora 138 kV bus and the 138 kV Gabbo bus. A second 250 kV circuit will join the new Gabbo 250 kV bus and Rancho. A single transformer, wye wye, 600 MVA, 3% 138/250 kV will energize the Aurora 250 kV bus from the 138 kV bus. Similarly Gabbo 250 and Gabbo 138 will be joined by an identical transformer. At Rancho, the 250 and 138 kV buses will be joined by a 350 MVA, 3%, 250/138 kV, wye wye transformer. A new power flow study shows the following prefault bus voltages:

	138 kV buses	250 kV buses
Aurora	$1.06 + j0.00$	$1.05 - j0.04$
Rawhide	$1.03 + j0.00$	–
Gabbo	$1.00 - j0.06$	$1.00 - j0.05$
Sparky	$0.99 - j0.06$	–
Rancho	$1.00 - j0.08$	$1.00 - j0.07$
Kahoka	$1.06 + j0.01$	–
Roger	$1.00 - j0.04$	–
Big Paul	$0.98 - j0.10$	–

Each of the new 250 kV circuits has a reactance of 0.9% on its own base (250 kV, 100 MVA).

a. All fault capacities at system buses (except for Aurora and Kahoka) are to be 155 GVA or less. Examine ways to accomplish this by use of series reactors in the 250 kV circuit. Estimate the reactor cost using an "index" obtained using $(\sqrt{3}/250\times10^3)^2x$, where x is the required series reactance.

b. Repeat part (a) using reactors in the 138 kV system. Estimate the cost using the index $(\sqrt{3}/138\times10^3)^2x$.

c. Let G be the highest GVA fault capacity in the system. Do not include the Aurora or Kahoka buses in this figure. Run several fault studies for reactors placed in the 250 kV system and plot the highest G versus reactor cost index given in part (a). Repeat these studies using 138 kV reactors and again plot G versus reactor cost index. These two plots should be on the same set of axes. Conclude from this study the circuit in which reactors should be sited to obtain the most economical fault capacity reduction.

7.11 The positive and negative sequence bus impedance matrices of the Rawhide Electric Company's 138 kV transmission system are identical. These matrices were formed using the subtransient reactance of all generators. Selected entries of Z_{bus}^{+} are

$$\begin{array}{l} \text{Rawhide E} \\ \text{Rawhide W} \\ \text{Airport} \\ \text{Luning} \end{array} \begin{bmatrix} 0.01 + j0.03 & 0.00 + j0.01 & 0.00 + j0.02 & 0.00 + j0.02 \\ & 0.00 + j0.04 & 0.00 + j0.02 & 0.00 + j0.02 \\ & & 0.01 + j0.04 & 0.00 + j0.02 \\ & & & 0.00 + j0.04 \end{bmatrix}.$$

The zero sequence impedance matrix using the same bus ordering is

$$\begin{bmatrix} 0.05 + j0.10 & 0.01 + j0.06 & 0.00 + j0.06 & 0.00 + j0.06 \\ & 0.03 + j0.10 & 0.01 + j0.01 & 0.00 + j0.01 \\ & & 0.03 + j0.01 & 0.01 + j0.01 \\ & & & 0.05 + j0.10 \end{bmatrix}.$$

These matrices are in per unit on a 138 kV, 100 MVA base. This problem concerns imbalanced faults at the Airport bus. Phase–to–ground faults are calculated for ϕA–g. There is no need to recalculate other phase–to–ground faults since they are readily obtained by symmetry. Similarly phase–phase–ground faults are studied for ϕA–ϕB–g and phase–phase faults for ϕA–ϕB.

a. Write a subroutine to set up the three simultaneous algebraic equations for a phase–to–ground fault. Call this FAN.

b. Repeat part (a) to set up the simultaneous equations for a phase–phase–ground fault. Call this subroutine FABN.

c. Repeat part (a) for a phase–phase fault. Call this subroutine FAB.

d. Write a subroutine that will perform an imbalanced fault study at bus k. Each of the three imbalanced faults cited above are to be included. Calculate and print the fault current both in sequence and phase variables.

e. Perform an imbalanced fault study at the Airport bus. Assume that the prefault conditions there are 137 kV line–to–line voltage.

f. Estimate the subtransient and transient region fault currents at the Airport bus. The only system generator is at Rawhide East, a 263 MVA machine with the following parameters on a 37 kV base:

	Positive Sequence	Negative Sequence	Zero Sequence
x	130%	140%	250%
x'	27%	30%	35%
x''	17%	18%	21%

Note that the unit transformer at Rawhide East is connected delta wye; this is very important to consider in your recalculation of Z^0_{bus}.

7.12 Figure P7.12 shows the Buffalo Nickel Power Company 345/138 kV power system for which your recommendations are sought concerning reactor siting. Circuit breakers in the system are sized for an interruption capacity of 110 GVA on 345 kV buses and 11 GVA on 138 kV buses. Due to system expansion, the company believes that the fault capacity at some buses is out of range. For this reason, they would like to place series line reactors in the system. Of course, cost is to be minimized.

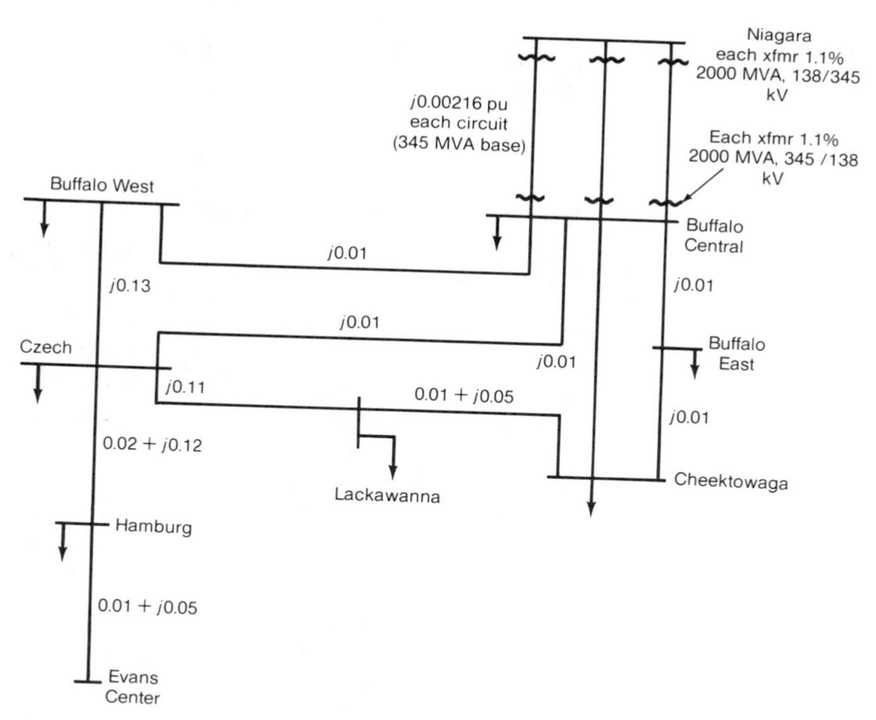

FIGURE P7.12 Buffalo Nickel Power Co. 138-kV system; per unit Z shown on 100-MVA base.

You are asked to run a base case fault study. Identify any problems. Run several cases with inserted line reactors and make a recommendation for reactor siting. Give the cost of your plan and modified fault capacities.

The Niagara bus is an infinite bus and the fault capacity there is not considered. The Buffalo Central bus fault capacity should be 110 GVA or below. The line reactor costs are as follows:

Eastinghouse "Super–Turn" Reactors

Ohms/Phase	Price	Catalog No.
	345 kV Reactors	
0.10	$120,000	1F-ST-10
0.30	140,000	1F-ST-30
0.70	180,000	1F-ST-70
1.00	210,000	1F-ST-100
1.50	250,000	1F-ST-150
	138 kV Reactors	
0.10	$ 25,000	1E-ST-10
0.30	35,000	1E-ST-30
0.50	45,000	1E-ST-50
0.70	55,000	1E-ST-70
1.00	75,000	1E-ST-100
1.50	120,000	1E-ST-150

Eastinghouse discounts: Under $100,000 none; $100,001 to $500,000 7%; over $500,000 10%.

Verrry-Round Co. "Happy Henry Reactors"

Ohms/Phase	Price	Catalog No.
	345 kV Reactors	
0.09	$114,000	XA-A
0.13	115,000	XA-B
0.15	117,000	XA-C
0.20	120,000	XA-D
	138 kV Reactors	
0.55	$68,250	YA-A
0.60	69,000	YA-B
0.65	70,000	YA-C
0.70	71,000	YA-D
	Small Sub Reactors*	
0.60	$71,000	YB-B
0.70	73,000	YB-C

*Suitable for compact substation ("Small Subs") design

Verrry-Round Discounts:

	Regular Reactors	Small Sub Reactors
0 – $122,000	-0.5%	-0.7%
$122,001 – $483,000	-5.0%	-7.0%
Above $483,000	-8.0%	-8.0%

The base case power flow study voltages are

Niagara	346 kV
Buffalo Central	345 kV
Buffalo East	137 kV
Buffalo West	137 kV
Czech	135 kV
Hamburg	127 kV
Evans Center	127 kV
Lackawana	131 kV
Cheektowaga	137 kV

7.13 In this problem, imbalanced faults are to be studied by preparing a computer program which will use as input data Z^{+}_{bus} and Z^{o}_{bus}. Prefault bus voltages are estimated as $V^{pre}_{+-o} = (\sqrt{3}\ 0\ 0)^t$. Only ϕA–g faults are studied (other phase–to–ground faults are found by symmetry).

a. Write a program that will read the Z^{+}_{bus} and Z^{o}_{bus} data and perform a phase–to–ground fault study at buses specified in another data image. This is done by repeatedly solving the sequence equations

$$V^{post}_{+-o} = V^{pre}_{+-o} + Z^{+-o}_{bus} I^{post}_{+-o} .$$

Examine the form of the sequence equations in Example 7.2.

b. Test your program by performing ϕA–g faults at buses 2, 3, and 4 for the following impedance data:

$$Z^{+}_{bus} = \begin{bmatrix} j0.050 & j0.049 & j0.049 & j0.048 & j0.041 \\ & j0.071 & j0.070 & j0.031 & j0.031 \\ & & j0.070 & j0.031 & j0.031 \\ & & & j0.092 & j0.031 \\ & & & & j0.092 \end{bmatrix}$$

$$Z^{o}_{bus} = \begin{bmatrix} j0.131 & j0.131 & j0 & j0 & j0 \\ & j0.031 & j0.050 & j0.050 & j0 \\ & & j0.712 & j0.050 & j0.050 \\ & & & j0.131 & j0.131 \\ & & & & j0.712 \end{bmatrix}.$$

7.14 Repeat Exercise 7.13 using Clarke's components. Examine Example 7.5 to determine the form of the component equations.

CHAPTER 8

Power System Stability

8.1

THE CONCEPT OF STABILITY

An electric power system is a dynamic, nonlinear system. The dynamics occur due to changes in demand, generation, line switching, lightning surges, and faults. These dynamics are often classified by the speed of occurrence: the high speed phenomena (less than 5/60 s) include switching and lighting surges; the intermediate speed occurrences (5/60 - 500/60 s) are primarily electromechanical transients of the synchronous machine rotors. Slower phenomena, such as changes in load and generation, are virtually steady state phenomena which are analyzed by power flow studies. The models needed to study these dynamics vary in detail depending on the speed of occurrence. The fastest lightning surges, for example, occur in the nanosecond range and analysis requires the use of distributed parameter models with detailed representation of the transmission lines. The partial differential equations describing these fast phenomena contain forcing functions of finite amplitude and damping due to transmission line resistance. The peak voltage on the line, and $v(t)$ and $i(t)$ are of considerable interest, but these quantities are bounded and, hence, stability, in the strict sense of the term, is not usually not at issue. At the other end of the time spectrum, slow load/generation changes are analyzed by sinusoidal steady state methods which are modeled by algebraic equations. Stability is assessed in such cases by observing whether the demand can be met: in practical terms, if the power flow study converges to an acceptable solution, questions of stability are not raised. In the case of electromechanical transients, it is necessary to consider stability more closely because disturbances may cause machine rotors to loose synchronism with the system power frequency. In this chapter attention is focused on electromechanical transients of synchronous machines in an interconnected power system.

An interconnected power system containing one or more synchronous generators is said to be stable if a disturbance does not result in loss of synchronism of any machine rotor with the power frequency. Because machine models are nonlinear and often of high dimension, it is common to resort to numerical solutions and analog simulations to examine stability. These simulations are carried on for a period sufficient to indicate whether synchronism is retained (typically 5 to 20 s). In some cases, it is possible to linearize the model and appeal to the solution of a linear, ordinary, vector differential equation. In such cases, certain parameters of the differential equation are examined to check stability. Numerical solutions and other simulations, when used to determine stability, are termed ***indirect methods***. These methods are indirect in the sense that the mathematics of the models are avoided in assessing the stability. The alternative to indirect methods is ***direct methods***, which entail examination of the mathematical properties of the functions present in the models. The most famous of these methods is the direct method of Liapunov, which is discussed near the end of this chapter.

The stability of power systems is further classified according to the intensity of the disturbance admitted. When large disturbances are considered, the term ***transient stability*** applies to systems that retain synchronism. For small disturbances, the terms ***small signal stability*** or ***small disturbance stability*** apply. Older literature contains the term ***dynamic stability***, which was used to denote small signal stability. The question of what constitutes a small disturbance is resolved by relegating small signal analysis to those problems in which linearization is allowable. When linearization produces inaccuracy sufficient to alter the stability study, it is necessary to employ nonlinear models. When nonlinear models are used, the term ***transient stability study*** applies.

For synchronous machine dynamics significantly longer than 5 cycles (1 cycle = 1/60 s at 60 Hz), the stator circuits and interconnecting network may be considered to be in steady state. The field circuit, however, is controlled by an automatic voltage regulator as well as other stabilizing controllers; these devices are not high speed controllers and the long time constants in the field itself usually preclude steady state analysis for this circuit. The most influential factor that determines the rotor dynamics, $\delta(t)$, is the electromechanical phenomena of the rotor inertia. The relationship between total rotor torque, $\Sigma\, T$, and rotor dynamics is Newton's second law,

$$\sum T = J\ddot{\delta}\,. \tag{8.1}$$

Equation (8.1) is nonlinear because the electrical rotor torques, the damping torques, and possibly the prime mover torques are nonlinear functions of δ and $\dot{\delta}$. In (8.1), J is the total rotor inertia and this term is occasionally rewritten as

$$J = \frac{2H}{\omega_r},$$

where H is termed the ***inertia constant***.

This chapter is necessarily only an introduction to power system stability analysis. Digital, numerical methods are highlighted. The reader interested in more information, particularly on more details on component models and interconnection modeling, is referred to one of the textbooks devoted to power system stability [36,40–42,48]. References on alternate numerical methods are cited in the sections below (references [5, 9, and 10] are useful for the fundamentals). The emphasis here is on electromechanical transients. The reader interested in fast transients (e.g., 5 to 50 μs), such as those associated with electromagnetic phenomena, is advised to examine Humpage [48] for an excellent summary of techniques as well as a thorough analysis based on Z-transforms.

8.2

MACHINE AND NETWORK MODELS

The torque term in (8.1) is composed of a prime mover torque, T_{pm}, an electrical torque, T_e, and several frictional torques, T_f. Using generator sign convention,

$$\sum T = T_{pm} - T_e - T_f \ .$$

Each of these terms is briefly discussed below followed by a description of an appropriate network model for transient and small signal stability studies. Often the mechanical torques are lumped and $T_{pm} - T_f$ is denoted T_m.

Prime Mover Torque

The prime mover of most synchronous generators is a series of steam turbines. Modern installations entail two or three such turbines on a shaft common to the machine rotor. The valves that operate these turbines typically have controllers which are significantly slower than the rotor inertia dynamics. For the main steam valve, a controller response in the order of over 10 s is not unusual. Apart from fast valving methods, usually the torques of the low pressure and high pressure turbines may be considered as fixed. In some special purpose studies, one or more of the turbine torques may be considered to be varying with time. This variation is assumed to correspond to the valve position and a linear variation is assumed. Figure 8.1 is a pictorial illustrating this point.

Hydroelectric installations comply with the indicated remarks for steam turbines. The prime mover torque for hydro installations is controlled by wastegates which regulate the flow to the water wheel. The response time of these wastegates is slower than the dynamics of the rotating elements and therefore T_{pm} is considered to be constant.

Magnetic Torques

The rotor torques that result from the magnetic field in the machine air gap are termed magnetic torques. These torques may also be termed electric torques since they effectively produce the generated electrical power. For a salient pole machine [1], the steady state electric torque is T_e,

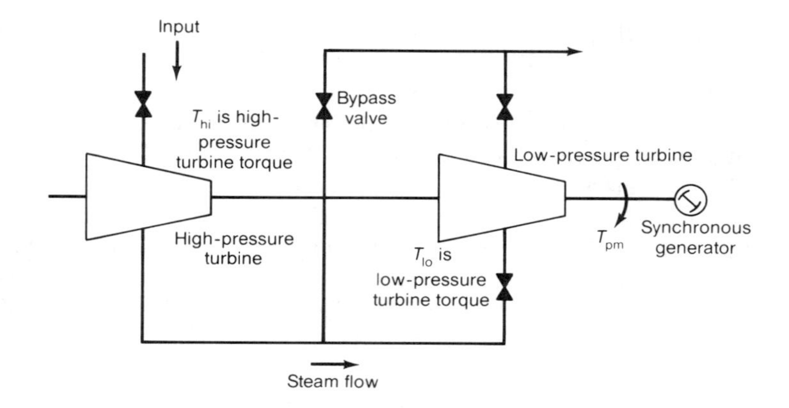

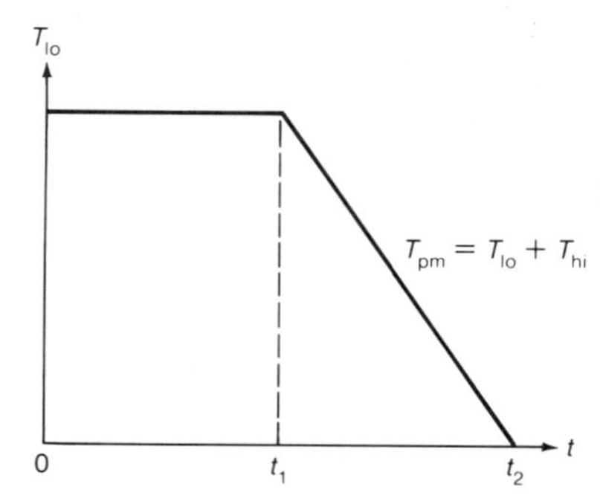

FIGURE 8.1 Prime-mover torque.

$$T_e = \frac{1}{\omega_r}\left(\frac{|E_{af}||V_t|}{x_d} \sin \delta + |V_t|^2 \frac{x_d - x_q}{2x_d x_q} \sin 2\delta\right) . \tag{8.2}$$

In this expression, ω_r is the mechanical rotor speed in rad/s and x_d and x_q denote the direct and quadrature axis reactances. The terminal voltage is V_t and the (open circuit) stator voltage produced by the rotating field is E_{af}. The first term in (8.2) depends on the sine of the torque angle δ, and this term is the ***synchronous torque.*** The second term depends on $\sin 2\delta$ and is termed the ***reluctance torque.*** For a round rotor machine, the reluctance torque vanishes since $x_d = x_q$. For a round rotor machine,

$$T_e = \frac{|E_{af}||V_t| \sin \delta}{\omega_r x_d} . \tag{8.3}$$

The remarks indicated here presuppose balanced 3ϕ operation. The imbalanced case is much more complex and is considered in [2–4].

Since most transient stability studies are performed in the transient period, transient reactances x'_d and x'_q should be used in (8.2). More accurately, transient reactances should be used during approximately the first 10 cycles (1 cycle = 1/60 s for a 60 Hz system), and x_d and x_q should be used thereafter.

Equation (8.2) is a rather simple model for a salient pole synchronous machine. Usually, this model suffices considering both the level of accuracy required by the study and the fact that most power system transients stability analysis are of the uplanned type. The description of these scenarios is not usually known to a high degree of accuracy. In some cases, Park's equations are used to model the synchronous machine; even these equations, however, are only a lumped approximation of a complicated, distributed parameter phenomenon. Finite element analysis of the magnetic circuits of the generator could be used to model the machine, but this degree of sophistication is rarely warrented.

Frictional Torques

The frictional torques in a synchronous generator occur due to windage losses, bearing friction, and slip ring friction. These frictional losses are usually approximated by a viscous friction model,

$$T_f \simeq B\dot{\delta} \,. \tag{8.4}$$

An advantage of this model beyond its simplicity is that rotor torques resulting from damper (amortisseur) bars are also approximately proportional to $\dot{\delta}$. Thus the viscous friction coefficient, B, is truly a compound parameter which models all frictional torques and certain magnetic torques. More complex, nonlinear models for T_f are occasionally used, but the sparsity of test data for many turbogenerator systems usually does not justify highly detailed models.

Coherency

The term *coherency* refers to a condition under which the dynamics of rotors of two or more synchronous generators are nearly the same. Under the assumption of coherency, the several state variables associated with the rotor angles and velocities of coherent machines are replaced by a single angle and a single velocity. For example, the coupled differential equations for the dynamics of two coherent machine become

$$\frac{2}{\omega_o}(H_{eq})\ddot{\delta}_{eq} = T_{m1} + T_{m2} - T_{e1} - T_{e2}$$

when H_{eq} and δ_{eq} is the equivalent inertia constant and rotor angle. Note that

$$H_{eq} = H_1 + H_2$$

$$\delta_{eq} = \delta_1 = \delta_2 .$$

The advantage of the coherency assumption is that of reduction of number of state variables (i.e., reduction of order). The disadvantages include degraded accuracy and the loss of identity of the individual machine states.

Network Models

In an interconnected electric power system, energy flows from bus to bus via the network. The dynamics of each machine, $\delta_i(t)$, are influenced by this energy flow through the T_e term in the differential equation of motion of the machine rotor. The specific terms in T_e which correspond to this coupling are δ and V_t. At machine i, energy is transferred to the network depending on the machine bus voltage magnitude and phase. These bus voltage parameters are network parameters and are essentially in the sinusoidal steady state. Usually, detailed data on the very short (i.e., few cycles) period directly following the disturbance are not required. Therefore, the assumption of instantaneous network response is adequate, and the usual assumptions of the sinusoidal steady state are made. Specifically, this assumption is

$$\Delta I_{bus} = Y_{bus} \Delta V_{bus} . \quad (8.5)$$

The disturbance is modeled as a ΔI term and post–disturbance ΔV terms are calculated. This calculation need be done only at generation buses to obtain machine terminal voltages. Subsequently, one or more $\delta(t)$ calculations are made by solution of the machine rotor differential equations. This calculation of $\delta(t)$ at the machines results in new injection currents at generation buses. These new currents are conveniently handled as ΔI_{bus} terms by subtracting the prefault values. The ΔI_{bus} terms are inserted in (8.5), and the network is assumed to respond instantaneously: a ΔV_{bus} results and the process is repeated at each integration step.

The assumption of instantaneous network response deserves additional comment: when the network is "tight" (i.e., low line impedances), the network time constants are very small, and they probably lie in the millisecond range. Thus response to signals at the power frequency is virtually instantaneous. When the system is large, time constants in the network can be correspondingly long. The assumption of instantaneous change in V_{bus} as the I_{bus} elements change becomes less valid. The increase in modeling complexity required for inclusion of network dynamics is considerable since an additional differential equation for each long line is required. This problem has been handled by simulation on an analog computer. Fortunately, it is not common to encounter cases of very long lines and requirements of high accuracy in the same study.

Automatic Voltage Regulators

The function of an automatic voltage regulator (AVR) is to hold the terminal voltage magnitude of a synchronous generator at a prescribed set point. This is accomplished by increasing the machine field current, I_f, when $|V_t|$ is to be increased. The increase in I_f causes an increase in air gap flux and corresponding increase in open circuit voltage $|E_{af}|$. When $|E_{af}|$ increases, elementary considerations indicate that the reactive power generated increases. The increased Q will usually cause $|V_t|$ to increase. With reference to Figure 8.2, the AVR is of interest since for long transient stability studies, the dynamics of the AVR itself must also be considered. The AVR operates by developing an error signal, ϵ, which is the difference between the set point and the terminal voltage. The terminal voltage is taken either from a potential transformer (PT) on one phase, or from a three phase PT. In either case, the PT output is rectified and compared to a dc setpoint signal. Usually, the PT and rectifier dynamics are ignored; thus

$$\epsilon = (|V_c| - |V_t|)K,$$

where $|V_c|$ is the set point. The error signal operates a controller, H. The controller is usually a nonlinear device, including limiter circuits and magnetic amplifiers. Also, linear amplifiers and compensators are usually part of the controller. The controller energizes the machine field. If H

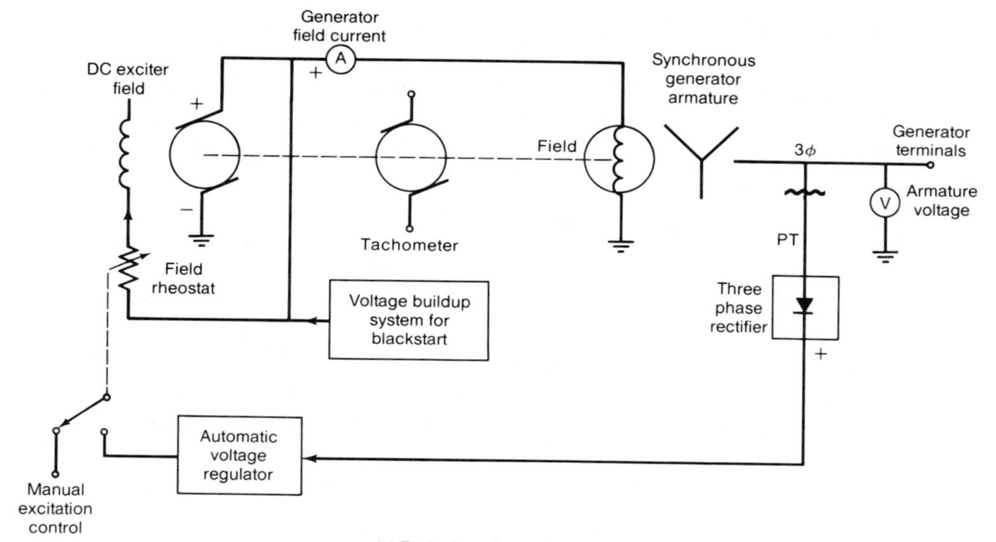

(a) Excitation of a synchronous generator showing location of automatic voltage regulator

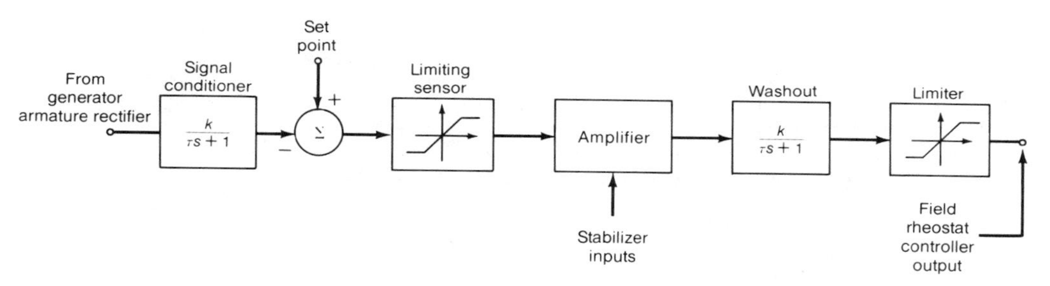

(b) Typical automatic voltage regulator

FIGURE 8.2 Automatic voltage regulator systems.

were linear, or if H may be so approximated, Laplace transforms may be used. Let the Laplace transform of the magnitude of the terminal voltage be $V_t(s)$. For no load conditions,

$$V_t(s) = \epsilon(s)H(s)\frac{K_f}{1+\tau_f s},$$

where K_f and τ_f are the field gain and time constant, respectively. Since $\epsilon(s)$ is $V_c(s) - V_t(s)$, the closed loop response is readily obtained,

$$\frac{V_t(s)}{V_c(s)} = \frac{H(s)K_f}{H(s)K_f + \tau_f s + 1}. \tag{8.6}$$

The time constants of the linearized AVR are the negative inverses of the roots of the characteristic equation in (8.6). If $H(s)$ had no dynamics, the time constant of V_t/V_c would be

$$\frac{\tau_f}{K_f K_h + 1},$$

where $K_h = H$. Similarly, when $H(s)$ has time constants of its own, the closed loop response is even longer than that indicated by $\tau_f/(K_f K_h + 1)$.

It is reasonable to conclude that the AVR is fairly slow in response, perhaps 5 to 30 s or longer. Although high speed AVRs exist, only in fairly long transient stability studies is ***detailed*** AVR modeling required. Eq. (8.6) is readily modeled using n_{avr} state variables where $n_{avr}+1$ is the order of the denominator of H(s). If (8.6) is rewritten as

$$\frac{V_t(s)}{V_c(s)} = \frac{N(s)}{D(s)},$$

one readily obtains

$$D(s)V_t(s) = N(s)V_c(s),$$

and the right and left hand sides are found to be differential equations in V_c and V_t, respectively. Appropriate definition of states will result in reduction to several simultaneous ordinary differential equations in first order form. Each is solved in the same way as the machine equations. It is also possible to include the simpler nonlinear phenomena such as

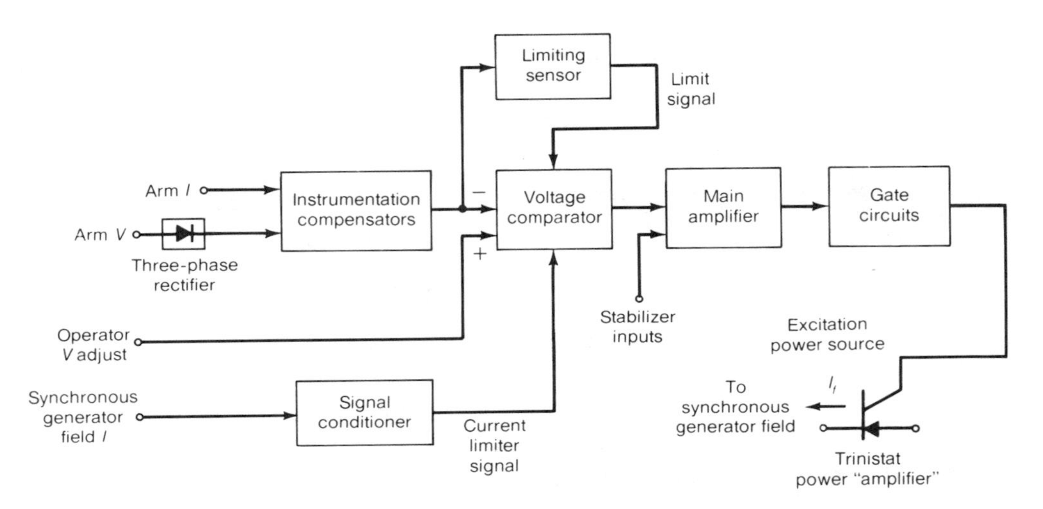

(a) Potential source rectifier exciter AVR using trinistat power amplifier

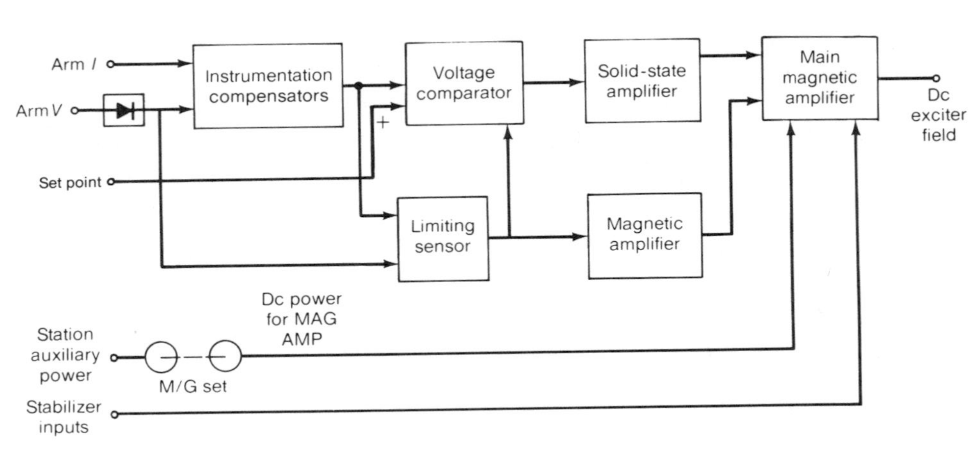

(b) Magnetic amplifier AVR

FIGURE 8.3 Two typical AVR configurations.

limiters and the magnetic saturation of the machine. More difficult is the inclusion of complex electronic processing of the error signal ϵ; for example, ϵ may be chopped and processed by power electronic semiconductors. These devices have advantageous control properties such as gain control by variation of SCR firing characteristics. The gross nonlinearity of such switching phenomena can be modeled only in approximate terms, and it is not unusual to perform this modeling by a rough estimate which is linear. The time constants of the linear estimate may be identified by matching actual controller response (obtained from field measurements) to computer results.

Figure 8.3 shows several typical AVRs suggested by IEEE, and Table 8.1 gives representative AVR parameters.

Power System Stabilizers

Many years ago, at the inception of interconnection of power systems, stability occupied a less important role in power engineering than at present. This is due to two distinct factors: early generators had very high inertia constants, and resulting angular accelerations were limited by these high values. Second, interconnection was not so extensive. High energy levels were unavailable from interties, and interaction between generators was correspondingly limited. The gradual appearance of low inertia machines (on a per unit basis) has resulted in increased importance of stability. This is reflected in the increasing importance of the control of energy available to machine rotors. This energy has the potential of causing inappropriate fluctuation of $\delta(t)$, or worse, instability. Since about 1965, most synchronous generator installations at central power stations have included power system stabilizers (PSSs). The function of the PSS is to use several local measurements and, perhaps, remote telemetered measurements to augment stability by supplementary excitation control.

Table 8.1
Typical AVR and PSS Parameters

Automatic Voltage Regulators	
Closed–loop gain	+14 to +30 dB
Bandwidth	0.3 Hz to 12 Hz
Rise time	0.2 to 3 s
Power System Stabilizers	
Closed–loop gain	-10 to +30 dB
Washout time constant [a]	1 to 50 s
Output signal limits	Typically 0.25 pu

[a] Typically much longer than the transducer and lead/lag time constants.

Typical inputs to the PSS are active and reactive power measurements at the machine terminals and Δf. The latter parameter, frequency deviation, is the input that distinguishes the PSS from other controllers. Since a PSS is a *system* controller, inputs that reflect system dynamics must be used. Frequency deviation is such an input. Figure 8.4 shows a typical configuration, and Table 8.1 shows representative PSS parameters.

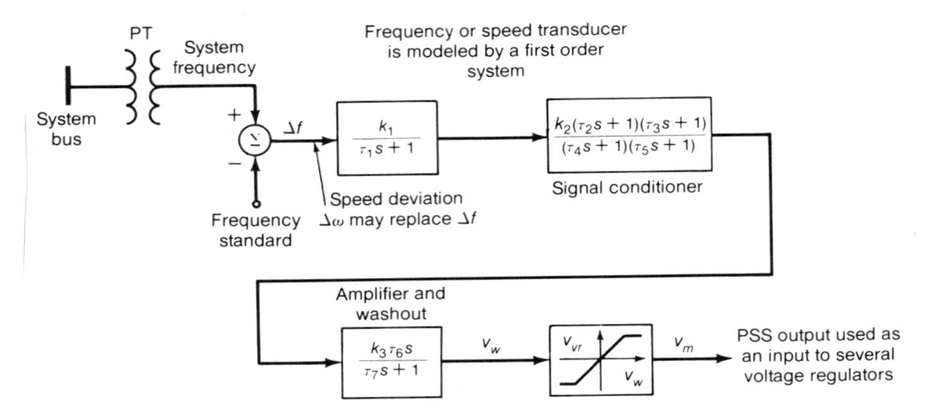

FIGURE 8.4 Typical power system stabilizer configuration and model.

The primary objective of a PSS is the enhanced stability of the machine under control and remote machines under a wide range of load schedules. The PSS control signal is used to either directly control the generator field or, more commonly, provide an auxiliary input to the AVR, which, in turn, controls the field. Since the PSS strategy is one of load following, it has relatively slow response. For this reason, the PSS action is often omitted from transient stability studies. If the AVR is modeled, the PSS output, V_{pss}, may usually be considered fixed. The excitation control, V_r, is related to the PSS and AVR inputs (V_{pss} and V_c) by

$$\dot{X} = A_{avr} X + b_{avr} \left[\begin{array}{c} V_c \\ \hline V_{pss} \end{array} \right] \tag{8.7a}$$

$$V_r = C_{avr} X + d_{avr} \left[\begin{array}{c} V_c \\ \hline V_{pss} \end{array} \right]. \tag{8.7b}$$

This is a linearized formulation. References [22,24,25] describe the calculation of the output equation matrix coefficients C_{avr} and d_{avr}. In some instances, it may be necessary to model PSS dynamics [22,25,26]: in such cases, the transfer function between the PSS inputs and V_{pss} must be written. A suitable number of state variables are then used to model the PSS and provide the needed data in (8.7). Aguilar [24] has suggested a way to modularize the machine–network–AVR–PSS equations, but it should be kept in mind that there is considerable variability in AVR and PSS configuration. This makes it difficult to write on all-encompassing production computer code to handle all designs. Commercial programs designed to handle specific configurations are readily available.

8.3

PREDICTOR FORMULAS FOR THE NUMERICAL SOLUTION OF THE TRANSIENT STABILITY PROBLEM

One form of the transient dynamic model for interconnected synchronous generators is

$$\frac{\omega_o}{2H_i}[(T_{pm})_i - B_i\dot{\delta}_i - \frac{|E_{afi}||V_i|}{\omega_{ri}x'_{di}} \sin \delta_i - \frac{|V_i|^2(x'_{di}-x'_{qi})}{2\omega_{ri}x'_{di}x'_{qi}} \sin 2\delta_i]$$

$$= \ddot{\delta}_i \qquad i = 1, 2, \ldots, g \tag{8.8}$$

$$V_{bus}(t) = \begin{bmatrix} V_1 \\ V_2 \\ \vdots \\ V_g \\ \hline V_{ng} \end{bmatrix} = V_{bus}(t_0) + Z_{bus}\,\Delta I_{bus} . \tag{8.9}$$

In (8.9), V_{ng} denotes a subvector of nongeneration buses and in (8.8), ω_o is the synchronous speed in rad/s. In the upper portion of V_{bus}, the generation bus voltages are found; these are used in the electrical torque terms in (8.8). Equation (8.9) represents the connection due to the electric network and ΔI_{bus} may represent the disturbance (e.g., fault). For the case of analysis of dynamics due to a fault, (8.9) is usually taken to be fixed V_{bus} at the postfault values. In such cases, only (8.8) must be solved. This is a set of g nonlinear, ordinary differential equations with $\delta_i(t_0)$ taken to be the prefault bus voltage phase angle at generator i and $\dot{\delta}_i(t_0)$ is such that $\dot{\delta}$ is initially zero.

In (8.8), the rotor speed of machine i, ω_{ri}, is essentially constant. Of course, δ_i must be expressed in radians. The inertia constant H_i is readily measured by methods relating to the kinetic energy stored in this rotating mass, W_i,

$$H_i = \frac{W_i}{\omega_o} . \tag{8.10}$$

Equation (8.8) is reduced to first order form using states $x_1 = \delta_1$, $x_2 = \delta_2, \ldots, x_g = \delta_g$, $x_{g+1} = \dot{\delta}_1, \ldots, x_{2g} = \dot{\delta}_g$. Let X be the state vector, then (8.8) is of the form

$$\dot{X} = F(X) . \tag{8.11}$$

A vector, nonlinear differential equation of the form (8.11) is solved numerically through the use of a predictor formula,

$$X(t + h) = \Gamma(X(t), X(t-h), \ldots, X(t_0)), \tag{8.12}$$

where Γ and F are $2g$ vector valued functions, and h is the time step size. A variety of predictor functions Γ are generated by the Taylor series expansion for $X(t + h)$. The most elementary of these is *Euler's method* [9,10],

$$X(t+h) = X(t) + \dot{X}(t)h + O(h^2)$$

where $O(h^2)$ is the remainder of the Taylor series and the argument, h^2, indicates that the lowest power of h present in that remainder is the square. When the remainder is ignored,

$$X(t+h) \simeq X(t) + \dot{X}(t)h = X(t) + F(X(t))h . \tag{8.13}$$

Equation (8.13) is Euler's predictor formula. In cases in which F is a function not only of $X(t)$, but also t itself, in other words,

$$\dot{X} = F(X(t), t) ,$$

Euler's predictor is

$$X(t+h) = X(t) + F(X(t), t)h \; . \tag{8.14}$$

Euler's predictor is not often used in power engineering applications because the computational burden is quite high. Also, truncation of the Taylor series at the $O(h)$ term results in undesirably large error at each step. Euler's predictor does not have good numerical stability properties, due to this early truncation. To hold the Taylor series remainder in (8.12) to two orders of magnitude below the $\dot{X}$ term,

$$h << 2||\dot{X}||/||\ddot{X}|| \; .$$

Since acceleration magnitudes can be an order of magnitude or more greater than velocities, a prudent choice of step size would be below 2 ms. For $h = 0.5$ ms, a 5 s transient stability study entails 10,000 predictor steps. The error at each step can compound this computational burden since error will propagate from step to step [6].

The remainder of this section is devoted to several other predictor methods. The most commonly used methods for the transient stability study of a power system are the Adams predictor and trapezoidal formulas discussed below. Many, if not most, of these predictor formulas are obtained by alternative forms and truncations of the Taylor series. These Adams and trapezoidal formula numerical integrations are usually employed in transient stability studies of 5 to 20 s. Numerical integration step sizes (h) depend on the system characteristics and needs of the study, but typical values are in the range 50 s to 1000 s.

Modified Euler Method

Consider the solution of

$$\dot{X} = F(X(t), t), \tag{8.15}$$

where X and F are either scalars or vectors of the same dimension. Expand $X(t + h)$ and $X(t - h)$ in Taylor series about $X(t)$ [5],

$$X(t+h) = X(t) + \frac{h}{1!}\dot{X}(t) + \frac{h^2}{2!}\ddot{X}(t) + \frac{h^3}{3!}\dddot{X}(t) + O(h^4)$$

$$X(t-h) = X(t) - \frac{h}{1!}\dot{X}(t) + \frac{h^2}{2!}\ddot{X}(t) - \frac{h^3}{3!}\dddot{X}(t) + O(h^4) \; .$$

Subtraction reveals that

$$X(t + h) - X(t - h) = 2hF(X(t), t) + \frac{h^3}{3}\dot{F}(X(t), t) + O(h^5) \; . \tag{8.16}$$

If $O(h^3)$ and higher terms are neglected, one obtains the *modified Euler formula,*

$$X(t + h) = X(t - h) + 2hF(X(t), t) \; . \tag{8.17}$$

The accuracy and stability of this predictor formula is better than the conventional Euler formula because a higher order truncation of the Taylor series is made. Note, however, that both $X(t)$ and $X(t - h)$ appear on the right–hand side of (8.17). If the notation X_n denotes the nth evaluated point of $X(t)$, and F_n denotes the nth evaluated point of F(X,t), then (8.17) is written

$$X_{n+1} = X_{n-1} + 2hF_n \; . \tag{8.18}$$

Such a predictor is termed ***not self starting,*** because (8.18) cannot be used for the first prediction. When the right hand side of a predictor formula involves X_{n-m} and/or F_{n-m}, $m > 0$, the predictor is not self starting. The Euler predictor

$$X_{n+1} = X_n + hF_n \tag{8.19}$$

is self starting since it may be used for all n including $n = 0$. To use the modified Euler predictor, Eq. (8.18), it is necessary to use the Euler predictor or some other self starting method for the $n = 0$ step.

Trapezoidal Rule

A predictor formula related to the modified Euler formula is the trapezoidal rule,

$$X_{n+1} = X_n + \frac{h}{2}(F_n + F_{n-1}) . \tag{8.20}$$

The trapezoidal rule is not self starting. The order of the truncation in the Taylor series to obtain (8.20) is $O(h^2)$. The trapezoidal formula is a very popular choice for transient stability studies primarily because of the case in programming coupled with reasonable accuracy. Sometimes computing efficiency of a numerical algorithm is based in part on the number of times F must be evaluated per integration step. The Euler and modified Euler methods require one evaluation of F per step. In (8.20), the trapezoidal rule appears to have two evaluations of F per step; with some care in programming, F_n at one stage may be used for F_{n-1} at the next stage. Because of the equal spacing of steps (i.e., h is constant), only one evaluation of F per step is required.

The origin of the term ***trapezoidal*** lies in the geometrical interpretation of (8.20) for scalar F: the right hand side of this formula is recognized as the area of a trapezoid with altitude h and bases F_n and F_{n-1}.

Adams Predictor Formulas

The Euler, modified Euler, and trapezoidal rule formulas each may be geometrically interpreted as a linear approximations of F in an interval at or near t. If a parabolic approximation is made, the second order Adams predictor formula results,

$$X(t) = X_n + \alpha(t - t_n) + \beta(t - t_n)^2$$

$$\frac{dX}{dt} = \alpha + 2\beta(t - t_n) .$$

Let $t = t_n$, revealing that

$$\alpha = \frac{dX}{dt}\Big| t_n = F_n .$$

Then let $t = t_{n-1}$, to obtain β:

$$F_{n-1} = \alpha + 2\beta(t_{n-1} - t_n)$$

$$\beta = \frac{F_{n-1} - \alpha}{2(-h)} .$$

The second order Adams predictor is therefore

$$X_{n+1} = X_n + \frac{h}{2}(3F_n - F_{n-1}) \, . \tag{8.21}$$

This is *not self starting.* With care in programming, only one function evaluation per step is required.

Higher order Adams predictors result from the expansion of $X(t)$ in higher order polynomials in $(t - t_n)$. A representative and popularly used formula is the Adams fifth order predictor,

$$X_{n+1} = X_n + \frac{h}{24}(55F_n - 59F_{n-1} + 37F_{n-2} - 9F_{n-3}) \, . \tag{8.22}$$

The form of all orders of Adams predictors is as shown in (8.22); they are all non self starting; they contain equal spacing of steps; they may be programmed with one function evaluation per step. For the latter reason and for experimentally observed numerical stability in power engineering problems, the Adams predictor formulas are widely used in transient stability studies. The fifth order Adams predictor, Eq. (8.22), has Taylor series truncation at $O(h^5)$. The reader is cautioned concerning terminology: depending on how a family of numerical algorithms is indexed, a fifth order method may be termed fourth order or sixth order. Further, names designating specific algorithms are often taken from the person who is credited with early development of the technique. These names are not completely standard. Thus the terminology used here may not be consistent with that found elsewhere.

The numerical stability of the Adams predictors has been examined by Hamming [6] and others [7], and [8] contains a discussion of starting procedures.

Runge – Kutta Methods

In this section attention turns to a group of well known predictor formulas known as the Runge–Kutta predictors. The primary applications of these predictors lie outside power engineering, although certain applications in the analysis of transients in rotating machines are known [13]. Runge–Kutta predictors are self starting, and for this reason they are used in commercially available numerical integration subroutines to start Adams or other nonself starting predictors. The numerical integration of

$$\dot{X} = F(X(t), t)$$

by a predictor generally involves calculation of X_{n+1} as a function of X_n, X_{n-1}, X_{n-2}, ..., F_n, F_{n-1}, F_{n-2}, . . . , and t_n. If the predictor is self starting, the formula must be free of X_{n-1}, X_{n-2}, . . . , F_{n-1}, F_{n-2} ... . Under this assumption,

$$X_{n+1} = \Gamma(X_n, F_n, t_n) \, . \tag{8.23}$$

Equation (8.23) gives difficulty in obtaining high accuracy since it appears that Taylor series expansions for X_{n+1} cannot be arranged/subtracted/processed to cancel low order (in h) terms. For methods such as the Adams formulas, values of X_{n-m} and F_{n-m}, $m > 0$ are used to obtain this cancellation at the expense of the self starting property.

The Runge–Kutta method involves the desired cancellation of low power terms in h in the Taylor series by several functional evaluations. By such an approach, the self starting property is retained at the expense of multiple evaluations of F. Consider several evaluations of the function F at points in X and t space. For the moment, let these points be located at X_n, t_n and a few other locations which are found in a way to be described below:

$$K^{(1)} = F(X_n, t_n)$$

$$K^{(2)} = F(X^{(2)}, t^{(2)})$$

$$K^{(3)} = F(X^{(3)}, t^{(3)})$$

$$\vdots$$

$$K^{(m)} = F(X^{(m)}, t^{(m)}) \,. \tag{8.24}$$

Let the point $X^{(2)}$ be found as a linear combination of the previously evaluated K terms and X_n. In this case

$$X^{(2)} = \alpha_{21}K^{(1)} + \beta_2 X_n \;;$$

and let $X^{(3)}$ be a linear combination of $K^{(2)}$, $K^{(1)}$, and X_n,

$$X^{(3)} = \alpha_{31}K^{(1)} + \alpha_{32}K^{(2)} + \beta_3 X_n \,.$$

The general point $X^{(i)}$ is

$$X^{(i)} = \sum_{j=1}^{i-1} \alpha_{ij}K^{(j)} + \beta_j X_n \,. \tag{8.25}$$

Further, let the time component of the arguments in (8.24) be of the form

$$t^{(i)} = t_n + \gamma_i h \,. \tag{8.26}$$

Thus we are suggesting that m functional evaluations of F be made, one at (X_n, t_n), and the other $m-1$ at points $(X^{(i)}, t^{(i)})$ where $X^{(i)}$ is a linear combination of X_n and the previously evaluated functions, and the $t^{(i)}$ are t_n advanced by some multiple or fraction of the step size, h. Now let a linear combination of the $K^{(i)}$ be the predictor,

$$X_{n+1} = X_n + \sum_{k=1}^{m} \varsigma_k K^{(k)} \,. \tag{8.27}$$

Equation (8.27) is the *Runge–Kutta predictor formula*. The question remains as to how the linear combination coefficients α, β, γ, and ς should be chosen. The answer to this question is quite involved and is omitted here except to say that (8.27) is expanded in a Taylor series and each $K^{(k)}$ term is also expanded in a Taylor series. Coefficients of like powers of h are forced to be equal in this expansion. It turns out that there are more linear combination coefficients than there are equations, and some of the α, β, γ, and ς terms must be chosen. Having chosen those terms, the others are calculated. The choice of α, β, γ, ς depends on the application, but several selections appear to work well in a wide range of applications: a few are tabulated below. The interested reader will find a "representative derivation" of the Runge–Kutta method in [9] and an excellent discussion of alternative rationales behind selection of α, β, γ, ς in [10,12]. Note that (8.27) is self starting but requires m evaluations of function F for each step. In the Adams predictor formulas, the right hand side evaluations of F were evenly spaced and if previously evaluated F's were stored, only one new evaluation of F per integration step was required. This is not the case for the Runge–Kutta predictor. The function evaluations are unevenly spaced, and m evaluations of F per integration step are required. Of course, X and F are vectors of identical dimension. The $K^{(i)}$ vectors are of the same dimension as X and F.

Second Order Runge–Kutta [11]

$$K^{(1)} = F(X_n, t_n)$$

$$K^{(2)} = F(X_n + \frac{h}{2}K^{(1)}, t_n + \frac{h}{2})$$

$$X_{n+1} = X_n + \frac{h}{2}(K^{(1)} + K^{(2)}) \tag{8.28}$$

Alternative Second Order Runge–Kutta

$$K^{(1)} = F(X_n, t_n)$$

$$K^{(2)} = F(X_n + \frac{h}{2}K^{(1)}, t_n + \frac{h}{2})$$

$$X_{n+1} = X_n + hK^{(2)} \tag{8.29}$$

Third Order Runge–Kutta

$$K^{(1)} = F(X_n, t_n)$$

$$K^{(2)} = F(X_n + \frac{h}{2}K^{(1)}, t_n + \frac{h}{2})$$

$$K^{(3)} = F(X_n - hK^{(1)} + 2hK^{(2)}, t_n + h)$$

$$X_{n+1} = X_n + \frac{h}{6}(K^{(1)} + 4K^{(2)} + K^{(3)}) \tag{8.30}$$

Alternative Third Order Runge–Kutta

$$K^{(1)} = F(X_n, t_n)$$

$$K^{(2)} = F(X_n + \frac{h}{3}K^{(1)}, t_n + \frac{h}{3})$$

$$K^{(3)} = F(X_n + \frac{2h}{3}K^{(2)}, t_n + \frac{2h}{3})$$

$$X_{n+1} = X_n + \frac{h}{4}(K^{(1)} + 3K^{(3)}) \tag{8.31}$$

Fourth Order Runge–Kutta

$$K^{(1)} = F(X_n, t_n)$$

$$K^{(2)} = F(X_n + \frac{h}{2}K^{(1)}, t_n + \frac{h}{2})$$

$$K^{(3)} = F(X_n + \frac{h}{2}K^{(2)}, t_n + \frac{h}{2})$$

$$K^{(4)} = F(X_n + hK^{(3)}, t_n + h)$$

$$X_{n+1} = X_n + \frac{h}{6}(K^{(1)} + 2K^{(2)} + 2K^{(3)} + K^{(4)}) \tag{8.32}$$

Alternative Fourth Order Runge–Kutta

$$K^{(1)} = F(X_n, t_n)$$

$$K^{(2)} = F(X_n + \frac{h}{3}K^{(1)}, t_n + \frac{h}{3})$$

$$K^{(3)} = F(X_n + \frac{h}{3}K^{(1)} + hK^{(2)}, t_n + \frac{2h}{3})$$

$$X_{n+1} = X_n + \frac{h}{8}(K^{(1)} + 3K^{(2)} + 3K^{(3)} + K^{(4)}) \tag{8.33}$$

Fifth Order Runge–Kutta [12]

$$K^{(1)} = F(X_n, t_n)$$

$$K^{(2)} = F(X_n + \frac{2h}{7}K^{(1)}, t_n + \frac{2h}{7})$$

$$K^{(3)} = F(X_n + \frac{4h}{7}K^{(2)}, t_n + \frac{4h}{7})$$

$$K^{(4)} = F(X_n + \frac{6h}{7}K^{(3)}, t_n + \frac{6h}{7})$$

$$K^{(5)} = F(X_n + hK^{(4)}, t_n + h)$$

$$X_{n+1} = X_n + h[\frac{11}{96}K^{(1)} + \frac{7}{24}K^{(2)} + \frac{35}{96}K^{(3)} + \frac{7}{48}K^{(4)} + \frac{1}{12}K^{(5)}]. \tag{8.34}$$

Principal Applications of Runge–Kutta Formulas

As stated earlier, the principal interest in Runge–Kutta formulas for power engineers lies in the solution of certain rotating machine transient problems. The method is not in more widespread use primarily because m function calculations are required per integration step. Also, the method is not quite as easily programmed as other predictor methods.

A great advantage of the Runge–Kutta formulas is that they are self starting. Also, in a wide range of numerical problems, even with jump discontinuities, the method is numerically stable (i.e., the propagation of error from integration step to integration step does not increase sharply). Owing to the self starting nature of the method, it is occasionally used for the starting points of an Adams predictor. For example, a fourth order Runge–Kutta predictor used to start a fourth order Adams predictor is termed a *fourth order* Adams/Runge–Kutta starter. When used as a starter for high order predictor formulas, the Runge–Kutta disadvantage of multiple F evaluations is not a serious disadvantage: after the starting procedure, the Runge–Kutta formula is not used and the multiple F evaluations are no longer necessary. Such an application is commonly found in general purpose commercial numerical integration packages. The original algorithm was described by Runge in 1895 [14].

Other Predictor Methods

Occasionally, the following formulas will prove useful. The references shown provide further details on the characteristics of these formulas, which are used primarily in research applications.

Gill's predictor [11]

$$K^{(1)} = hF(X_n, t_n)$$

$$K^{(2)} = hF(X_n + \frac{1}{2}K^{(1)}, t_n + \frac{h}{2})$$

$$K^{(3)} = hF(X_n + (-\frac{1}{2} + \sqrt{\frac{1}{2}})K^{(1)}, t_n + \frac{h}{2})$$

$$K^{(4)} = hF(X_n - \sqrt{\frac{1}{2}}K^{(2)} + (1 + \sqrt{\frac{1}{2}})K^{(3)}, t_n + h)$$

$$X_{n+1} = X_n + \frac{1}{6}(K^{(1)} + 2(1 - \sqrt{\frac{1}{2}})K^{(2)} + 2(1 + \sqrt{\frac{1}{2}})K^{(3)} + K^{(4)}). \quad (8.35)$$

The Taylor series is truncated at $O(h^5)$. Gill's predictor is a variation of the Runge–Kutta method. Its principal use is a starter for non self starting methods.

Nystrom Extrapolation Formula [13]

$$X_{n+1} = X_{n-1} + h(2F_n + \frac{\nabla^2}{3} F_n + \frac{\nabla^3}{3} F_n + \frac{29}{30} \nabla^4 F_n + \frac{14}{25} \nabla^5 F_n + \cdots) \quad (8.36)$$

where ∇ is the backward difference operator

$$\nabla F_n = F_n - F_{n-1}$$

$$\nabla^2 F_n = F_n - 2F_{n-1} + F_{n-2}$$

$$\nabla^3 F_n = F_n - 3F_{n-1} + 3F_{n-2} - F_{n-3}$$

and so on (using Pascal's triangle as coefficients for F_{n-k}).

Example 8.1

In this example, several predictor formulas are compared for the numerical solution of (8.8) for a single synchronous machine on an infinite bus. The bus voltage is initially at rating, but a fault in the system depresses the bus voltage to 60% for $1^+ \leq t \leq 6$ cycles. Machine parameters are

$$\begin{array}{llllll} H & = 4.0 & |E_{af}| & = 1.20 & x_d' & = 0.50 \\ x_q' & = 0.30 & T_{pm} & = 1.085676 & \omega_r & = 1.0 \\ B & = 0.10 & \omega_o & = 120\pi & |V_t| & = 1.0 \end{array}$$

with all quantities expressed in per unit except ω_o, which is the mechanical rotor speed rating in rad/s.

Solution

Figure 8.5 and Table 8.2 show the numerical simulation results using the Euler, fourth order Adams (Euler starter), and fourth order Runge–Kutta methods. In each of the three cases, the integration step size (Δt) illustrated is 0.02 cycle. Note that 1 cycle is 1/60 s.

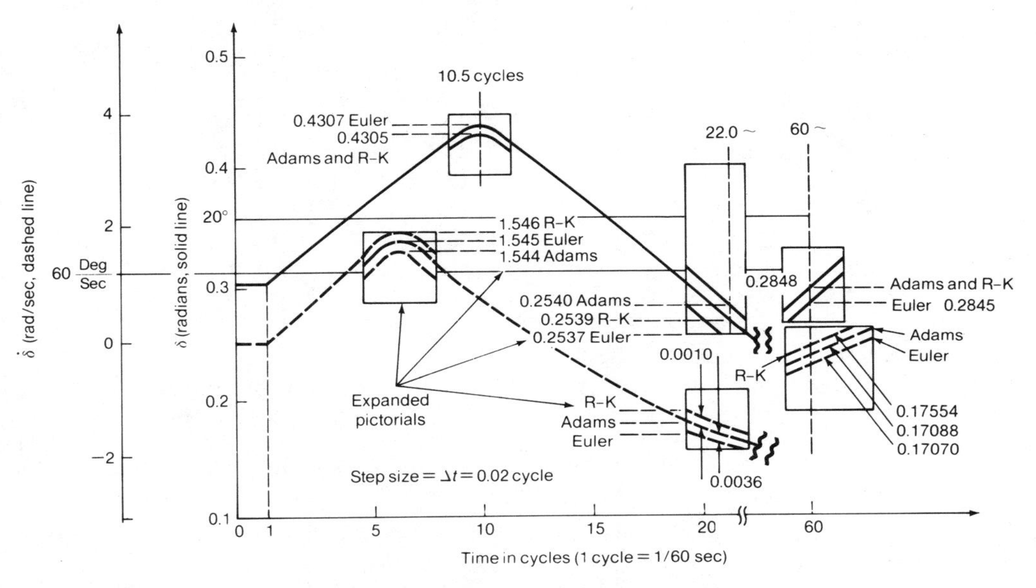

FIGURE 8.5 Comparison of Euler, Adams (fourth order), and Runge-Kutta (R-K, fourth order) methods for Example 8.1.

Examination of the figure for $t << 1$ s (i.e., $t << 60$ cycles) reveals little distinction between the three methods/two step sizes. Expanded depictions in the figure help distinguish the curves, but agreement in $\delta(t)$ within about 0.004 radian/second for the same step size. Also, the values of t for which peaks in $\delta(t)$ and $\dot{\delta}(t)$ occur are in very close agreement (less than 0.01 cycle in each case).

Later in the integration period (e.g., $t = 1.0$ s) there is very reasonable agreement and numerical stability in $\delta(t)$. The $\dot{\delta}(t)$ characteristic, however, begins to show greater discrepancy between the Euler method and the higher order methods. The Euler method shows higher rotor velocity by about 0.005 rad/s at $t = 1.0$ second.

Very short integration steps, exemplified by $\Delta t = 0.02$ cycle in Figure 8.5, reveal very little disagreement between the low–order and high order methods depicted. At three times this resolution, $\Delta t = 0.06$ cycle (i.e., 1.0 ms), the comparison depicted in Table 8.2 results. A step size of

Table 8.2
Comparison Between the Euler and Fourth Order Adams and Runge–Kutta Methods in Example 8.1

Parameter [a]		$\Delta t = 0.02$ cycle	$\Delta t = 0.06$ cycle
$\delta(9.6)$			
	E	0.4284	0.4297
	A	0.4282	0.4293
	R	0.4283	0.4282
$\dot{\delta}(9.6)$			
	E	0.3093	0.3245
	A	0.3038	0.3078
	R	0.3015	0.3038
$\delta(60)$			
	E	0.2845	0.2837
	A	0.2848	0.2846
	R	0.2848	0.2848
$\dot{\delta}(60)$			
	E	0.1755	0.1869
	A	0.1770	0.1719
	R	0.1709	0.1707

[a] Times in cycles (1 cycle = 1/60 s);
E, Euler; A, Fourth order Adams; R, Fourth order Runge–Kutta.

0.06 cycle is considered to be a very small step size in a transient stability study. Examination of this table reveals:

1. For small t, calculated $\delta(t)$ values by the three methods agree between the $\Delta t = 0.02$ and $\Delta t = 0.06$ studies.

2. For small t, calculated $\dot{\delta}(t)$ values of the higher order methods (i.e., Runge–Kutta and Adams) agree between the larger and smaller step sizes.

3. Euler calculation of $\dot{\delta}$ for small t shows degradation at triple step size. This degradation is manifest in large $\dot{\delta}(t)$ values at large Δt.

4. For large values of t (e.g., t = 60 cycles), $\delta(t)$ calculated by the three methods is stable between the large and small step size. Suspect $\delta(60)$ is observed, however, for the Euler method at $\Delta t = 0.06$. This suspicion is revealed by an apparent accumulation of error in the order of 0.001 rad.

5. For large values of t, the Adams and Runge–Kutta methods show stable $\dot{\delta}(t)$ for large step size, but the Euler method exhibits about 0.01 rad/s discrepancy.

There are certain cautions which warrant mention with regard to the generalization of results of this example. The large value of inertia constant and modestly short fault duration do not severely test the numerical methods illustrated. If the dynamics were faster and of greater intensity, such as those in a lower inertia system with prolonged fault, the discrepancies in the coarse step size Euler solution would be greater. Also, for longer simulations, the accumulation of error becomes unstable and intolerable in the Euler method. For these reasons, the higher order solution methods are used in power engineering applications. Further, historical trends indicate that the inertia constant of synchronous generators is likely to decrease, thereby reinforcing these observations.

In this example, Δt may be increased considerably for the Adams and Runge–Kutta methods. In fact, for short simulation times, step sizes up to a second are not unusual when using high order integration methods. The corresponding savings in central processing time is significant.

8.4 PREDICTOR-CORRECTOR METHODS

A principal difficulty of predictor methods of the numerical solution of

$$\dot{X} = F(X(t))$$

is that the error at each prediction step,

$$X(t+h) = \Gamma(X(t), X(t-h), .., X(t_0))$$

propagates into the next step. It is possible to control this error through the use of a correction applied after each predictor step. This is pictorially illustrated in Figure 8.6; a step is taken at $t = t_1$ and x_1^p is obtained from the predictor formula

$$x_1^p = \gamma(x_0) .$$

At this point, x_1^p and x_0 are used to obtain a correction which is applied to x_1^p,

$$x_1^c = x_1^p + \gamma_c(x_0, x_1^p) . \tag{8.37}$$

The function γ_c is termed a corrector formula and x_1^c is the corrected value. A further step is shown in Figure 8.6.

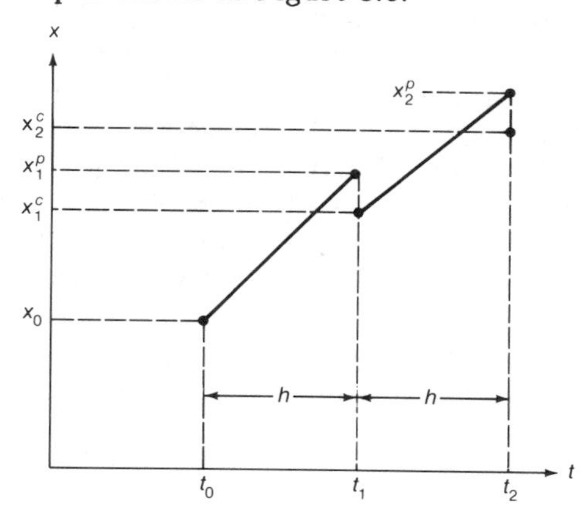

FIGURE 8.6 Predictor-corrector solution.

At this point it is important to reflect on the tradeoff between numerical error control and increased execution time required by the additional function evaluations (namely, the evaluation of the corrector formula). In most power engineering applications, predictor–corrector methods are marginally of value. This is the case since synchronous machine dynamics occur with rather slowly varying second derivatives of the torque angle. Thus the error at each predictor step is low. This may not be the case in all instances, particularly when the inertia constant is very low. As an illustration, consider a superconducting generator in the 10 MVA class. Such a machine would probably have a non–ferrous rotor assembly (perhaps plastic parts are used), and the rotor inertia is expected to be at least an order of magnitude lower than a conventional 10 MVA machine. For such a machine, the turbine is likely to be the principal element in the total rotor inertia. The decreased rotor mass causes much faster rotor dynamics, and $\ddot{X}$ can change from the predicted step much more than for a conventional machine. The remainder term in the Taylor series has the potential of being an order of magnitude or more greater than what might be expected. Hence truncation of the Taylor series could cause such serious error that correction is required. A more complete mathematical discussion of this point maybe found in [5,6,16].

An additional point that should be considered in assessing the appropriateness of the use of a predictor–corrector method is related to the tradeoff between accuracy and number of evaluations of F. In many predictor–corrector algorithms, both γ and γ_c are linear combinations of hF evaluated at various points. The points are typically equally spaced. For example, the predictor

$$X_{n+1} = X_{n-3} + \frac{4h}{3}(2F_n - F_{n-1} + 2F_{n-2})$$

truly involves only one evaluation of F per integration step. It is necessary to store F_n, F_{n-1}, F_{n-2} and use these values as required. When n is incremented, the several F's are "pushed down" and only one new F is added to the storage register. The typical corrector is similarly a linear combination of F's and the same storage register is used. Usually, the predicted value of X_{n+1} is used to evaluate F_{n+1}, and since the predicted value is subsequently corrected, this F_{n+1} can not be used further. Thus typically, two function evaluations are required for each integration step.

As stated above, the principal reason for the lack of popularity of predictor–corrector methods for conventional transient stability solutions in power engineering is that the correction is quite small. Even when the step size is of the order of magnitude of the period of the power frequency (e.g., 1/60 s for a 60 Hz system), the magnitude of the correction obtained is typically smaller than $10^{-6}n$, where the state vector is of order n. The amount of the correction is highly system dependent, as well as algorithm dependent. As a general guideline, high order problems involving low rotor inertias in which large step size is to be used offer the most reasonable potential for the predictor–corrector methods.

Milne's Methods

The Milne methods are predictor–corrector methods which are based on estimators (predictors) which are subsequently used to obtain a better estimate of the derivative, $\dot{X}$, in the interval being integrated. The formula in which the "better estimate" of $\dot{X}$ is used is Simpson's rule,

$$\int_q^{q+h} f(\lambda)\, d\lambda = \frac{1}{6}(f(q) + 4f(q+\frac{h}{2}) + f(q+h)) - \frac{h^5 f^{(4)}(\theta)}{2880}, \qquad (8.38)$$

where $f^{(4)}$ denotes the fourth derivative of f and θ is in the interval $q \leq \theta \leq q+h$. For $h = 2$, a well known form of Simpson's rule is found

$$\int_{-1}^{1} f(\lambda)\, d\lambda \simeq \frac{1}{3}[f(-1) + 4f(0) + f(1)] \; . \tag{8.39}$$

Kuo [10] shows that (8.38) is a special form of the fourth order Runge–Kutta predictor. A few forms of the Milne formulas are given below, and the general form is given by Kelly in [15]. The value of these methods lies in the very low truncation error per step. However, some long simulation applications may exhibit increasing error (i.e., numerical instability) late in the integration period. The predictor formulas are labeled "P" and the correctors are labeled "C."

Fifth order Milne predictor–corrector

$$P{:}\; X_{n+1} = X_{n-3} + \frac{4h}{3}(2F_n - F_{n-1} + 2F_{n-2}) \tag{8.40}$$

$$C{:}\; X_{n+1} = X_{n-1} + \frac{h}{3}(F_{n-1} + 4F_n + F_{n+1}) \tag{8.41}$$

Seventh order Milne predictor–corrector

$$P{:}\; X_{n+1} = X_{n-5} + \frac{3h}{10}(11F_n - 14F_{n-1} + 26F_{n-2} - 14F_{n-3} + 11F_{n-4}) \tag{8.42}$$

$$C{:}\; X_{n+1} = X_{n-3} + \frac{2h}{45}(7F_{n+1} + 32F_n + 12F_{n-1} + 32F_{n-2} + 7F_{n-3}) \tag{8.43}$$

Modified Adams or Adams-Bashforth method

$$P{:}\; X_{n+1} = X_n + \frac{h}{24}(55F_n - 59F_{n-1} + 37F_{n-2} - 9F_{n-3}) \tag{8.44}$$

$$C{:}\; X_{n+1} = X_n + \frac{h}{24}(9F_{n+1} + 19F_n - 5F_{n-1} + F_{n-2}) \tag{8.45}$$

Example 8.2

In this example, a selection of predictor and predictor–corrector methods are used to solve

$$\ddot{y} = -2t\dot{y} - y + 0.02e^{-t^2}(t \sin t - \cos t)$$

$$y(0) = 0.01 \quad \dot{y}(0) = 0 \; .$$

Solution

The exact solution of this nonlinear, time varying equation is

$$y(t) = 0.01e^{-t^2} \cos t.$$

The following methods will be used to illustrate predictor–corrector methods:

1. Euler's method.
2. Fifth order Milne predictor–corrector
3. Adams–Bashforth predictor–corrector.

To use these methods, the equation is placed in first–order form,

$$x_1(t) \equiv y(t)$$

$$x_2(t) \equiv \dot{y}(t) ,$$

which gives

$$\dot{x}_1 = x_2$$

$$\dot{x}_2 = -x_1 - 2tx_2 + 0.02e^{-t^2}(t \ sin \ t - cos \ t)$$

$$X(0) = \begin{bmatrix} 0.01 \\ 0.00 \end{bmatrix} .$$

Results of the comparison are indicated in Table 8.3. In this table, the error between the calculated and exact value of $x(t)$ is indicated for selected values of t in the period $0 \leq t \leq 3.0$. Note that in this example, the *exact* value of $x(t)$ is known; obviously, this is not usually the case. In Table 8.3, note that the columns of errors have different powers of ten as scale factors; these are indicated in the column headings. The following observations are made in this example:

Table 8.3
Comparison of Euler, Milne, and Adams–Bashforth Methods

Time (s)	Error in Euler's method ($\times 10^{-6}$)	Error in Milne's method ($\times 10^{-8}$)	Error in Adams–Bashforth method ($\times 10^{-8}$)
		$\Delta t = 0.001$	
0.5	3.008	2.574	3.970
1.0	5.569	1.740	2.940
1.5	10.576	1.121	2.110
2.0	9.498	0.883	1.661
2.5	8.032	0.976	1.428
3.0	7.193	1.284	1.284
		$\Delta t = 0.010$	
0.5	30.901	244.000	381.584
1.0	55.154	152.670	270.320
1.5	106.129	86.939	188.575
2.0	95.286	67.975	146.335
2.5	80.480	54.087	125.418
3.0	72.055	73.269	112.725

1. The error in both Milne's and the Adams–Bashforth method are generally more than two orders of magnitude lower than Euler's method.

2. The Adams–Bashforth method exhibits excellent stability with decreasing error with t [the effect is especially noted if the tabulated errors are normalized by $x(t)$].

Additional observations are made by examining the magnitude of the correction vector,

$$||x^c(t) - x^p(t)||.$$

For this example, these corrections are tabulated in Table 8.4. It is observed that:

3. For Milne's method, the error in $x(t)$ is greater than the order of the correction.

4. For the Adams–Bashforth method, the error in $x(t)$ is much greater than the order of the correction.

5. The correction for the Adams–Bashforth method can become very small and in the range of computer accuracy.

Table 8.4
Magnitude of the Corrections in the Milne and Adams–Bashforth Methods

Time (s)	Correction vector Milne's method ($\times 10^{-8}$)	Correction vector Adams–Bashforth ($\times 10^{-13}$)
	$\Delta t = 0.001$	
0.5	0.0046	0.0051
1.0	0.0070	0.0046
1.5	0.0130	0.0034
2.0	0.0281	0.0027
2.5	0.0701	0.0037
3.0	0.2005	0.000009
	$\Delta t = 0.010$	
0.5	3.94	502.65
1.0	5.19	500.57
1.5	8.79	350.35
2.0	17.73	29.85
2.5	40.55	39.55
3.0	103.08	1.23

8.5 PRACTICAL TRANSIENT STABILITY STUDIES

The foregoing sections provide the majority of the programming and modeling techniques to perform a transient stability study. In this section, attention turns to questions of how these studies are used, who uses them, and concluding remarks on the practical implementation of a transient stability study.

Applications in Protective Relay Design

A power system protective relay is a subsystem that senses system parameters (e.g., voltage, current, power, frequency, sequence, phase), processes those data, and derives a decision as to whether corrective action is required (e.g., circuit breaker trip). The numerous details of relay design and function are far too voluminous to present here, but suffices it to say that in some instances, the relay speed and thresholds are designed so that in cases of anomalous system operation, a corrective action will be taken to retain system stability. An example is the case of an overcurrent relay design; consider a simple radial circuit subject to a bus fault at the receiving terminal. The sending terminal is an infinite bus. At the inception of the bus fault, the delivered active power drops to zero and the line current increases sharply. Consideration of balance of power principles in system supply generators indicate that these machines will experience net accelerating torque on their rotors. This is the case since the electrical demand has dropped, but the turbine-governor has not reacted instantaneously and the mechanical torques delivered to the rotor by the turbines are unchanged. If the machines overspeed too long, the torque angles, $\delta_i(t)$, could increase so much that synchronism will be lost. The relay, therefore, must open the fault and attempt reclosure at a speed which will permit retention of synchronism. Many faults are of the type that deenergization of the bus will result in clearing the fault; reclosure speed, then, should be sufficiently rapid so that $\delta_i(t)$ is less than the critical value. A stability study is used to determine not only the critical $\delta_i(t)$, but also the critical clearing time required. Transmission circuit relays close–in to generator buses, or in key (i.e., heavily loaded) lines are designed in conjunction with transient stability studies using several studies to assess tradeoffs of costs associated with high speed breakers/relays and enhance probability of transient stability. In an application to a strictly radial transmission circuit, stability considerations are not likely to use transient stability studies by computer since elementary considerations using the equal area criterion (discussed below) are less elaborate and more easily done. For networks, however, a transient stability study must be done.

Relay design at the transmission level is often, at least in part, heuristic. Typical transmission networks are so complex that exact design of zones of protection, impedance relay trip regions, and time dial settings, to name a few parameters, are virtually impossible. Usually, a standard design is used in conjunction with a transient stability study to verify the adequacy of the thresholds, time dial settings, and sensitivities. In addition, the appropriateness of the zones of protection is obtained from analysis of the standard design and multiple transient stability runs with faults placed at alternate points in the system.

Some types of protective relaying design do not involve stability or stability studies, these applications include relays for the protection of individual transmission system components and many differential relays. For example, a differential relay for a large power transformer is designed with the protection of the transformer in view. Probably, threshold settings and relay breaker speeds will be dictated by the transformer capabilities. In this instance, stability phenomena are not primary considerations. On the other hand, most relays at a generating station switchyard are designed with stability considered. A bus fault in such a switchyard could interrupt a considerable power level and cause unacceptable machine rotor acceleration. In such an application, in addition to consideration of loss of synchronism, generator protection may be taken into account through the calculation of net rotor torque, ΔT,

$$\Delta T = T_{mech} - T_{elect}. \tag{8.46}$$

Applications in Switching Strategy Evaluation

Any redistribution or interruption of power flow in a power system has a concommitant potential for stability difficulties. Most switching procedures will not strongly affect machine torque angles, since there is typically little accompanying change of machine power generated. In some cases, however, a transient stability study is done to assess the impact of a given switching sequence.

Postfault Scenario Evaluation

Many unplanned events in power system operation are evaluated after the fact in order to assess whether relaying or operating procedures require modification. For such evaluations, transient stability studies are invaluable. Sometimes, the event will be partially unknown, and the transient stability curves will be matched with system recordings in order to reconstruct accurately the circumstances of the event. Many substations are equipped with voltage, current, active power, reactive power, and frequency recording capability (often, paper charts). These recordings, as well as those obtained from large industrial customers (who often monitor parameters similar to those cited above), are used for postfault scenario evaluation.

AVR and PSS Design

Since the automatic voltage regulator (AVR) and power system stabilizer (PSS) are generation controllers designed, in part, to enhance system stability, it is common to use stability studies to evaluate alternative designs. Automatic voltage regulators and power system stabilizers are designed primarily with nominal operating regimes in mind. For these design considerations, the appropriate stability study methodology is probably small signal methodology. Small signal stability assessment methods are described later in this chapter. When large disturbances are considered, AVR and PSS design procedures include transient stability studies. Since AVR and PSS dynamics are long, applications in this area usually entail studies with long run times.

Typical Run Time

When a numerical method is used to assess transient stability, the question of how long a solution simulation time is required becomes an issue since a stable solution for 1 or 2 s is no guarantee that stability will be retained in the subsequent second. Transient stability studies are regularly run for at least a few seconds, and 10 to 30 s runs are occasionally required. In the linearized system, all time constants are usually less than 5 s, and the cited transient stability simulation times will generally reveal a machine $\delta(t)$ that is loosing synchronism. In this regard, it may be useful to use the concept of center of angle *(COA)*,

$$COA(t) = \frac{1}{g} \sum_{i} \delta_i(t),$$

where the sum is carried over g machines. Then machine i has a rotor angle $\Delta\delta_i(t)$ relative to the *COA*,

$$\Delta\delta_i(t) = \delta_i(t) - COA.$$

A plot of $\Delta\delta_i(t)$ may more readily reveal a machine "departing from the system" (i.e., loosing synchronism). If there is doubt, the simulation time of a transient stability study should be increased to reveal the outcome of a potential loss of synchronism.

Figure 8.7 shows a few actual results of a transient stability study.

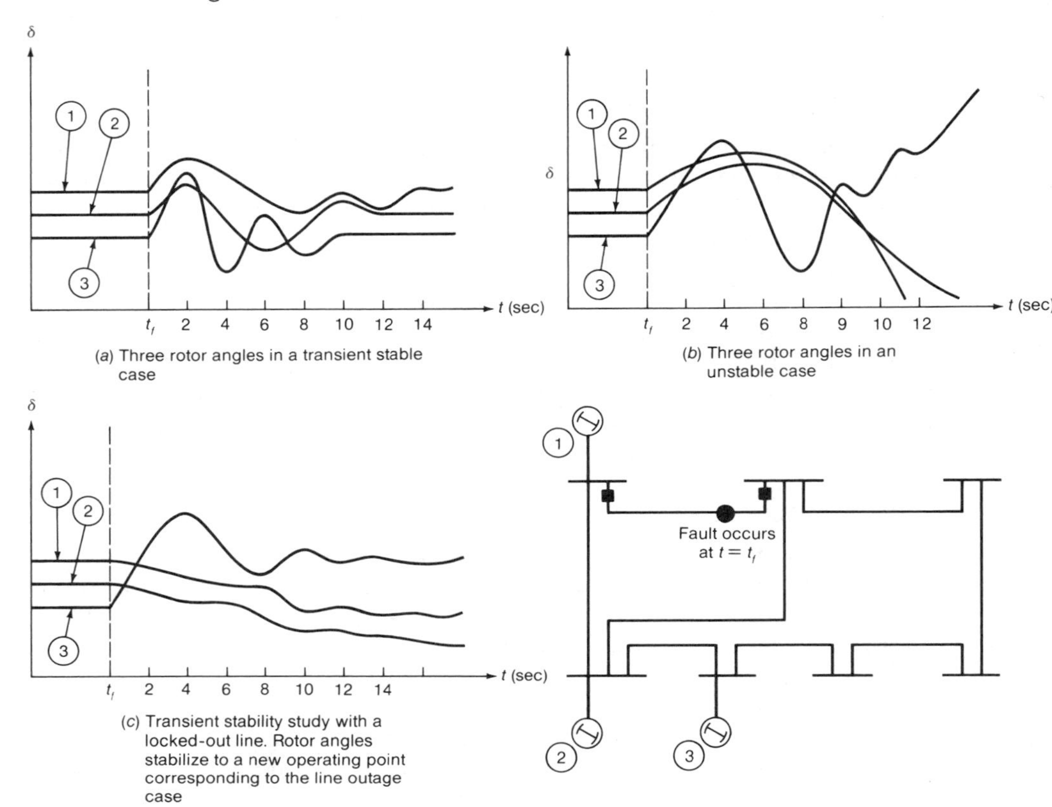

FIGURE 8.7 Typical transient stability study results.

Debugging Transient Stability Study Programs

The following remarks may be useful for debugging transient stability study programs:

1. *Removing the disturbance:* The initialization point and certain aspects of the numerical integration may be checked by removing the disturbance from the problem. In other words, the value of the state vector X and other parameters that are passed to the numerical integration subroutine are initialized at points which are expected to be the steady state ($t < 0$) values. For such initialization, $X(t)$ should remain fixed.

2. *Integrate a known system:* To separate errors in the power system model from the numerical integration routines, integrate a known set of equations. For example,

$$\dot{X} = AX$$

$$A = diag\ (-1\ -2\ -3\ \cdots)$$

$$X(0) = (1\ 1\ 1\ \cdots)^t\ .$$

The solution is

$$X(t) = (e^{-t} \; e^{-2t} \; e^{-3t} \cdots)^t .$$

3. *Round-off and faulty initialization arising from approximation of certain parameters:* For many pencil-and-paper applications, certain approximations involving π, $\sqrt{2}$, $\sqrt{3}$, ϵ, and other quantities are used. For example, the rotor speed of a two pole machine (60 Hz) is 120π rad/s, which is commonly written as 377. The error in this approximation, while small, may propagate in an undesirable way. It is recommended to use the full computer accuracy in setting these parameters. For example,

```
PI = 4. * ATAN(1.)
OMEGAR = 120. * PI
```

4. *Faulty sign of torque terms:* When studied in isolation, the sign of electrical, frictional, and prime mover torques are not important. The total rotor torque, $\Sigma\ T$, however, must be composed of these terms with proper sign. A machine terminal fault should cause the rotor angle to accelerate (δ increasing, $\dot{\delta}$ positive). If this is not observed, check the sign of the torque terms.

5. *Faulty perunitization:* Perunitization in transient stability problems is not quite as elementary as in sinusoidal steady state applications. The principal difficulty arises in the fact that time is not perunitized. One obvious solution is to work completely in *actual* values. This is not always satisfactory since most electrical parameters (e.g., machine reactances) are given in per unit. When solving the transient stability equations in per unit, most, if not all, AVR and PSS equations will be in per unit. The machine field equation

$$\frac{V_f}{r_f} \frac{1}{\tau_f s + 1} = | E_{af} |$$

should be perunitized so that when the field dc voltage V_f is at rating, $| E_{af} |$ at $t \rightarrow \infty$ is also at rating. The field time constant τ_f is simply in seconds.

The principal difficulty arises in the rotor equation of dynamic motion

$$\sum T = J \ddot{\delta} .$$

$$= \frac{2H}{\omega_o} \ddot{\delta} .$$

If $\sum T$ is perunitized (i.e., at $P = P_{base}$ and $\omega_r \simeq$ synchronous speed, $| T_{pm} | = | T_{elect} | = 1.00$ pu), it is necessary to perunitize the inertia constant H. The correct per unit base in this regard is such that 1 rad/s^2 rotor acceleration occurs when $2H/\omega_o$ is numerically equal to the base torque. For example, in a 100 MVA, 60 Hz system with rotor synchronous speed of 120π rad/s, the logical selection of base torque is

$$T_{base} = \frac{100 \times 10^6}{120\pi} N \cdot m.$$

Therefore, when

$$\frac{2H}{\omega_o} = \frac{100 \times 10^6}{120\pi}$$

the rotor acceleration will be 1 rad/s^2. Let ω_o *not* be perunitized, and let ω_o be retained as 120π rad/s; then the logical choice of per unit base for H is

$$H_{base} = \frac{100 \times 10^6}{120\pi} .$$

Assuming that ω_o is written as the actual rad/s synchronous rotor speed (i.e., 120π for a 60 Hz two pole machine) and assuming that $\ddot{\delta}$ is expressed in rad/s^2, then

$$H_{base} = T_{base} \text{ (numerically)}.$$

This result is readily obtained from the simultaneous solution of

$$\sum T_{actual} = \frac{2\, H_{actual}}{\omega_o} \ddot{\delta}_{\text{actual}}$$

$$\sum T_{pu} = \frac{2\, H_{pu}}{\omega_o} \ddot{\delta}_{\text{actual}},$$

with

$$\sum T_{pu} = \frac{\sum T_{\text{actual}}}{T_{base}} \qquad H_{pu} = \frac{H_{\text{actual}}}{H_{\text{base}}} .$$

Typical values of H_{pu} are in the range 1 to 3.

6. *Excessive output:* In many applications, it is inadvisable to print out (or graph) each calculated point, x_n. It is useful to use a parameter such as NPRINT in a way that only when N is evenly divisible by NPRINT will a state be printed.

7. *Discontinuous fluctuations of states:* If some states are fluctuating widely from integration step to integration step, check for too large a value for step size h.

Connection with the Equal Area Criterion

The fundamental relation describing the dynamics of a synchronous machine is

$$\sum T = \frac{2H}{\omega_0}\ddot{\delta} . \tag{8.47}$$

In the absence of frictional torques,

$$T_{pm} - T_e = \frac{2H}{\omega_0}\ddot{\delta}, \tag{8.48}$$

where T_e is the decelerating electrical torque given by a model of appropriate complexity to the problem at hand. Consider the power associated with the total torque $\Sigma\, T$,

$$P = \omega \sum T . \tag{8.49}$$

If this power is integrated with respect to torque angle δ, recognizing a one-to-one correspondence between $d\delta/dt$ and mechanical rotor speed, one obtains

$$\int P \, d\delta = \int P \frac{d\delta}{dt} \, dt \tag{8.50}$$

$$= \int P\omega \, dt \, .$$

Therefore,

$$\int P \, d\delta = \int P\omega \, dt = \int \omega \sum T \, d\delta \, . \tag{8.51}$$

The $\int P\omega \, dt$ term is recognized as proportional to $\int P dt$ when the rotor speed is nearly synchronous speed. This term is proportional to the accelerating energy imparted to the rotor. One concludes that the areas under the T_{pm} and T_e curves plotted versus δ are proportional to the accelerating and decelerating energy imparted to the machine rotor.

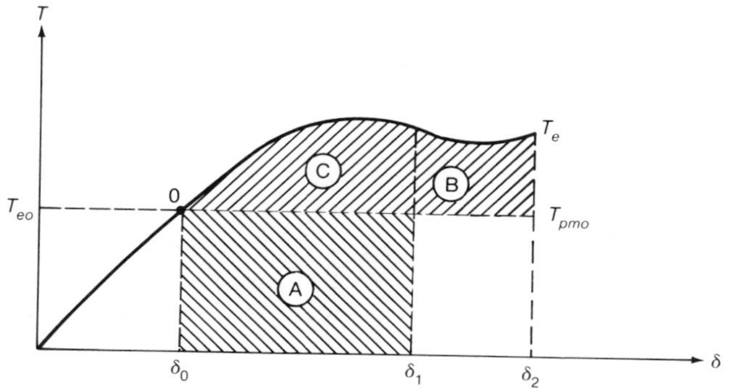

FIGURE 8.8 Torque versus δ characteristic used in equal-area criterion.

Figure 8.8 illustrates a typical $T_e(\delta)$ characteristic with initial operating point O marked at $T_e = T_{eo}$. The prime mover torque is T_{pmo} such that $|T_{eo}| = |T_{pmo}|$. If T_e and/or T_{pm} change, the equilibrium is disturbed and acceleration or deceleration may result. If T_e drops to zero, for example, the machine will accelerate. This is the case for a fault at the machine terminals. If acceleration persists until $\delta(t)$ reaches δ_1 (see Figure 8.8), the rotor will have accepted accelerating kinetic energy W_{acc}, which is proportional to the area A in Figure 8.8. If the fault now clears, the torque angle will continue to "open" (i.e., increase) to δ_2. However, the rotor will decelerate since

$$|T_e| > |T_{pm}| \, .$$

The angle δ_2 is readily calculated since decelerating area B must just balance (equal) A. When $\delta(t)$ reaches δ_2, the torque angle will decrease toward the equilibrium point O. In the process of doing this, the rotor will accept decelerating kinetic energy W_{dec}. Kinetic energy W_{dec} is proportional to areas B plus C in Figure 8.8. Having arrived at $\delta(t) = \delta_0$, the rotor angle will continue to decrease since the net accelerating/decelerating kinetic energy will be proportional to area C. Under the terms of this frictionless model, the rotor will oscillate indefinitely about O.

Note that angle δ_2 is readily located [even if rather complex formulas for $T_e(\delta)$ and T_{pm} are used]. Also, it is possible to identify whether synchronism will be retained since excessively large δ_2 will result in conditions in which accelerating area A is so large that decelerating area B may never counterbalance this rotor acceleration. Thus quick assessments of transient stability and maximum rotor swing are possible. The assessment has the following limitations, however:

1. It is difficult to incorporate effects of friction. The stability determination is conservative in this regard.

2. Variation of machine terminal voltage and/or field excitation results in changes in the $T_e(\delta)$ characteristic. These variations, although slow, are not readily incorporated into the method.

3. Although δ is readily calculable by this method, $\delta(t)$ versus t is *not.* This is the case because $\delta(t)$ does not appear explicitly in the model. There are techniques to alleviate this limitation, but they are approximate.

This method is termed the ***equal area method,*** and criteria that result in retention of stability are termed ***equal area criteria.*** The principal advantage lies in its simplicity. Complex models for torque versus torque angle are easily included, but the technique is largely relegated to single machine applications.

For single machine applications, or for multimachine cases in which W_{acc} and W_{dec} are an order of magnitude or more greater than the energy obtained from other generators via the network, the equal area method offers a handy approximate check for numerical solutions. In these cases, the equal area method may be adequate to assess stability and calculate critical clearing times.

8.6

SMALL SIGNAL STABILITY OF ELECTRIC POWER SYSTEMS

The term ***small signal stability*** applies to stability studies in which the machine rotor swings are sufficiently small to allow the linearization of the dynamic equation. Typical small signal stability studies entail rotor swings in the range of ± 0.1 rad around the operating point. For such swings, round rotor machine torques are nearly linear in δ and the torque expressions should be expanded in a Taylor series and truncated at the term that is linear in δ. For salient pole machines, when the reluctance torque is much smaller than the synchronous torque, or the rotor swing is smaller than that indicated above, or both, the same linearization procedure applies.

The dynamic equations of an electric power system are of the form

$$\dot{X} = F(X) \ . \tag{8.52}$$

The *sin* δ and *sin* 2δ terms in vector F are linearized by expanding this vector in a Taylor series. Sometimes, frictional torque terms in F must also be linearized. In small signal studies, effects of the automatic voltage regulator (AVR) and power system stabilizer (PSS) are usually included. Further discussion at the end of this section relates to these field circuit controllers. Expansion of the vector function F around the operating point X_0 results in

$$F(X) = F(X_0) + (\frac{dF}{dX}|_{X_0})(X - X_0) + O[(X - X_0)^t(X - X_0)] \approx AX + b, \tag{8.53}$$

where the high order terms $O[(X - X_0)^t(X - X_0)]$ have been dropped and

$$A = (\frac{dF}{dX}|_{X_0}) . \tag{8.54}$$

The A matrix in (8.53) is the ***associate matrix*** of (8.52) and this matrix is the Jacobian matrix of partial derivatives of rows of F with respect to rows of the state vector X. The stability of (8.53) is determined by the eigenvalues of A: if the eigenvalues of A have only negative real parts, small signal stability is insured,

$$det\ (A - \lambda I) = 0$$

$$Re\ (\lambda_i) < 0 \text{ for } \textit{all } i \rightarrow \textit{small signal stability.}$$

The associate matrix A is not symmetric, the eigenvalues of A are generally complex, and the process of evaluating these eigenvalues is computationally burdensome for large A. For a system of g generators, A is at least $2g$ by $2g$; when auxiliary controllers are included, the size of A may be quite large. There are numerous alternatives to determine the signs of $Re\ (\lambda_i)$. These include the Routh method [17] and others [18,19].

8.7

PRACTICAL APPLICATIONS OF SMALL SIGNAL STABILITY STUDIES

There are two principal application areas for small signal stability studies: AVR and PSS design, and system operating procedure analysis. In the area of AVR and PSS design, as well as of other boiler–turbine–generator controllers, small signal analysis is very convenient, particularly for a quick assessment of design adequacy and parameter (e.g., gain) selection. Although small signal instability does not imply transient instability, in machine controller design, small signal instability is not permitted. This is the case because good operating procedure does not permit oscillation of voltages, currents, or power flows. Production stability analysis programs often contain a "quick look" option for stability assessment in which linear analysis is used.

When small signal instability is suspected by operators, the operating procedures that produce the suspicious conditions are analyzed using small signal stability studies. Unstable conditions are occasionally the result of interaction of controllers of machines at different generating stations. Some rather unusual, if not spectacular cases of small signal instability have resulted by such interaction between controllers which were separated by hundreds of kilometers [20,21]. Undamped, persistent oscillation of system parameters is not an acceptable operating condition. A small signal stability study will help identify such a condition, and it is indispensable in alleviating the problem.

Small signal methods have been used in certain other theoretical applications [22,23] in which the principal motivation for the application is the simple, linear mathematical formulation.

In each of these application areas, care must be taken to justify the use of small signal methods. The linearization used is seldom valid for fluctuations of active power beyond 300 MW in large systems. For this reason, most analyses of system dynamics are done using transient (i.e., large signal) methods.

8.8 LIAPUNOV'S DIRECT METHOD

Numerical simulations to assess stability are *indirect methods* since they do not exploit properties of the mathematical model to determine inherent properties directly. Small signal methods rely on eigenvalue analysis to determine stability, and in this regard these methods directly exploit the mathematical model to answer questions of system stability. A methodology such as small signal analysis is termed a *direct method.* Unfortunately, the eigenvalue analysis used in small signal methods implies linearized behavior; it is natural to ask whether there is a direct method which is applicable for nonlinear systems. There are several such techniques, but the term *direct method* is usually associated with Liapunov, who examined several methods of stability analysis. Liapunov's first method is an indirect method that is, in essence, numerical evaluation. Liapunov's second method is a direct method that is both celebrated as a broadly applicable stability technique, and criticized as a method which is difficult to apply and as a technique which has a limited number of practical applications. The reader will have to assess the effectiveness of the method on its own merits, but it is fair to say that despite the considerable interest in Liapunov's direct method applied to power system stability, there have been very few actual applications. The reason for this will be discussed at the end of this section.

Liapunov's original thesis was written in the Russian language and many libraries retain French translations [27]. References [28–32] will serve to document both the theory of the method and engineering applications. The student interested in the basic method is advised to examine the literature of automatic control theory for elementary applications; reference [5] contains a particularly readable description of the method.

The focus of this article is Liapunov's *second* or *direct* method. The system of nonlinear equations,

$$\dot{X} = F(X, t) \, , \tag{8.55}$$

are examined for stability. The essence of Liapunov's indirect method relates to properties of an energy function, $V(X)$, which is a scalar valued function of the state vector X. The first order form in (8.55) does not restrict applications since all lumped systems may be modeled in this form. The energy function V does not have a specific form, but it does broadly represent a measure of energy contained in the system. In this context, the term *energy* is used as in communications theory, where the energy in a signal $x(t)$ is W,

$$W = \int_0^T x^2(t)\, dt \ .$$

For dynamic systems in which the state vector represents positions and velocities of system components, V may be proportional to the kinetic plus potential energy of the system. Usually, high order systems have state variables which are not readily identifiable as distinct positions, velocities, accelerations, or other physical parameters. In these cases, V must assume a generalized interpretation which measures the energy content of all the states. As an example, the general quadratic form

$$V(X) = X^t QX \tag{8.56}$$

may be viewed as an energy function when Q is a positive definite matrix. Momentarily, let Q be symmetric; then the positive definite quality of Q implies that V is *always* a positive scalar except at $X = 0$, where V is obviously also zero. An example of a suitable energy function of the form in (8.56) occurs for

$$Q = diag\ (\lambda_1, \lambda_2, \ldots),$$

where the λ_i are all real and positive. For the purpose of studying candidate $V(X)$ functions, the following definitions are useful:

1. A scalar function $V(X)$ of vector argument X is ***positive definite*** if $V(X)$ is positive for all X except $X = 0$, where $V(0) = 0$.

2. A scalar function $V(X)$ of vector argument X is ***positive semidefinite*** if $V(X)$ is nonnegative for all X and at $X = 0$, $V(0) = 0$.

3. A scalar function $V(X)$ of vector argument X is ***negative definite*** if $-V(X)$ is positive definite.

4. A scalar function $V(X)$ of vector argument X is ***negative semidefinite*** if $-V(X)$ is positive semidefinite.

Examples of some of these functional classifications are:

Positive Definite

$$V\left(\begin{bmatrix} x_1 \\ x_2 \end{bmatrix}\right) = x_1^2 + x_2^2$$

$$V\left(\begin{bmatrix} x_1 \\ x_2 \\ x_3 \end{bmatrix}\right) = X^t \begin{bmatrix} 1 & 0.1 & 0 \\ 0.1 & 1 & 0 \\ 0 & 0 & 2 \end{bmatrix} X$$

Positive Semidefinite

$$V\left(\begin{bmatrix} x_1 \\ x_2 \end{bmatrix}\right) = x_1^2$$

$$V\left(\begin{bmatrix} x_1 \\ x_2 \\ x_3 \end{bmatrix}\right) = x_1^2 + (x_2 + x_3)^2.$$

The use of energy functions to assess stability is given by the Liapunov theorems:

Liapunov's Asymptotic Stability Theorem

The system

$$\dot{X} = F(X)$$

is asymptotically stable in the vicinity of $X = 0$ if for positive definite $V(X)$, the function $\dot{V}(X)$ is negative definite.

Liapunov's Bounded State Stability Theorem

The system

$$\dot{X} = F(X)$$

is bounded state stable [i.e., all elements of the state vector $X(t)$ have finite bounds in the vicinity of $X=0$] if for positive definite $V(X)$, the function $\dot{V}(X)$ is negative semidefinite. The term ***stability in the sense of Liapunov (ISL)*** refers to bounded state stability.

Liapunov's Instability Theorem

The system

$$\dot{X} = F(X)$$

is unstable if for $\dot{V}(X)$ negative definite, the function $V(X)$ is not globally positive semidefinite.

General Remarks on Liapunov Theory

The great advantage of Liapunov theory lies in its broad range of applicability, including nonlinear systems. Unfortunately, the asymptotic stability theorem and bounded state stability theorem do not give any information on stability if the requirement on $V(X)$ and $\dot{V}(X)$ are *not* fulfilled. In other words, in practical applications, if a suitable $V(X)$ can not be found, no general conclusion on stability can be drawn. Also, the specific form of $V(X)$ is unknown, and aside from the mathematical acuity of the user, there are few methods to generate forms of the energy function which apply in a wide range of applications. These criticisms have been debated in detail in the literature, and each criticism has been, in part, alleviated.

Example 8.3

In this example, Liapunov's direct method is illustrated for a small system taken from outside power engineering. Consider the nonlinear system

$$\ddot{x} + x^2(2 + \sin x) = 0 .$$

Examine this system for bounded state stability.

Solution

In first order form, the system is

$$\dot{X} = \begin{bmatrix} x_2 \\ -x_1^2(2 + \sin x_1) \end{bmatrix},$$

where $X^t = [x_1, x_2]$ and $x_1 = x$, $x_2 = \dot{x}$. Select the energy function V motivated by consideration of the kinetic energy plus potential energy of physical systems,

$$V(X) = \frac{1}{2}x_2^2 + \int_0^{x_1} \xi^2(2 + \sin\,\xi)\, d\xi \tag{8.57}$$

$$= \frac{1}{2}x_2^2 + \frac{2}{3}x_1^2 + 2x_1 \sin x_1 - x_2^2 \cos x_1 + 2(\cos x_1 - 1)\,.$$

The $V(X)$ function is positive definite since the integral term is always positive except at $x_1 = 0$, where this term is zero. This integral is positive since the integrand is positive or zero for all ξ. The $\dot{V}(X)$ function is readily shown to be

$$\dot{V}(X) = 0,$$

and hence is negative semidefinite. The system is therefore bounded state stable for all X.

Applications in Power Engineering

In this part of the section, Liapunov theory as applied to the transient stability problem is presented. El-Abiad and others are generally credited with early applications [38,39]. At machine i, the swing equation is

$$J_i \frac{d^2\delta_i}{dt^2} = T_{mi} - T_{ei},$$

where T_m and T_e are mechanical and electrical torques, respectively. In terms of the inertia constant, H_i, the swing equation for machine i is

$$\frac{2H_i}{\omega_o} \frac{d^2\delta_i}{dt^2} = T_{mi} - T_{ei}\,. \tag{8.58}$$

The electrical torques, T_{ei}, are readily written in terms of the electric power,

$$\omega_o T_{ei} = P_{ei} = |v_i|^2 G_{ii} + \sum_{\substack{j=1 \\ \neq i}}^{n} |v_i||v_j|\, Y_{ij} \sin\,(\delta_{ij} - \alpha_{ij}), \tag{8.59}$$

where $|v_i|$ denotes the bus voltage magnitude at i and $Y_{ij} \underline{/\theta_{ij}}$ is the ij entry of Y_{bus}, G_{ii} is the real part of Y_{ii}, and

$$\delta_{ij} = \delta_i - \delta_j \tag{8.60}$$

$$\alpha_{ij} = \theta_{ij} - \frac{\pi}{2}\,.$$

The bus phase angles, δ_i, appear in (8.59) through the "line phase angle," δ_{ij}, defined in (8.60). Equation (8.58) is therefore

$$\dot{x}_{2i-1} = x_{2i}$$

$$2H_i \dot{x}_{2i} = P_{mi} - |v_i|^2 G_{ii} - \sum_{\substack{j=1 \\ \neq i}}^{n} |v_i||v_j|\, Y_{ij} \sin\,(\delta_{ij} - \alpha_{ij})\,. \tag{8.61}$$

where x_{2i-1} is δ_i and x_{2i} is $\dot{\delta}_i$. Note that the states x generally appear on the right hand side of (8.61) in the δ_{ij} terms. Equation (8.61) is of the form

$$\dot{X} = F(X),$$

where X is a $2n$ vector in the case of an n machine system.

At this point, a suitable Liapunov function is sought to demonstrate stability. References [33–36] contains several suggested forms of $V(X)$, each with its own merits and drawbacks. One form of $V(X)$ will be produced here to illustrate the approach. This illustrative alternative is Fouad's form [35], in which the transfer conductances, G_{ij}, are zero. In this case, the proposed $V(X)$ is

$$V(X) = \sum_{i=1}^{n-1} \sum_{k=i+1}^{n} \left[\frac{H_i H_k}{4\omega_o^2} \omega_{ik}^2 - a_{ik}\delta_{ik} - \frac{2b_{ik}}{\omega_o} (H_i + H_k) \cos \delta_{ik} \right] - \sum_{i=1}^{n} \sum_{\substack{j=1 \\ \neq i}}^{n-1} \sum_{\substack{k=j+1 \\ \neq i}}^{n} \frac{H_i b_{jk}}{2\omega_o} \cos \delta_{jk} + K_1, \tag{8.62}$$

where

$$\omega_{ik} = \dot{\delta}_i - \dot{\delta}_k$$

$$= x_{2i} - x_{2k}$$

$$\delta_{ik} = \delta_i - \delta_k$$

$$= x_{2i-1} - x_{2k-1}$$

and the a's and b's and K_1 are constants selected such that $V(X)$ is positive definite and $\dot{V}(X)$ is negative definite. The values of a, b, K_1 are dependent on the operating region around which stability is evaluated, in other words, where

$$T_{mi} - T_{ei} = 0$$

for all machines. Both the operating points and the values of a, b, K_1 are found *numerically*. This usually involves calculation of the zeros of $V(X)$ – this is a serious disadvantage of the method. In the region in X space where the selected a, b, K_1 give the desired properties of $V(X)$, the power system is asymptotically stable. Outside this region, nothing is known about stability. The drawback to this method is the numerical burden of calculating a, b, K_1 and the extent of the region in which the desired properties of V(X) and $\dot{V}(X)$ are obtained. The former of these drawbacks can be assessed only by actual implementation. The latter is somewhat an inherent feature of Liapunov's direct method. A brief discussion is found at the conclusion of this article. Also note that $V(X)$ in (8.62) uses $G_{ij} = 0$. These transfer conductances are, in fact, nonzero and when they are neglected, the damping that is obtained from the system active power losses is absent from the model. This neglect of system damping results in exacerbation of the conservatism of the region of stability (i.e., the region of stability is, in fact, larger than that obtained). The reason for neglecting the transfer conductances is strictly to reduce the mathematical complexity. A further computational burden is the calculation of all values of machine torque angles for which T_{mi} minus T_{ei} is zero.

Other Forms of V(X)

The key to Liapunov's direct method is the selection of $V(X)$. The numerical burden, extent of conservatism of the region of stability, and complexity of the model depend on the choice of $V(X)$. The complexity of the selection of this function is reduced, in part, by appealing to properties of the physics of dynamics, for example, (8.62) is not as arbitrary as it may first appear. The first (double summation) term is of the form of the sums of kinetic energies. The second term may be viewed as potential energy. The cited form is

$$v(X) = X^t QX + \int_0^{C^t X} f^t(\sigma)R \, d\sigma,$$

where Q and R are matrices, C is a vector selected to generate the required forms, and $f(\sigma)$ is a vector of vector argument σ which contains the nonlinearities present in the system. An alternative is Zubov's method, which is an iterative method to generate $V(X)$ and $\dot{V}(X)$ with successively larger regions of X to give the required properties.

Conervatism of Liapunov's Second Method

The preceding discussion reveals that Liapunov's second method results in a conservative calculation of the region in X-space in which a system is stable. For system which is asymptotically stable near $X = 0$, in which F is linearized to the system

$$\dot{X} = AX ,$$

the Liapunov function

$$V(X) = X^t QX \tag{8.63}$$

applies. For this case

$$\begin{aligned} \dot{V}(X) &= \dot{X}^t QX + X^t Q\dot{X} \\ &= X^t A^t QX + X^t QAX \\ &= X^t RX, \end{aligned} \tag{8.64}$$

where $R = A^t Q + QA$. If A were diagonal, Q is readily selected to be A^{-1} and when elements of A (i.e., the eigenvalues) have negative real parts, R is diagonal with elements possessing positive real parts. Hence $\dot{V}$ is negative definite and V is positive definite. Even if A is not diagonal, a change of variables is readily used to obtain the same result. This exercise implies that the quadratic form (8.63) yields no new information over the linearized analysis. The system may be asymptotically stable well outside the region where linearization is valid, but (8.63) and (8.64) do not reveal this region.

Of course, the reply to this criticism is that the quadratic form (8.63) must be augmented by nonlinear functions to increase the region in which stability is shown. There is no general method to perform this augmentation, and there is never a guarantee that all points in the region of asymptotic stability will be identified.

PATTERN RECOGNITION METHODS

The concluding technique considered in this chapter bears numerous similarities to the direct method of Liapunov. The method of pattern recognition was developed in connection with information processing techniques [37] and represents an attempt to automate human attributes. This approach uses features, ϕ_i, of a process to determine whether the process is classified into one or another classes. In the case of power system stability, the classes are "stable" or "unstable." The features are arranged in a feature vector Φ and the scalar function $c(\Phi)$ is used to classify the system as stable or unstable,

$$\textit{If } c(\Phi) \geq 0 \rightarrow \textit{stable.}$$

$$\textit{If } c(\Phi) < 0 \rightarrow \textit{unstable.}$$

The obvious questions relate to the selection of c and Φ. The function c is known as a classifier.

It is logical to select machine angles and velocities as features, ϕ_i. Further, it logical to choose kinetic energy minus a threshold level as a classifier; this is the case since systems with high kinetic energy will tend to be unstable, while those with low kinetic energy will tend to be stable.

The pattern recognition approach is similar to Liapunov's direct method in that the classifier function bears a remarkable resemblance to the energy function. The feature vector resembles the state vector. There is a further nonmathematical resemblance between the methods: both have been proposed for power system transient stability studies but neither has significantly penetrated this application area. The advantage of both the Liapunov and pattern recognition approaches lies in the fact that on-line numerical simulation is avoided.

Selection of Classifier

The essence of the pattern recognition approach is the relegation of most of the calculation process to off-line. This is done by designing a classifier function, $c(\Phi)$, which is able to distinguish between stable and unstable samples. For example, if n_s stable cases and n_u unstable cases are used to "train" the classifier (i.e., to design the classifier), the $n_s + n_u$ cases are studies in transient stability studies off line. The training set corresponds to not only processing time because of the many required transient stability studies, but also to calculation time because the classifier function must be adjusted. A two dimensional feature vector pictorial is given in Figure 8.9; in this pictorial, "U" denotes unstable cases and "S" denotes stable cases. Each point depicted entails an off-line transient stability study. The dashed line $c_1(\Phi)$ denotes a linear classifier,

$$c_1(\Phi) = a\Phi + b, \tag{8.65}$$

in which parameters a and b are chosen to separate the samples in the training set. In the pictorial illustrated, the nonlinear classifier $c_2(\Phi)$ appears to do a better job.

To design a classifier $c(\Phi)$, consider feature vector Φ as a vector of n

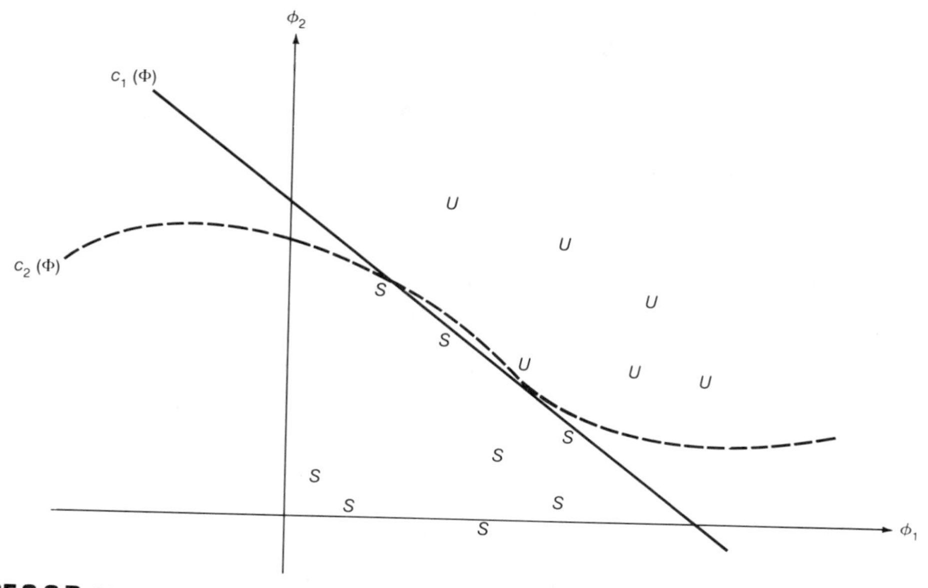

FIGURE 8.9 Pattern recognition method for transient stability analysis.

"primary" variables $\phi_1, \phi_2, \ldots, \phi_n$. Denote the stable and unstable classes as Ω_s and Ω_u. The Bayes' maximum likelihood test is

$$\text{if } f(\Omega_s/\Phi) > f(\Omega_u/\Phi) \rightarrow \Phi \text{ is in } \Omega_s$$

$$\text{if } f(\Omega_s/\Phi) < f(\Omega_u/\Phi) \rightarrow \Phi \text{ is in } \Omega_u,$$

where $f(\Omega_s/\Phi)$ and $f(\Omega_u/\Phi)$ are the aposteriori probability density functions for the stable and unstable classes. These densities are related to the conditional density $f(\Phi/\Omega_k)$, $k = s, u$, by Bayes' theorem,

$$f(\Omega_k/\Phi) = \frac{f(\Phi/\Omega_k)P\{\Omega_k\}}{f(\Phi)} . \tag{8.66}$$

Thus a probabilistic model is used for the distribution of elements of Φ. It is an easy matter to manipulate (8.66) to obtain a classifier function $c(\Phi)$,

$$c(\Phi) = \frac{f(\Phi/\Omega_s)}{f(\Phi/\Omega_u)} - \frac{P\{\Omega_u\}}{P\{\Omega_s\}} . \tag{8.67}$$

This is the maximum likelihood classifier. When $c(\Phi)$ is greater than zero, the sample is in Ω_s (stable). When $c(\Phi)$ is less than zero, the sample is in Ω_u (unstable). The notation $P\{\Omega_u\}$ and $P\{\Omega_s\}$ refers to the sample probability of unstable and stable cases in the training set. Equation (8.67) is based on maximum likelihood theory and is similar to minimizing the probability of misclassification. It is necessary to model $f(\Phi/\Omega_s)$ and $f(\Phi/\Omega_u)$ statistically – this is a serious drawback since any specific assumption of form of f (e.g., Gaussian) is subject to questions of accuracy, stationarity, and ease in calculation of defining parameters. There are some indications that the Gaussian assumption for f gives a reliable classifier [43,44]; this method results in the following form of $c(\Phi)$:

$$c(\Phi) = \frac{1}{2}(\Phi - M_s)^t \textstyle\sum_s^{-1} (\Phi - M_s) - \frac{1}{2}(\Phi - M_u)^t \sum_u^{-1} (\Phi - M_u)$$

$$+ \frac{1}{2} ln \frac{det \sum_s}{det \sum_u} - \frac{P\{\Omega_s\}}{P\{\Omega_u\}} . \qquad (8.68)$$

In (8.68), M_s and M_u are the mean (expected) values of vector Φ in the stable and unstable classes,

$$M_s = E(\Phi) \ \Phi \in \Omega_s$$

$$M_u = E(\Phi) \ \Phi \in \Omega_u$$

(in the training set), and $\sum_s$ and $\sum_u$ denote the covariance matrices of Φ in the stable and unstable classes. The terms M and $\sum$ are sample statistics. Equation (8.68) is recognized as a multivariate Gaussian form, a point that is discussed in Chapter 9. In (8.68), if $c(\Phi)$ is positive, Φ belongs to Ω_u; if it is negative, Φ belongs to Ω_s.

Other forms of $c(\Phi)$ have been suggested and tested in transient stability analysis for power systems. These alternatives include the "K nearest neighbor decision rule" and the polynomial classifier. The former is based on a functional approximation for $f(\Phi/\Omega)$:

$$f(\Phi/\Omega) \simeq K - \frac{1}{N} \frac{1}{A} . \qquad (8.69)$$

In (8.69), Ω is either Ω_s or Ω_u, N is the number of samples from the training set, and A is the volume of a sphere (in multidimensional space) such that K elements of the N samples lie in the sphere. Evidently,

$$A = A(K, N, \Phi)$$

$$K \leq N .$$

Equation (8.69) must be used for Ω_s and Ω_u. In the Ω_s case, only stable samples are considered (i.e., all N training set elements are stable), and in the Ω_u case, the training set consists of only unstable cases. The advantage of the K nearest neighbor classifier is that accuracy superior to the Bayes' maximum likelihood is obtained; it is necessary to store all samples, however, to construct $c(\Phi)$. The polynomial classifier is based on a polynomial approximation of $f(\Phi/\Omega)$. The method is similar to interpolation formulas used to estimate values of f between available samples of Φ. By this method,

$$f(\Phi/\Omega_k) = \frac{1}{N_k} \sum_{i=1}^{N_k} \gamma(\Phi, \Phi_i) \qquad (8.70)$$

where k is either s or u, and $\gamma(\Phi, \Phi_i)$ is the contribution of the ith sample in the training set to f. The functional form of the polynomial classifier is

$$c(\Phi) = \frac{1}{N_s} \sum_{i=1}^{N_s} \gamma(\Phi, \Phi_s) - P\{\Omega_u\} \frac{1}{N_u} \sum_{j=1}^{N_u} \gamma(\Phi, \Phi_j) . \qquad (8.71)$$

In (8.70) and (8.71), $\gamma(\Phi, \Phi_i)$ is a polynomial in $\phi_1, \phi_2, \ldots, \phi_n$.

Selection of Feature Vector

A selection of the features to be used, Φ, is a key selection concomitant to the selection of the classifier function, c. The strategy in making the selection is obviously to render the stable and unstable classes separable by the classifier.

In transient stability assessment, there is a heuristic choice of features as the kinetic energy of each machine rotor,

$$\phi_i = H_i \frac{\omega_i^2}{\omega_o} .$$

Also, the active power dispatched at each generator, P_{gi}, has been found to be effective in Φ. Note that in an economically dispatched system, the P_{gi} are related by the dispatch algorithm and it is resonable to use

$$\phi_j = \sum_i P_{gi}$$

as one feature rather than each P_{gi} as a feature.

Practical Applications

This section opened with a remark that the pattern recognition approach shares numerous similarities to the Liapunov methods. Perhaps the most striking similarity is that the methods have been "proven" in limited laboratory applications in which problems of dimensionality were not at issue because small systems were studied. The pattern recognition approach suffers from two undesirable characteristics which have impeded its practical application: ***false dismissals*** may result (these are cases in which unstable systems are not so recognized) or ***false alarms*** may occur (these are cases in which stable systems are classified as unstable). Figure 8.10 illustrates the case in which a false alarm may occur if linear classifier c_1 is used. On the other hand, classifier c_2 may result in false dismissals. From an operating point of view, false alarms are tolerable, but false dismissals are not.

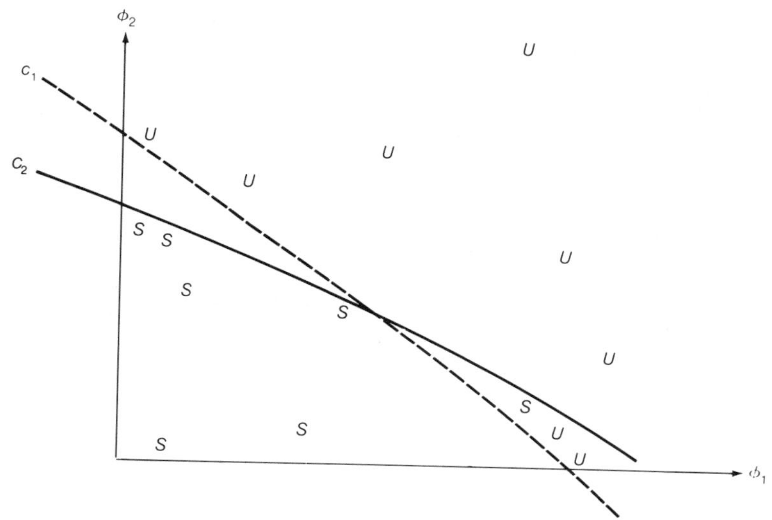

FIGURE 8.10 Separability problems in pattern recognition method.

The great advantage of the pattern recognition method is that long computation is precalculated (i.e., done off-line). Stability is rapidly assessed by evaluating one function, $c(\Phi)$.

References [43,44] discuss application methods in detail. There have been limited applications of the technique in practical power systems; these are described in [45,46].

Bibliography

[1] A. Fitzgerald, L. Kingsley, and S. Umans, *Electric Machinery,* McGraw–Hill, New York, 1983.

[2] J. Chassande, E. Pillet, M. Poloujadoff, and L. Pierrat, "Transient Low Frequency Unsymmetrical Operation of Synchronous Machines," *IEEE Trans. Power Apparatus and Systems,* v. PAS-99, no. 3, May–June 1980, pp. 1298–1305.

[3] C. Concordia, *Synchronous Machines,* Chapman & Hall, London, 1951.

[4] A. Blaylock, H. Hindmarsh, and K. Foster, "Some Critical Aspects of Generator Capability Under Unbalanced Operating Conditions," *IEEE Trans. Power Apparatus and Systems,* v. PAS-96, no. 5, September–October 1977, pp. 1470–1478.

[5] J. Gibson, *Nonlinear Automatic Control,* McGraw–Hill, New York, 1963.

[6] R. W. Hamming, *Numerical Methods for Scientists and Engineers,* McGraw–Hill, New York, 1962.

[7] A. Nordsieck, "On Numerical Integration of Ordinary Differential Equations," *Mathematics of Computation,* v. 16, no. 77, January 1962, pp. 22–49.

[8] R. Alonso, "A Starting Method for the Three Point Adams Predictor–Corrector Method," *J. Association for Computing Machinery,* v. 7, 1960, pp. 176–180.

[9] Z. Kopal, *Numerical Analysis,* Wiley, New York, 1955.

[10] S. S. Kuo, *Numerical Methods and Computers,* Addison-Wesley, Reading, Mass., 1965.

[11] U.S. Department of Commerce, National Bureau of Standards, *Handbook of Mathematical Functions,* Applied Mathematics Series, no. 55, U.S. Government Printing Office, Washington, D.C., May 1968.

[12] G. Hall and J. Watt, *Modern Numerical Methods for Ordinary Differential Equations,* Clarendon Press, Oxford, 1976.

[13] R. Bellman and R. Kalaba, *Modern Analytic and Computational Methods in Science and Mathematics,* American Elsevier, New York, 1967.

[14] C. Runge, "Über die numerische Auflösung totaler Differentialgleichungen," *Mathematical Annals,* v. 46, 1895, pp. 167–178.

[15] L. Kelly, *Handbook of Numerical Methods and Applications,* Addison–Wesley, Reading, Mass., 1967.

[16] B. Carnahan, H. Luther, and J. Wilkes, Applied Numerical Methods, Wiley, New York, 1969.

[17] R. Schwarz and B. Friedlaend, *Linear Systems,* McGraw–Hill, New York, 1965.

[18] R. W. Brockett, "The Status of Stability Theory for Deterministic Systems," *IEEE Trans. Automatic Control,* v. AC-11, July 1966, pp. 546–606.

[19] N. Rao and H. Rao, "Phase Plane Technique for the Solution of Transient Stability Problems," *Proc. IEE,* v. 10, 1963, pp. 1451–1461.

[20] O. Hanson, C. Goodwin, and P. Dandeno, "Influence of Excitation and Speed Control Parameters in Stabilizing Signal and Computer Program Verification," IEEE Trans. Power Apparatus and Systems, v. PAS-87, February 1968, pp. 315–322.

[21] H. Moussa and Y. Yu, "Dynamic Interaction of Multi-machine Power Systems and Excitation Control," *IEEE Trans. Power Apparatus and Systems,* v. PAS-93, July–August 1974, pp. 1150-1158.

[22] G. T. Heydt and R. C. Burchett, "Probabilistic Methods for Power System Dynamic Stability Studies," *IEEE Trans. Power Apparatus and Systems,* v. PAS-97, no. 3, May–June 1978, pp. 695–702.

[23] R. Alden and H. El-Din, "Multi-machine Dynamic Stability," *IEEE Trans. on Power Apparatus and Systems,* v. PAS-95, September 1976, pp. 1529–1534.

[24] A. Aguilar, "Power System Dynamics Under Small Parturbations," Purdue University Technical Report TR-EE 74-48, West Lafayette, Ind., December 1974.

[25] M. O. Mansour, "Hierarchical Control of Interconnected Systems," Purdue University Technical Report TR-EE 75-34, Purdue University, West Lafayette, Ind., 1975.

[26] F. Keay and W. South, "Design of a Power System Stabilizer Sensing Frequency Deviation," *IEEE Trans. on Power Apparatus and Systems,* v. PAS-90, March–April 1971, pp. 707–713.

[27] M. Liapunov, "Le Problème général de la stabilité du mouvement," *Annales de la Faculté de Science de Toulouse,* v. 9, 1907, pp. 203–474. The original was published in Russian in 1892. An English translation appears in *Annals of Mathematical Studies,* no. 17, Princeton University Press, Princeton, N.J., 1947.

[28] J. LaSalle and S. Lefschetz, *"Stability by Liapunov's Direct Method with Applications,"* Academic Press, New York, 1961.

[29] A. I. Luré, *Some Nonlinear Problems in the Theory of Automatic Control* (original in Russian), Her Majesty's Stationery Office, London, 1957.

[30] A. Letov, *Stability in Nonlinear Control Systems,* translated by J. Adashko, Princeton University Press, Princeton, N.J., 1961.

[31] W. Hahn, *Theory and Application of Liapunov's Direct Method,* translated by H. Hosenthien, Prentice-Hall, Englewood Cliffs, N.J., 1963.

[32] J. Willems, "Improved Liapunov Function for Transient Power System Stability," *PIEE,* v. 115, no. 9, Sept. 1968, pp. 1315–1317.

[33] M. Ribbens–Pavella and B. Lemal, "Fast Determination of Stability Regions for On Line Transient Power Systems Studies," *Proc. IEE,* v. 123, no. 7, July 1976, pp. 689–696.

[34] T. Athay, R. Podmore, and S. Virmani, "A Practical Method for the Direct Analyses of Transient Stability," *IEEE Trans. on Power Apparatus and Systems,* v. PAS-98, no. 2, March–April 1979, pp. 573–581.

[35] A. A. Fouad, "Stability Theory–Criteria for Transient Stability," Proc., Conference on System Engineering for Power, Henniker, N.H., 1975, pp. 421–450. These proceedings also contain numerous articles and bibliographies on power system stability.

[36] P. M. Anderson, and A. A. Fouad, *Power System Control and Stability,*" Iowa State University Press, Ames, 1977.

[37] K. Fukunaga, *Introduction to Statistical Pattern Recognition,* Academic Press, New York, 1972.

[38] A. H. El-Abiad, and K. Nagappan, "Transient Stability Regions of Multimachine Power Systems," *IEEE Trans. Power Apparatus and Systems,* v. PAS-85, February 1966, pp. 169–179.

[39] P. Aylett, "The Energy–Integral Criterion of Transient Stability Limits of Power Systems," *PIEE,* v. 105(C), 1958, pp. 527–536.

[40] M. A. Pai, Power System Stability Analysis by the Direct Method of Liapunov, North Holland, Amsterdam, 1981.

[41] J. H. Chow, *Time Scale Modeling of Dynamic Networks with Applications to* Power Systems, Springer-Verlag, New York, 1982.

[42] V. Venikov, *Transient Processes in Electric Power Systems,* Government Printing Office (Mir), Moscow, 1977.

[43] H. Hakkimmashhadi, "Fast Transient Security Assessment," Ph.D. thesis, Purdue University, West Lafayette, Ind., August 1982.

[44] H. Hakkimmashhadi and G. T. Heydt, "Fast Transient Security Assessment," *IEEE Trans. Power Apparatus and Systems,* v. PAS-102, no. 12, December, 1983, pp. 3816–3824.

[45] T. Sakaguchi and K. Matsumoto, "Development of a Knowledge Based System for Power System Restoration," *IEEE Trans. Power Apparatus and Systems,* v. PAS-102, February 1983, pp. 320–329.

[46] C. L. Gupta and A. H. El-Abiad, "Transient Security Assessment of Power Systems by Pattern Recognition," *1977 IEEE Summer Power Meeting,* San Francisco, July 1977.

[47] E. Kimbark, *Power System Stability: Synchronous Machines,* Dover, New York, 1956.

[48] W. D. Humpage, *Z-Transform Electromagnetic Transient Analysis in High Voltage Networks,* Peregrinus, London, UK, 1982.

Exercises

Exercises that do not require computer programming

8.1 In this exercise, the simple scalar differential equation

$$\dot{x} = -2x \qquad x(0) = 1$$

will be used to illustrate a few predictor and predictor–corrector methods. The exact solution is

$$x(t) = e^{-2t} .$$

a. Euler's method. Noting that $x(0) = 1$, write an expression for $x(h)$ in terms of only h. Use Euler's method to obtain this relation. Use $x(h)$ to obtain $x(2h)$ in terms of only h. Continue the process and find $x(3h)$, $x(4h)$, ... By inspection of these points, deduce the general form of $x(kh)$ in terms of only k (hint: examine Pascal's triangle in row $k - 1$.) Having found $x(kh)$, it is possible to find the error in using numerical integration, ϵ,

$$\epsilon_k = e^{-2kh} - x(kh) .$$

It would be nice to know the largest h permissible yet holding

$$\epsilon_k < \bar{\epsilon}$$

where $\bar{\epsilon}$ is the largest tolerable error. This is not easy to do since the expression for ϵ_k is transcendental. Using appropriate approximations, obtain an estimate of the largest value of h for a given $\bar{\epsilon}$. Does your expression hold for all k?

b. Modified Euler method. For the same differential equation, find an expression for $x(kh)$ and compare the result numerically to result of part (a).

c. Solve the given differential equation using the second order Runge–Kutta method, $h = 0.01$. Compare the first few steps to the exact solution.

8.2 The fifth order Adams method is derived by fitting a polynomial in $(t - t_n)$ to $X(t)$. Derive this predictor formula and state whether the method is self-starting.

8.3 In this exercise, the Adams predictor is investigated in some detail. One way to write the general form of this predictor is

$$X_{n+1} = X_n + h(1 + \sum_{i=1}^{q} \alpha_i \nabla^i)F_n$$

$$\alpha_{\ell+1} = \int_0^1 \frac{\lambda(\lambda+1)(\lambda+2)(\cdots)(\lambda+\ell)}{(\ell+1)!} d\lambda,$$

where ℓ goes from zero to $q - 1$ inclusive and the notation ∇^i denotes the ith backward difference

$$\nabla^1 F_n = F_n - F_{n-1}$$

$$\nabla^2 F_n = F_n - 2F_{n-1} + F_{n-2}$$

$$\nabla^3 F_n = F_n - 3F_{n-1} + 3F_{n-2} - F_{n-3}$$

and so on. The coefficients of F_{n-i} in $\nabla^q F_n$ are taken from Pascal's triangle.

a. Verify the formulas (8.21) and (8.22) using truncation $q = 1$ and $q = 3$, respectively

b. Show that the error is of order ϵ,

$$\epsilon = \alpha_{q+1} h^{q+2} X^{(q+2)},$$

where the superscript on X denotes the order of the derivative of X.

c. Verify that for the Adams predictor formula at truncation q, the sum of the coefficients of F_n, F_{n-1}, F_{n-2}, ..., is h.

8.4 Using the notation introduced in Exercise 8.3, the Adams–Bashforth predictor corrector formulas are

$$P: X_{n+1} = X_n + h(1 + \frac{1}{2}\nabla + \frac{5}{12}\nabla^2 + \frac{3}{8}\nabla^3 + \cdots)F_n$$

$$C: X_{n+1} = X_n + h(1 - \frac{1}{2}\nabla - \frac{1}{12}\nabla^2 - \frac{1}{24}\nabla^3 - \cdots)F_{n+1}.$$

The coefficients of hF_n in the predictor are α_i given in Exercise 8.3. The coefficients of hF_{n+1} in the corrector are β_i.

a. Find an expression for β_i in terms of only i. Note that $\beta_4 = -19/720$.

b. Write three representative Adams–Bashforth predictor corrector algorithms (i.e., use different truncation points).

c. When this predictor corrector algorithm is truncated at the ∇^1 term, a simple numerical integration algorithm is obtained. Show that the error in this case is

$$\epsilon = -\frac{h^3}{12}X^{(3)},$$

where the superscript on X denotes the third derivative.

Exercises that require computer programming

8.5 Using Euler's predictor, solve the following equations numerically for $0 \le t \le 5$ seconds. (Note: you will have to estimate the step size in each case and examine your solution to determine whether the choice was appropriate.)

a. $\dot{x} = -x + e^{-2t}$, $x(0) = 1.0$

b. $\ddot{x} + 2\dot{x} + x = 2e^{-t} - e^{-5t}$, $x(0) = 1.0$, $\dot{x}(0) = 0.0$.

c. $$\dot{X} = \begin{bmatrix} 0 & 1 & 0 \\ 0 & 0 & 1 \\ 1 + 0.01\sin t & 1 + 0.01\cos t & 1 \end{bmatrix} X + \begin{bmatrix} 0 \\ 0 \\ 1 \end{bmatrix} u$$

$$u(t) = e^{-t}$$

$$X^t(0) = [1.0\ 0.0\ 0.0].$$

8.6 In this problem, the solution of the swing equation

$$\frac{2H}{\omega_o}\ddot{\delta} = T_m - T_e$$

will be performed using the fourth order Runge–Kutta method. For this purpose, the synchronous speed will be 120π rad/s and the inertia constant such that the kinetic energy of the rotor at synchronous speed is 10000 J. At $t = 0$, the machine is in sinusoidal steady state generating 100 MW. The electrical torque is of the form

$$T_e = K_1 \sin\delta + K_2 \sin 2\delta$$

and the reluctance torque term is 10% of the synchronous torque term.

a. Solve the swing equation if the mechanical (driving) torque is suddenly removed at $t = 0$ and replaced at $t = 1$ s. Neglect triction. Use the Runge–Kutta method.

b. Solve the swing equation if T_e is zero for $0 \le t \le 1$ s and T_e is restored at 1 s. Is synchronism retained.

c. Find the maximum time for which a bus fault at the machine terminals may be tolerated yet retain synchronism.

8.7 Compare the fourth order Adams predictor to the Euler predictor for the numerical solution of

$$\ddot{\delta} = -0.01\dot{\delta} - \delta + F$$

for the conditions

$$\begin{array}{llll} \delta(0^-) & = 1 \text{ radian} & \delta(0^+) & = \delta(0^+) \\ \dot{\delta}(0^-) & = 0 & \dot{\delta}(0^+) & = \dot{\delta}(0^-) \\ F(0^-) & = 1 \text{ radian} & F(0^+) & = 0 \end{array}$$

for the time interval $0 \le t \le 5$ s. Explain the discrepancy between the two numerical methods for the step size selected.

8.8 Repeat Exercise 8.7 for

$$\ddot{\delta} = -0.01\dot{\delta} - 2 \sin \delta + 1$$

$$\begin{array}{llll} \delta(0^-) & = \frac{\pi}{6} \text{ radian} & \delta(0^+) & = \delta(0^+) \\ \dot{\delta}(0^-) & = 0 & \dot{\delta}(0^+) & = \dot{\delta}(0^-) \\ F(0^-) & = 1 \text{ radian} & F(0^+) & = 0. \end{array}$$

8.9 In this problem, the seventh order Milne predictor corrector formula is examined for the solution of

$$\ddot{\delta} = -0.01\dot{\delta} - 1.55 \sin \delta - 0.01 \sin 2\delta,$$

where $\delta(0)$ is such that $\dot{\delta}(0)$ and $\ddot{\delta}(0)$ are both zero.

a. Find $\delta(0)$.

b. Solve for $\delta(t)$ for $0 \le t \le 5$ s using $h = 0.001$ s. Print $\delta(t)$ in steps of 20 ms. Use the seventh order Milne predictor corrector.

c. In this part, the corrector formula will be used to estimate the accuracy. This is done by defining

$$\epsilon_x = || x^p_{n+1} - x^c_{n+1} || \,.$$

If ϵ_x is less than 0.01, the step size will be taken as adequate; if ϵ_x is greater than 0.01, it will be assumed that h must be increased. Write a computer code to perform this adaptive step size. Note that every time the step is adapted (changed), the non self starting formulas must be restarted. You may wish to use a fairly simple starter for this purpose, although a Runge–Kutta starter will yield more meaningful results

d. In part (c), ϵ_z was used to decrease h as required. Develop an algorithm to increase and decrease h depending on ϵ_z. Program and test your method. Comment on execution time.

8.10 Use the Adams–Bashforth method to solve the swing equation given in Exercise 8.9. Note that the truncation in the Taylor series used to obtain this predictor corrector is $0(h^5)$.

8.11 In this exercise, a coupled multimachine system is studied. The differential equations (in per unit and radian measure) are

$$\ddot{\delta}_1 = -0.01\dot{\delta}_1 - 2.00 \sin \delta_1 - 0.02 \sin 2\delta_1 - 0.05 \sin \delta_2 + F_1$$

$$\ddot{\delta}_2 = -0.03\dot{\delta}_2 - 1.91 \sin \delta_2 - 0.01 \sin 2\delta_2 - 0.05 \sin \delta_1 + F_2$$

and the system is in sinusoidal steady state at $t = 0$, where mechanical power terms F_1 and F_2 are 1.042 and 0.985, respectively.

a. Find the steady state ($t = 0$) operating conditions.

b. A fault occurs for T_{f1} seconds at machine 1. The $\sin \delta_1$ and $\sin 2\delta_1$ terms in both the first and second equations are replaced by zeros for

$$0 \leq t \leq T_{f1} .$$

Also, the $\sin \delta_2$ term in the $\ddot{\delta}$ expression is zero in this interval. After $t = T_{f1}$, the equations revert to the form shown (i.e., the fault clears). Solve this simultaneous set for representative values of T_{f1} and find the maximum value of T_{f1} to retain synchronism.

c. Repeat with a fault at machine 2.

CHAPTER 9

Stochastic Methods in Power System Analysis

9.1

RANDOM VARIABLES AND STOCHASTIC PROCESSES

In preceding chapters, the focus of attention has been primarily in two areas: steady state analysis (e.g., power flow studies, fault studies) and transient analysis (e.g., transient stability studies). The principal difference in these areas is the way time is treated in the models. In all cases, however, the variables and parameters have been deterministic. Actual power systems exhibit numerous parameters and phenomena which are either nondeterministic or so complex and dependent on so many diverse processes that they may readily be regarded as nondeterministic. Power system loads, for example, are nondeterministic because the active and reactive power levels are functions of many loads which are activated and deactivated depending on time parameters (hour, day of the week, season, etc.), weather (temperature, humidity), socio-economic factors, and numerous other factors. Also, there is an arbitrary character to some loads. Taken in unison, these factors result in load levels which are stochastic in nature. Sampled values or measurements of such a stochastic process are random variables. Examples of stochastic processes and random variables in power engineering are the system demand at a specified time, forced outages of transmission and generation components, sparkover characteristics of high voltage dielectrics, load growth, and fuel availability and cost. This chapter contains an introduction to stochastic methods in power system analysis. Since many of these methods are rather new, their application is not widespread. References document more detailed discussions of these methods. This chapter is necessarily simply an introduction to stochastic methodologies in power engineering since the topic is extensive in theory and scope. Whole books have been written on this subject [7,8]. In keeping with the emphasis in this book, only discussions of computer analysis techniques are presented here. References [1–3] contain the fundamentals of the theory of probability and stochastic processes, and [4] contains an exhaustive description of the Gaussian probability distribution which is often used to model power engineering phenomena.

A few fundamental relations are indicated here to establish the notation used in this chapter. The probability density function of random variable x, $f_x(x)$, is related to its probability distribution function, $F_x(x)$, by

$$F_x(x) = \int_{-\infty}^{x} f_x(\lambda)\, d\lambda \,. \tag{9.1}$$

The expectation of x is given by

$$E(x) = \int_{-\infty}^{\infty} x f_x(x)\, dx \tag{9.2}$$

and $E(x)$ is also termed the *mean* or *first moment* of x,

$$\mu_x = E(x) \,.$$

The kth central moment of x is $\mu_x^{(k)}$

$$\mu_x^{(k)} = E((x - \mu_x)^k) = \int_{-\infty}^{\infty} (x - \mu_x)^k f_x(x)\, dx \,, \tag{9.3}$$

and the kth raw moment of x is $m_x^{(k)}$

$$m_x^{(k)} = E(x^k) = \int_{-\infty}^{\infty} x^k f_x(x)\, dx \,. \tag{9.4}$$

The term "variance of x" refers to the second central moment of x. The random variable is said to be of order k if all its raw moments up to (and including) the kth exist. Also, if x is a scaler, the relations used and the variable itself are termed univariate.

If X is an n-vector, the several univariate entries form a multivariate or n-variate random vector. Bivariate and trivariate vectors are the $n = 2$ and $n = 3$ cases, respectively. The probability density function of vector X is denoted $f_X(X)$ and is a scalar valued function of vector valued argument. The density is related to the distribution of X by

$$F_X(X) = \int_{-\infty}^{x_1} \int_{-\infty}^{x_2} \cdots \int_{-\infty}^{x_n} f_X(\Lambda) d\lambda_1 d\lambda_2 \cdots d\lambda_n \tag{9.5}$$

where F_X(X) is a scalar valued function of a vector valued argument, x_1 through x_n are the n components of vector X, Λ is a dummy n-vector of integration, and λ_1 through λ_n are components of Λ. Equation (9.5) is written in condensed notation as

$$F_X(X) = \int_{-\infty}^{x} f_X(\Lambda)\, d\Lambda \tag{9.6}$$

where the notation $d\Lambda$ is a product of the n differentials indicated in (9.5) and n-dimensional integration is required. The expectation of vector X is termed the mean,

$$\begin{aligned} \mu_X &= E(X) = \int_{-\infty}^{\infty} \int_{-\infty}^{\infty} \cdots \int_{-\infty}^{\infty} X f_X(X)\, dx_1,\, dx_2 \cdots dx_n \\ &= \int_{-\infty}^{\infty} X f_X(X)\, dX \,. \end{aligned} \tag{9.7}$$

The mean, μ_X, is an n-vector. Equation (9.7) is readily shown to be

$$\mu_X = \begin{bmatrix} \mu_{x1} \\ \mu_{x2} \\ \vdots \\ \mu_{xn} \end{bmatrix} .$$

The analog of the variance in univariate theory is the covariance matrix in multivariate theory,

$$\sum_X = \int_{-\infty}^{\infty} \int_{-\infty}^{\infty} (X - \mu_X)(X - \mu_X)^t f_X(X) dx_1, dx_2 \cdots dx_n \; . \quad (9.8)$$

The notation $\sum_X$ refers to the covariance matrix of X and, hopefully, will not be confused with summation. Equation (9.8) may be rewritten as

$$\sum_X = E((X - \mu_X)(X - \mu_X)^t) \; . \quad (9.9)$$

Note that (9.8) and (9.9) apply for real X; in the case of complex X, the transposition in (9.8) and (9.9) becomes the Hermitian operation.

Higher (k>2) central moments of multivariate X are tensors of order k. Thus the third central moment of X is a three dimensional array of parameters each of whose entries is $\sigma_{x_i x_j x_k}$,

$$\sigma_{x_i x_j x_k} = E((x_i - \mu_{x_i})(x_j - \mu_{x_j})(x_k - \mu_{x_k})). \quad (9.10)$$

Raw moments of X are readily defined without resorting to subtraction of the mean.

The characteristic function of univariate x is the Fourier transform of its density function

$$\phi_x(\omega) = \mathcal{F}(f_x(x))$$
$$= \int_{-\infty}^{\infty} f_x(x) e^{-j\omega x} \, dx \; .$$

The natural logarithm of the characteristic function is the second characteristic function,

$$\psi_x(\omega) = \ell n \; (\phi_x(\omega)) \; . \quad (9.11)$$

In the n-variate case, the characteristic function is the n-dimensional Fourier transform of the scalar $f_X(X)$,

$$\phi_X(\Omega) = \mathcal{F}(f_X(X))$$
$$= \int_{-\infty}^{\infty} \int_{-\infty}^{\infty} \cdots \int_{-\infty}^{\infty} f_X(X) \; exp \; (-jX^t \Omega) \; d\omega_1, \, d\omega_2 \cdots d\omega_n$$

or, more compactly,

$$\phi_X(\Omega) = \int_{-\infty}^{\infty} f_X(X) \; exp \; (-jX^t \Omega) \; d\Omega \; . \quad (9.12)$$

Note that $\phi_X(\Omega)$ is a scalar valued function of a vector valued argument. The second characteristic function of X is the natural logarithm of ϕ_X,

$$\psi_X(\Omega) = ln(\phi_X(\Omega)) \; . \quad (9.13)$$

Finally, note that when the central moments of univariate x exist, the integrand in (9.11) may be expanded in a Taylor series about $x = 0$ to yield the following after term-by-term integration:

$$\phi_x(\omega) = \sum_{k=0}^{\infty} \frac{-1}{k!} m_x^{(k)} \omega^k \; . \quad (9.14)$$

This is the Taylor series for $\phi_x(\omega)$ around $\omega = 0$. Thus the coefficients of the Taylor series for ϕ_x are proportional to the raw moments of x. When x has zero mean and unit variance, the variate is termed "standard measure," and for standard measure variables, the Taylor series of ϕ_x has coefficients proportional to the central moments of x. The multivariate analog of (9.14) is left as an exercise. The Taylor series of the second characteristic function of univariate x, $\psi_x(\omega)$, about $\omega = 0$ is

$$\psi_x(\omega) = \sum_{k=0}^{\infty} \frac{(-1)^k}{k!} \tau_x^{(k)} \omega^k \tag{9.15}$$

where $\tau_x^{(k)}$ is the kth cumulant of standard measure x. References [1,5,6] give an extensive discussion of the second characteristic function and cumulants.

9.2 LOAD FORECASTING

Perhaps the most extensive application of probability theory in power engineering lies in the area of load forecasting. This generic term applies to a wide range of methods, including:

1. *Long term forecasting:* the forecasting of systems load and/or bulk power interchange levels for periods years into the future. Usually, peak loads are forecast.

2. *Medium term forecasting:* the forecasting of system and/or bulk power intercharge levels six months to one year into the future. Usually, peak loads are forecast, but occasionally off-peak values are also calculated.

3. *Short term forecasting:* refers to much less than one-year-ahead forecasts, and occasionally to one-day-ahead forecasts. The short term forecasts differ from their longer term counterparts in that peak forecasts assume less importance and forecasts at specific times (e.g., hour-by-hour forecasts) assume greater importance.

Each of these methods is described in this section. Load forecasting, particularly short term forecasting, contains numerous heuristic techniques which are frequently system dependent. As an example, short term forecasting in the northern parts of North America are highly weather dependent. Daily temperature has a significant impact on peak demand since many residences use electrical energy for heating and cooling. On the other hand, the medium term forecasting of demand in a steel producing region is highly dependent on economic factors.

The decomposition of the load into *residential, commercial,* and *industrial* will help decouple some of the load variation phenomena cited. For example, residential demand in northern temperate regions is highly sensitive to temperature; industrial loads are sensitive to economic factors; commercial demand is sensitive to day-of-the-week and the occurrence of holidays.

The remainder of this section is devoted to the details of a selection of load forecasting techniques. A great deal of further detail is available from Sullivan [34] and, for time series methods, Box and Jenkins [20].

Reference [21] contains a compact summary of short term forecasting methods, and [22,23] is a representative sampling of the literature on medium and long term forecasting.

Long Term Peak Forecasting

Long term peak forecasting must incorporate virtually all the physical and socio-economic phenomena which influence electrical demand. It is rather evident that long term forecasting cannot model weather phenomena except in the broadest of trends; previous peak data are extrapolated to obtain the forecast. Consider peak demand $d(t)$ which is known for several time intervals, for example, the past n_s years. For convenience, let $t = 0$ denote the present and negative t denote the past [for which $d(t)$ are known]. The forecast peak is $\hat{d}(t)$,

$$\hat{d}(t) = A^t f(t) , \tag{9.16}$$

where A is an n_f vector of constants and $f(t)$ is an n_f vector of fitting functions. For example, if straight line fits to peak data are used, $f(t)$ will contain a straight line [i.e., $row_1(f(t))$ will be simply t]; if exponential trends are expected, a row of $f(t)$ will contain a suitable exponential. Now consider the n_s samples of known peak data to be arranged in the n_s vector D. One obtains the forecast $\hat{D}$ using

$$\hat{D} = FA, \tag{9.17}$$

where

$$F = \begin{bmatrix} f^t(t_1) \\ f^t(t_2) \\ \vdots \\ f^t(t_{n_s}) \end{bmatrix} .$$

Equation (9.17) contains n_s rows and there are n_f unknown elements of A. Thus the mean square fit is obtained using the Moore–Penrose pseudoinverse of matrix F [18] (see Appendix B):

$$A = F^+ D . \tag{9.18}$$

Equation (9.18) is equivalent to minimizing the mean square error, ϵ, between $d(t)$ and $\hat{d}(t)$ over the n_s samples,

$$\epsilon = \frac{1}{n_s} \sum_{i=1}^{n_s} (d(t_i) - \hat{d}(t_i))^2$$

$$= \frac{1}{n_s}(D - \hat{D})^t (D - \hat{D}) . \tag{9.19}$$

Equation (9.18) should be used when all data in D are assumed to influence the selection of A equally. In long term forecasting in which the forecast is much longer than a few years, it is appropriate to consider *discounting* the earliest (i.e., most remote) elements of D by incorporating weighting factors. For example, if the contribution to the mean square error of the $t = t_i$ term is to be $w_i^2(d(t_i) - \hat{d}(t_i))^2$, the total mean square error is ϵ_w,

$$\epsilon_w = \frac{1}{n_s}(D - \hat{D})^t W^t W(D - \hat{D}), \tag{9.20}$$

where W is the diagonal n_s by n_s matrix of weights or discounting factors,

$$W = diag\ (w_i)\ .$$

Then (9.20) is readily rewritten in terms of WD and $W\hat{D}$ instead of D and $\hat{D}$. This reveals that

$$A = (WF)^+ WD\ . \tag{9.21}$$

Use (9.18) when discounting is not required, and use (9.21) when discounting is required. A typical form of W is

$$W = diag\ (\sqrt{\beta^{k-1}}) \qquad k = 1, 2, \ldots, n_s,$$

where the upper rows of D are the most recent (least discounted) data and β is termed the *discount factor.* Suitable discount factors lie in the range $0.25 \leq \beta \leq 0.90$.

The general problem of assessing the expected accuracy of the forecast is very difficult since the statistics of $d(t)$ for $t > 0$ are difficult to estimate. If the peak demand in the period before the forecast of the form

$$d(t) = A^t f(t) + \xi(t),$$

where Ξ is the vector of $\xi(t_1)$, $\xi(t_2)$, ..., $\xi(t_{n_s})$ terms, the expected value of ϵ is

$$E(\epsilon) = \frac{1}{n_s} E(\Xi^t \Xi)$$

$$= \frac{1}{n_s} \sigma_\xi^2, \tag{9.22}$$

where σ_ξ^2 denotes the variance of ξ. If the ξ is stationary, the statistics of ϵ will not change as t progresses from the sampling interval, $t \leq 0$, to the forecast interval, $t > 0$. Therefore, for nondiscounted applications, the expected mean square error is proportional to the variance of ξ and inversely proportional to n_s.

Fitting Functions

Suitable fitting functions for peak load forecasting include straight lines, quadratic forms, and exponentials. Special trends may be approximated by higher order polynomials, in which cases rows of vector $f(t)$ simply contain t^i for rows $i = 1, 2, 3, \ldots$. If too high an order of t is used, numerical instability may result; this instability is high sensitivity of elements of A with respect to non-discounted samples of $d(t)$. The use of a simple time fitting function presupposes that weather, socio-demographics, and economic conditions do not appear in the forecast explicitly. For long term forecasts, this is a reasonable approach.

Weather Sensitivity of Forecasts

The weather sensitivity of peak load, off peak load, and hourly load forecasts are primarily a function of temperature, humidity, and wind speed. In addition, gross weather phenomena may significantly influence the load; for example, a heavy snowfall or rainfall may cause commercial loads to drop. Most methods of inclusion of weather phenomena do not consider such gross phenomena; they are best considered by visual inspection of weather radar and (in the United States) National Weather Service reports. The cited weather parameters are usually included in peak load forecasts as piecewise linear functions,

$$P_{peak} = k_\theta(\theta)\theta + k_h(h)h + k_{ws}(ws)ws, \tag{9.23}$$

where P_{peak} denotes the system daily peak, and θ, h, and ws denote the forecast temperature peak (in summer), humidity, and midday wind speed, respectively. In the winter, θ may be the forecast low temperature. The constants in (9.23) are obtained from tables which are system and region dependent. For summer peak forecasts in northern parts of the United States, k_θ typically lies in the range 0.5 to 1.5% of system load per degree (Farenheit) for θ near 90 °F.

It is possible to isolate weather phenomena from other portions of the forecast in simply an additive way. Thus, (9.17) is used for the part of the load that is not weather sensitive and (9.23) is used for the weather sensitive part. By such a dichotomy, historical data are used in (9.18) by simple subtraction

$$A = F^+(D - k_\theta(\theta)\theta - k_h(h)h - k_{ws}(ws)ws).$$

In this formula, the weather sensitive terms are vectors of dimension commensurate with D. In this dichotomy, it is assumed that the weather data are orthogonal. If weather sensitivity is subtracted from D, it must be added to $\hat{D}$ in (9.17)

$$\hat{D} = FA + k_\theta(\theta)\theta + k_h(h)h + k_{ws}(ws)ws.$$

Seasonal Forecasts

Seasonal forecasts are typical intermediate term forecasts and are used primarily to estimate peak system loads. Sometimes peak bus demands at principal load centers are forecast. The principal method used is the time series

$$d_{peak}(t) = \sum_{i=1}^{n} \alpha_i d_{peak}(t-i) + \sum_{j=1}^{m} \gamma_j u(t-j) + \beta(t), \tag{9.24}$$

where d_{peak} is the peak demand on day t, u represents weather sensitive data, and β models long term trends. The usual simplification is to delete the u and β terms, thus accommodating weather and long term phenomena as stochastic elements of d. The coefficients α_i are termed *coefficients of the population model.* In many applications, the daily peak d_{peak} is *not* used, but a *deviation* from the previous annual peak, Δd_{peak}, is used. In either case,

$$d_{peak}(t) \simeq \sum_{i=1}^{n} \alpha_i d_{peak}(t-i).$$

The α_i may be identified by a mean square error fit of historical data; the method using pseudoinverses has been described above. Alternatively, a probability density function or elements of a stochastic model for $d_{peak}(t)$ may be assumed. Useful in these regards are the following concepts and parameters.

Autocovariance of d_{peak}

$$\gamma_j = autocovariance\ of\ d_{peak}(t)$$

$$= E\{(d_{peak}(t) - u)(d_{peak}(t-j) - u)\}$$

$$\mu = E\{d_{peak}(t)\}$$

Autocorrelation coefficient of d_{peak}

$$\rho_i = autocorrelation\ coefficient\ of\ d_{peak}(t)$$

$$= \frac{\gamma_j}{\gamma_o}$$

Estimate of ρ_j

$$\hat{\rho}_j = estimate\ of\ \rho_j$$

$$= \hat{c}_j(d_t)/\hat{c}_o(d_t)$$

$$\hat{c}_j(d_t) = \frac{1}{n}\sum_{t=1}^{n-j}(d_t - E(d_t))(d_{t+j} - E(d_t))$$

$$E(d_t) = \frac{1}{n}\sum_{t=1}^{n} d_t$$

$$d_t = d(t)$$

Typical errors for seasonal forecasts lie in the range 2 to 4%. Box and Jenkins [20] have written a definative book on the topic, and Uri [23] has prepared a useful condensed summary of the subject.

Short Term Forecasting

The terminology *short term* in this context implies forecasts which are minutes to days ahead. Usually, the forecast value is the total system load, although energy forecasts are also made. In some instances, individual bus loads at bulk transmission points are forecast. On the short end of the forecast time range cited, one has applications of economic generation dispatch, while on the long end, unit commitment is the primary application. With regard to the latter, generators are *committed* or scheduled for use with the load accommodated and spinning reserve allocated to comply with system reliability requirements.

Short term forecasting of the one-day-ahead class is perhaps the most heuristic of all the forecast methods discussed thus far – the system operator usually is a key element in this process. The operator essentially extrapolates previous load versus time characteristics. These remarks are less valid as the forecast is made at shorter time periods into the future. For example, the operator usually cannot be occupied with hour ahead forecasts.

The two most often used mathematical methods of short term forecasting are the time series model and the autoregressive moving average (ARMA) technique. The time series model is of the form

$$d(t) = \sum_{i=1}^{n} \alpha_i f_i(t) + \beta(t) \,. \tag{9.25}$$

In (9.25), $\beta(t)$ represents the longer term phenomena, such as seasonal or annual trends. The functions $f_1, f_2, \ldots, f_n$ are usually sinusoids of the form

$$f_i(t) = sin\ (\omega_i t + \gamma_i)$$

and the ω_i are chosen to represent the physical process under consideration. Equation (9.25) is essentially a truncated Fourier series representation of $d(t)$. For week-ahead forecasts, the ω_i are multiples of $2\pi/(168\times3600)$ rad/s. (Note: there are 168 hours in a week.) For day ahead forecasts, the ω_i are multiples of $2\pi/(24\times3600)$ rad/s. The period of f_i chosen in this way is T_i,

$$T_i = \frac{2\pi}{\omega_i} \,.$$

When the autocorrelation of the quantity $\varsigma(t)$,

$$\varsigma(t) = d(t) - \beta(t)$$

$$= \sum_{i=1}^{n} \alpha_i f_i(t) \tag{9.26}$$

is maximized, (9.26) is termed a Karhunen–Loeve expansion. It is possible to select other forms of $f_i(t)$ in (9.25) to produce desired short term effects (e.g., polynomials in t, exponentials in t).

The auto regressive moving average model is a dynamic or adaptive model in that model parameters adapt or are updated as the input data characteristics change. The ARMA method may be viewed as a dynamic time series,

$$d(t) = \sum_{i=1}^{n} \alpha_i d(t-i) + \sum_{j=1}^{m} \gamma_j u(t-j) + \beta(t) \,. \tag{9.27}$$

In (9.27), the sum over i represents the autocorrelated effects, and the sum over j is used to model short term weather effects. An expression such as (9.23) may be used for u

$$u(t) = k_\theta \theta + k_h h + k_{ws}(ws) \,.$$

The function $\beta(t)$, as before, represents long term trends.

In both the time series and ARMA models, initialization is a problem to be considered. Usually, historical data or other off line methods are used to initialize model parameters. Identification of model parameters is a mean square error approach assuming the probability density function of $d(t)$ (e.g., Gaussian, uniform). The usual assumption of short term stationarity of $d(t)$ is made.

An excellent survey of short term forecasting methods is given by Galiana [19] and specialized aspects of short term techniques are considered in [20–22].

Forecasting Accuracy

Load forecasting accuracy is highly system, forecast horizon, and event specific. If the forecast error is $\epsilon(t)$,

$$\epsilon(t) = | d(t) - \hat{d}(t) | ,$$

where d and $\hat{d}$ are the demand and forecast demand, one should assume that the mean of $\epsilon(t)$ is approximately zero. The variance of $\epsilon(t)$,

$$\sigma^2 = E(\epsilon^2(t)),$$

is a reasonable measure of the error, but not necessarily definitive of the bounds of $\epsilon(t)$. If a $\pm 5\sigma$ rule is used, a reasonable estimate of the error bound is attained with probability Pr,

$$Pr = \int_{-5\sigma}^{+5\sigma} f_{\epsilon}(\lambda)\, d\lambda ,$$

which, for Gaussian $f_{\epsilon}(\epsilon)$ is approximately 0.999.

From a practical point of view, $\epsilon(t)$ is usually less critical at off peak periods compared to peak periods. When weather sensitive models are used, the accuracy of weather forecasts influences demand forecast accuracy. Weather forecasts are usually issued for 12 hour periods (twice early in the day and once in early evening). Late in the 12 hour forecast interval, there is a degradation of accuracy. If certain criteria are met which result in a specially issued weather forecast (at other than the regular time, usually due to a considerable change in the forecast), the accuracy late in the forecast period is improved. Longer term forecasts, known as *area forecasts,* are also issued for the day ahead (and longer). Accuracy attainable by commercially available short term load forecasting programs is in the range 2 to 5%. Galiana [19] reports that cases of

$$\sigma \leq 0.02$$

with probability 0.5 (or more) is considered "good" performance, while

$$0.02 < \sigma \leq 0.05$$

with probability 0.5 (or more) is considered "fair to poor."

9.3 STOCHASTIC POWER FLOW STUDIES

The power flow problem as presented in Chapters 4 and 5 is a sinusoidal steady state problem with deterministic input data. Actual power studies contain uncertainty in several areas:

1. *Line data.* Transmission line data are usually estimated at their 50 °C values, and often circuit geometry is used to calculate self and mutual impedances. The ambient temperature is a random process and the line loading may be viewed as a random process. This uncertainty in line data can significantly influence line loading calculations.

2. *Load data.* As discussed in the preceding section, all forecast load data contain uncertainty. The further into the future the forecast, the larger the data variance. Even present data (telemetered to the processing point) may contain significant uncertainty due to bad data [9]. This point is considered in detail at the conclusion of this chapter.

3. *Generation data.* Forced outages result in uncertainty in generation data. At this point, attention turns to a reformulation of the

power flow study problem considering several of these uncertainties. A study in which line/load/generation statistics are used to calculate bus voltage/line loading statistics is termed a ***stochastic power flow study.*** A stochastic power flow study is a calculation of the statistics of variates that are functions of many random variables. Several representative approaches are presented. The reader is referred to the references for details and examples. These methods are not in widespread use because the algorithms are not fully proven and there is a general reluctance to use stochastic studies for planning and operations purposes. The latter is partially due to the requirement that power systems must be sized for extremal conditions. The reader is advised to take careful note of what information is attainable from the stochastic power flow study; a discussion of potential applications will conclude the section.

Stochastic Power Flow Study Based on the Jacobian Matrix

The active and reactive bus mismatches ΔP, ΔQ, are related to deviations of bus angle and voltage magnitude, $\Delta\delta$, ΔV by

$$\begin{bmatrix}\Delta P\\ \Delta Q\end{bmatrix} = J\begin{bmatrix}\Delta\delta\\ \Delta V\end{bmatrix}. \tag{9.28}$$

The Jacobian matrix, J, may be viewed as a transformation of variables. If ΔP, ΔQ is a random vector of probability density f_{PQ}, the density of $\Delta\delta$ and ΔV is obtained using the fact that $\Delta\delta$, ΔV is a function of several random variables,

$$\begin{bmatrix}\Delta\delta\\ \Delta V\end{bmatrix} = J^{-1}\begin{bmatrix}\Delta P\\ \Delta Q\end{bmatrix}. \tag{9.29}$$

For the general case of arbitrary statistics of ΔP, ΔQ, this is not an easy chore. If ΔP, ΔQ are Gaussian,

$$f_{PQ}(\begin{bmatrix}\Delta P\\ \Delta Q\end{bmatrix}) = G(M_{PQ}, \textstyle\sum_{PQ}), \tag{9.30}$$

where M_{PQ} is the mean of $[\Delta P\ \Delta Q]^t$ and $\sum_{PQ}$ is the covariance matrix of $[\Delta P\ \Delta Q]^t$. Let J^{-1} be constant in (9.29), and recognize that a linear transformation of a Gaussian random vector is also Gaussian. The density of $[\Delta\delta\ \Delta V]^t$ is $f_{\delta V}$,

$$f_{\delta V}(\begin{bmatrix}\Delta\delta\\ \Delta V\end{bmatrix}) = G(M_{\delta V}, \textstyle\sum_{\delta V}) \tag{9.31}$$

$$M_{\delta V} = J^{-1}M_{PQ} \tag{9.32}$$

$$\textstyle\sum_{\delta V} = J^{-1}\sum_{PQ}(J^{-1})^t . \tag{9.33}$$

In (9.30) and (9.31), the notation $G(\cdot, \cdot)$ denotes the Gaussian probability density function,

$$G(M,\textstyle\sum) = \dfrac{exp\ (-\frac{1}{2}(X - M)^t\sum^{-1} (X - M))}{(2\pi)^{n/2}\sqrt{det\ (\sum)}}, \tag{9.34}$$

provided that $\sum^{-1}$ exists. In (9.34), X is the multivariate random vector of dimension n. If $\sum^{-1}$ does not exist, (9.32) and (9.33) nonetheless are correct, but the density function does not exist. The characteristic function and second characteristic function exist and either $\phi_X(\Omega)$ or $\psi_X(\Omega)$ are used to define the nature of X.

Thus one formulation of the stochastic power flow problem requires a knowledge of the statistics of P, Q. Using a best Gaussian fit, M_{PQ} and $\sum_{PQ}$ are used as the mean and covariance of $[P\ Q]^t$. Then (9.32) and (9.33) are used to find the statistics of $[\delta\ V]^t$. The limitations are primarily:

1. Constant J is assumed.
2. Gaussian P, Q is assumed.

The first of these limitations is valid when the load variances are is small (e.g., diagonal entries are less than 0.1), and the second is valid under some loading conditions and at some load points. Bulk transmission loads tend to be rather Gaussian in distribution while local distribution points may assume arbitrary distribution.

Stochastic Power Flow Study Based on Distribution Factors

Line loads, $\bar{S}_{line}$, are related to bus injections, S_{bus}, in an approximate way by

$$\bar{S}_{line} = \rho S_{bus} \ . \tag{9.35}$$

Equation (9.35) is valid under the usual assumptions used in distribution factor methodology: nearly flat bus voltage profile and small fluctuations in S_{bus}. Let S_{bus}, $\bar{S}_{line}$, and ρ be resolved into real and imaginary parts,

$$\bar{P}_{line} + j\bar{Q}_{line} = (\rho_r + j\rho_i)(P_{bus} + jQ_{bus}) \ .$$

Thus

$$\bar{P}_{line} = \rho_r P_{bus} - \rho_i Q_{bus}$$

$$\bar{Q}_{line} = \rho_i P_{bus} + \rho_r Q_{bus} \ .$$

When P_{bus}, Q_{bus} are arranged in a vector $[P_{bus}\ Q_{bus}]^t$, it is possible to use the technique noted in (9.31) through (9.33) when $[P_{bus}\ Q_{bus}]^t$ is multivariate Gaussian with mean M_{PQ} and covariance $\sum_{PQ}$,

$$T = \begin{bmatrix} \rho_r & -\rho_i \\ \rho_i & \rho_r \end{bmatrix} \tag{9.36}$$

$$M_{line} = TM_{PQ} \tag{9.37}$$

$$\textstyle\sum_{line} = T \sum_{PQ} T^t , \tag{9.38}$$

where M_{line} and $\sum_{line}$ are the statistics of the line flow $[\bar{P}_{line}\ \bar{Q}_{line}]^t$.

The limitations of this technique are similar to those cited for the stochastic power flow study based on the Jacobian matrix. Note that ρ_r and ρ_i are generally rectangular submatrices, and T is generally rectangular. Therefore, even if $\sum_{PQ}$ is nonsingular, $\sum_{line}$ will be singular if 2 $\times$ (number of lines) $>$ 2 $\times$ (number of buses). This observation is a consequence of the fact that the rank of the product of several matrices is the minimum of the rank of any one of the matrices in the product. The rank of $\sum_{PQ}$ [see Eq. (9.38)] is 2 $\times$ (number of buses) if $\sum_{PQ}$ is nonsingular. Therefore, the rank of $\sum_{line}$ is equal to the rank of $\sum_{PQ}$ when there are more lines than buses. Thus the nullity of $\sum_{line}$ is nonzero, and $\sum_{line}$ is singular. Thus, it is not possible to write the joint density of $[P_{line} \; Q_{line}]^t$, but it is nonetheless possible to write its characteristic function.

Generalized Approach to Calculating Line and Bus Voltage Statistics

If the inverse Jacobian matrix, $J^{-1} = \Xi$, is nearly constant over the range of bus voltages expected in a stochastic power flow study,

$$\begin{bmatrix} \Delta\delta \\ \Delta V \end{bmatrix} = \Xi \begin{bmatrix} \Delta P \\ \Delta Q \end{bmatrix} .$$

Now consider $[\Delta P \mid \Delta Q]^t$ of arbitrary statistics. Denote this vector as

$$X = \begin{bmatrix} \Delta P \\ \Delta Q \end{bmatrix}$$

and recognize that the sample statistics of X are calculable from past data. Consider standard measure X without loss of generality,

$$E(X) = 0 \qquad E(XX^t) = I .$$

Then the kth order moment of X is

$$\sigma^{(k)}_{a_1 a_2 a_3 \cdots a_k} = \frac{1}{N_s} \sum_{i=1}^{N_s} (X)_{a_1 i} (X)_{a_2 i} \ldots (X)_{a_k i} . \tag{9.39}$$

In (9.39), the superscript (k) on σ denotes the moment order, a_1 through a_k denote the variates, N_s is the number of samples in the ensemble, and $(X)_{a_q i}$ denotes the a_q element of vector X for sample i. Proceed to calculate the statistics of the state vector $[\Delta\delta \;\; \Delta V]^t$,

$$Y = \begin{bmatrix} \Delta\delta \\ \Delta V \end{bmatrix} .$$

Then the qth order moment of Y is $s^{(q)}_{a_1 a_2 a_3 \cdots a_q}$,

$$s^{(q)}_{a_1 a_2 a_3 \cdots a_q} = E\Big(\prod_{i=1}^{q} (Y)_{a_i}\Big)$$

$$= E\Big(\prod_{i=1}^{q} (\Xi X)_{a_i}\Big) .$$

The term $(\Xi X)_{a_i}$ is the a_i row of the product ΞX,

$$(\Xi X)_{a_i} = \sum_{j=1}^{n_x} (\Xi)_{a_i j} (X)_j ,$$

where n_x is the dimension of X. One concludes that the first few moments of Y are readily calculable,

$$E(Y) = 0$$

$$\textstyle\sum_Y = \Xi \Xi^t$$

$$s^{(3)}_{a_1 a_2 a_3} = \sum_j \sum_k \sum_\ell (\Xi)_{a_1 j} (\Xi)_{a_2 k} (\Xi)_{a_3 \ell} \, \sigma^{(3)}_{jk\ell} . \tag{9.40}$$

The notation $\sum_j$ denotes $\sum_{j=1}^{n_x}$ in (9.40). Inspection of (9.10) reveals the general procedure to obtain $s^{(q)}_{a_1 a_2 \cdots a_q}$. Thus the moments of Y are known to *any* desired order.

Although the moments of a random vector do *not* determine its probability density function, they do characterize it, and there are methods to generate $f_Y(Y)$ such that the moments of Y (i.e., the $s^{(q)}$ values) equal the appropriate coefficients of the characteristic function of Y [see Eq. (9.14)]. Sauer and Heydt [10–11] describe how f_Y is constructed using the Gram–Charlier series type A. This series is, in effect, a multivariate form of (9.14) and details of the series are given in [12].

Applications of the Stochastic Power Flow Study

The statistics of $[\delta\ V]^t$ or $[\overline{P}_{line}\ \overline{Q}_{line}]^t$ may be used to determine such quantities as

1. The probability of bus voltage out of range.

2. The probability of line loading out of range.

3. Conditional probabilities of specific events (e.g., given the condition that bus voltage $|v_1|$ is within range, calculate the probability that any one $|v_i|$, $i \neq 1$, is out of range).

4. Expected values of bus voltage magnitude, angle, or line flow.

5. An "index of confidence" for the load flow study in general or for specific expected values of V_{bus} or S_{line}.

These applications reflect how uncertainty in input data is reflected in bus voltage and line load data. Also, it is possible to assess alternative designs by comparing probabilities of out-of-range $|V_{bus}|$ and/or $|\overline{S}_{line}|$. For several of these applications, it is necessary to integrate a multivariate density to obtain a probability. This may be done by numerical integration. If f_Y is Gaussian, there are some specialized formulas which may be applicable. For bivariate Y, the probability of an event,

$$Pr\{y_1 \leq \overline{y}_1,\ y_2 \leq \overline{y}_2\} = F_Y\left(\begin{bmatrix}\overline{y}_1\\ \overline{y}_2\end{bmatrix}\right)$$

is given by the tetrachoric series [13]. For the higher dimensional case, the corresponding distribution function, $F_Y(\overline{Y})$, is given by the generalized tetrachoric series [14].

Sasson [15] has reported some applications of the stochastic power flow studies, and there have been proposed variations in formulation to give such parameters as the statistics of operating fuel costs (i.e., a stochastic optimal power flow study) [16,17].

9.4 UNCERTAINTY IN GENERATION

As stated in the previous section, apart from the load level, a principal source of uncertainty in electric power systems is the available generation level. The available generation is generally less than the installed capacity due to *forced* outages of generators (i.e., unplanned outages) and *planned* outages (e.g., scheduled outages for maintenance). The adequacy of the available generation to serve the demand depends not only the level of the demand, but also the duration of each level of demand. For the assessment of generation adequacy, the concept of a load duration curve (LDC) is useful. An LDC is a cumulative histogram of load demand during a specified period. For example, a typical daily LDC is shown in Figure 9.1; the load is less than P_1 for t_1 hours per day. The demand is less than P_{peak}, for 24 hours.

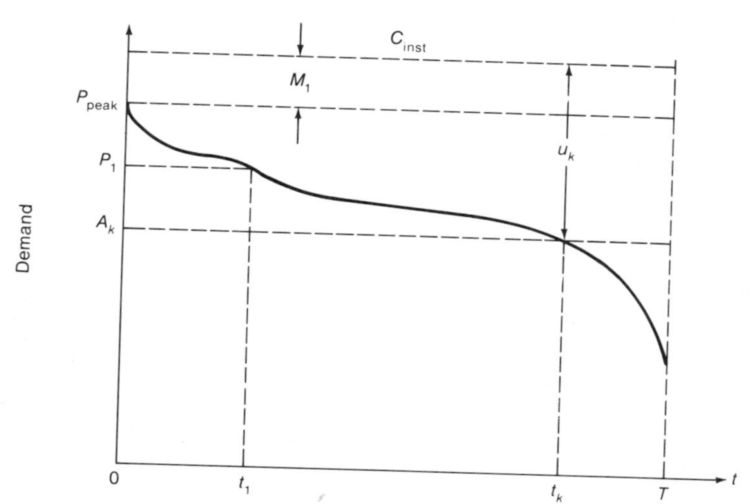

FIGURE 9.1 Load duration curve.

Obviously, for very large available capacity, for example far greater than P_{peak}, the probability of serving the load will be high. The installed capacity, C_{inst}, and capacity margin, M_1, are indicated in Figure 9.1. For the purpose of calculating the probability of serving the load, it is convenient to introduce the concept of a capacity outage table (COT) to be used with the LDC. A capacity outage table is a table listing the amount of generation on forced outage versus the probability of that operating state. Usually, only forced outages are considered since planned outages are scheduled at appropriate times. Let the probability of a specified generator being in service be denoted as p. Then the forced outage rate (FOR) is $1-p$. To illustrate the process of constructing a COT, consider a three machine system with each generator having the same FOR (denoted as $q = 1 - p$). Let the units be identical with capacity C megawatts each. Then the condition of zero megawatts on outage occurs as follows:

Available	On Outage	Probability
$3C$	0	p^3

It is assumed that the generators are forced outaged as independent probabilistic phenomena. The case of C megawatts on outage can occur three different ways. Therefore, the next line in the COT is

Available	On Outage	Probability
$3C$	0	p^3
$2C$	C	$3p^2q$

The remaining two lines of the table are readily calculated. Usually, a cumulative probability column is also calculated:

Available	On Outage	Probability	Cumulative Probability
$3C$	0	p^3	1
$2C$	C	$3p^2q$	$1 - p^3$
C	$2C$	$3q^2p$	$1 - p^3 - 3p^2q$
0	$3C$	q^3	$1 - p^3 - 3p^2q - 3q^2p$

In this COT, the probability column is denoted f. If f is indexed by the available capacity ([e.g., $f(A = 3C)$ is p^3 in this case], $f(A)$ is the individual state probability of available capacity A. Alternatively, f may be indexed by the amount of capacity on outage: thus $f(U)$ is the individual state probability of U megawatts on outage. The cumulative probability is F and this parameter may be indexed by A or U. Cumulative probability $F(A = C_1)$ is the probability that the available capacity is C_1 or less. In the example shown, $F(A = 3C)$ is unity since the total installed capacity is $3C$. It is possible to index F according to the unavailable capacity: $F(U = C_1)$ is the probability that the capacity on outage is C_1 megawatts or more.

This capacity outage table, when used in connection with the system LDC, gives information on the system generation adequacy to meet the load. The remainder of this article focuses on the computer algorithms used to generate the COT, approximations for the state probability, and the methodology of using the COT and LDC for generation adequacy assessment.

Digital Computation of the Capacity Outage Table

As described above, the COT is a tabulation of available capacity, capacity on outage, state probability (i.e, probability of the specific outage state), and cumulative probability. Denote these parameters as A, U, f, and F, respectively. Also, denote the total installed capacity as C. Then

$$A + U = C \tag{9.41}$$

$$\sum_{A \geq a_1} f(A) = F(a_1), \tag{9.42}$$

where $f(a)$ is the probability of available capacity $A = a$ (exactly), $F(a_1)$ is the cumulative probability of available capacity $A \leq a_1$, and $\sum_{A \leq a_1}$ refers to summation over all states in which the available capacity is less than or equal to a_1. At this point, consider the digital computer algorithm to construct the COT. A "building algorithm" approach is used here whereby the input generation list is read one unit at a time and the COT is constructed as the units are read. The allotted (dimensioned) space for the COT is first cleared to zero. The first generator is read:

Capacity	C_1
FOR	q_1
Availability	$p_1 = 1 - q$

Obviously, only C_1 and q_1 need to be read. The table is simply two lines.

A	U	f	F
C_1	0	p_1	1
0	C_1	q_1	$1-p_1$

The next unit in the generation list has capacity C_2 and FOR equal to q_2. Each line in the existing COT may give rise to two lines in the new COT because the second generator may be operative or outaged. For example, the existing first line in the COT had available capacity entry $A = C_1$; this line now becomes

A	
$C_1 + C_2$	(unit 2 "up")
C_1	(unit 2 outaged)

At this point, the question arises as to where the two new lines are placed in the new COT. One resolution of this question is to place the added line at the bottom of the table, do not calculate F, then reorder the table according to larger A's at the top. This is illustrated as follows for the two unit illustration (let $C_1 > C_2$ for this illustration):

Old Table

A	U	f
C_1	0	p_1
0	C_1	q_1

New Table (Not Reordered)

A	U	f
C_1	.	.
0	.	.
$C_1 + C_2$	.	.
$0 + C_2$	.	.

The f entry for lines C_1 and $C_1 + C_2$ is equal to q_2 and p_2 times the old entries in those positions, respectively:

A	U	f
C_1	.	p_1q_2
0	.	.
$C_1 + C_2$	.	p_1p_2
$0 + C_2$	.	.

and the f entry for lines 0 and $0 + C_2$ is q_2 and p_2 times the old entry, respectively:

A	U	f
C_1	.	p_1q_2
0	.	q_1q_2
$C_1 + C_2$	.	p_1p_2
$0 + C_2$	.	q_1p_2

At this point, the table is reordered with the larger A entries at the top. The unavailable capacity could also be calculated if desired, but this is best done after the last generation unit is read and the final COT is calculated. The same remark applies to cumulative probabilities F. There are two difficulties with this procedure: (1) the addition of a new generator may give rise to a value of A already in the table (i.e., a "new" row of the COT is not generated), and (2) the number of allotted rows of the table may already be filled, and it will be necessary to use an interpolation procedure to obtain an approximate COT within the available allotted memory.

The first of these difficulties is readily solved by noting that when an entry of the COT is generated (due to a new unit added) such that an existing value of A is replicated ($\pm\epsilon$, where ϵ is a small resolution parameter), a new line is *not* generated, but instead, the existing value f is increased by the probability of the new state. An example will illustrate the process.

The second cited difficulty relates to generation of more lines in the COT in instances of a filled COT. When the available space allotted for the COT fills, it is necessary to first order the table such that descending values of A occur with ascending row number. This numerical ordering is equivalent to numerical sorting, and a "bubble sort" is useful. Having reordered the table, the problem now is to insert a new entry in a filled, ordered COT. Pictorially, this is illustrated as

$$
\begin{array}{lll}
 & A_i & f_i \\
A_{new}\ f_{new} \rightarrow & & \\
 & A_{i+1} & f_{i+1} \\
 & A_{i+2} & f_{i+2} \\
 & A_{i+3} & f_{i+3}
\end{array}
$$

Let

$$A_i < A_{new} < A_{i+1};$$

then the new state is entered, with the following result:

A	f
A_i	$f_i + \dfrac{A_{i+1} - A_{new}}{A_{i+1} - A_i} f_{new}$
A_{i+1}	$f_{i+1} + \dfrac{A_{new} - A_i}{A_{i+1} - A_i} f_{new}$

If the A_{new} entry is greater than A_1 or smaller than A_n (n = number of rows allotted for the COT), the new A value should replace the existing A, and the interpolated value of the state probabilities should replace the f entry.

After the A versus f table is constructed, the U column is readily generated,

$$U_i = C - A_i$$

and the F entries are

$$F_i - f_i = F_{i+1}.$$

Example 9.1

Consider a generation unit list as follows:

Generator Capacity (MW)	Forced Outage Rate
800	0.010
800	0.011
231	0.005
400	0.020
400	0.020

Construct a capacity outage table with a maximum of 10 rows.

Solution

The first unit is read and the COT is

A	f
800	0.990
0	0.010

The remaining eight rows of the table are zeros. The next unit is read and two new rows are generated for each existing row (where $A_{new} = A_{old} + C_{new}$ and $A_{new} = A_{old}$):

A	f
800	(0.990) (0.011)
0	(0.010) (0.011)
1600	(0.990) (0.989)
800	(0.010) (0.989)

and the table is scanned for duplicate entries. In instances of duplicate entries, the rows are consolidated by summing the f's,

A	f
800	0.02078
0	0.00011
1600	0.97911

The next unit is read (231 MW) and each existing value of A gives rise to a new row of the COT:

A	f
800	0.00010390
0	0.00000055
1600	0.00489555
1031	0.02067610
231	0.00010945
1831	0.97421445

At this point, there are six rows in the COT. In the process of incorporating the next generating unit ($C_5 = 400$, $q_5 = 0.020$), the table fills. At the point where the table fills, the $A_{old} = 1031$ entry has just been processed. Before proceeding to the $A = 631$ and $A = 2231$ entries, the fact that the table fills is recognized, and the table is ordered according to descending values of A,

A	f
2000	0.004797639
1831	0.019484289
1600	0.000097911
1431	0.020262578
1200	0.000101822
1031	0.000413522
800	0.000002078
400	0.000000539
231	0.000002189
0	0.000000011

There are two additional entries to be inserted: $A = 231 + 400$ and $A = 1831 + 400$. The first of these is inserted by the interpolation-like procedure:

$$\begin{array}{ll} 800 & x \\ 631 & (0.00010945)(0.98) \\ 400 & y \end{array}$$

The method cited earlier is used and x and y are found

$$x = 0.000066166 \quad y = 0.000043712$$

As the second is found to be greater than 2000, 2231 replaces 2000 and the old value of $f(2000)$ must be proportioned between the $A = 2231$ and $A = 1831$ entries:

A	f
2231	$(0.97421445)(0.98) + \frac{169}{400}(0.004797639)$
1831	$0.019484289 + \frac{231}{400}(0.004797639)$

The last generator in the list is a 400 MW unit with FOR equal to 0.020. Interpolation is required for each new entry. The U and F entries are completed at the end of the process. The result is shown below:

A	U	f	F
2631	0	0.958115977	1.000000000
1831	800	0.040849394	0.041884022
1610	1021	0.000096169	0.001034627
1431	1200	0.000816078	0.000938456
1200	1431	0.000066879	0.000122378
1031	1600	0.000008270	0.000055498
800	1831	0.000045400	0.000047227
400	2231	0.000001781	0.000001827
231	2400	0.000000044	0.000000045
0	2631	0.000000001	0.000000001

The interpolation procedure introduces some error into the final COT. If the original generation list is ordered with high capacity units first, the required interpolations that occur late in the formulation of the table will be associated with small capacity generators. Thus the error associated with interpolation will be small.

Effects of Capacity Addition and Removal on the COT

It is not difficult to generalize the cited procedure. For the modification of the COT upon addition of a generating unit of capacity C_{add}, and FOR equal to q_{add},

$$F_{new}(U_1) = F_{old}(U_1 - C_{add})q_{add} + F_{old}(U_1)p_{add} \tag{9.43a}$$

$$p_{add} = 1 - q_{add} \, . \tag{9.43b}$$

The COT building algorithm is essentially the use of (9.43), allowing the unavailable capacity U_1 to take on values from zero up to the new installed capacity (in that order). In (9.43), the subscripts *new* and *old* refer to after the installation of C_{add} and before the installation, respectively. Use of (9.43) requires calculation of an F column in the COT. If the argument of $F_{old}(U_1 - C_{add})$ is nonpositive in (9.43), use unity for this term. Equation (9.43) is generally applied to a rounded (interpolated) COT.

Equation (9.43) may also be used upon outage of a generating unit. For example, a unit retirement may require modification of the existing COT. In such an application, (9.43) is reformulated as follows:

$$F_{after\ out}(U_1) = p^{-1}(F_{before\ out}(U_1) - F_{after\ out}(U_1 - C)q),$$

where $F_{after\ out}$ and $F_{before\ out}$ refer to the post- and preoutage cumulative probability, and a unit of capacity C and FOR equal to q is outaged. Billinton [8] gives examples of the use of the capacity addition and capacity outage formulas.

Practical Considerations on COT Construction; Debugging

The algorithm cited above works well when the allotted table size for the COT is large. The long table length avoids inaccuracies in modeling due to interpolation. Also, if the largest units are ordered early in the generation list, the required interpolations after the table fills will be for small ΔA values; this results in small error due to interpolation.

The most common errors in programming the COT building algorithm relate to the interpolation procedure. Extremal values of the table and rows of the table which are almost identical must be considered. Note that

$$\textstyle\sum_A f(A) = 1, \tag{9.44}$$

where $\sum_A$ refers to summation over all outage states. Equation (9.44) is useful as a check at each stage of construction of the COT.

Modeling the Cumulative Probability Function

It is possible to form mathematical models for $f(A)$ and $F(A)$. Usually, these models, which are in fact statistical models, assume that the available generation is a continuous parameter. For small power systems, the latter is not a reasonable assumption, but for systems of n machines, where n is large, there will be as many as 2^n different outage states (not considering overlapping). For typical systems with 10 to 50 generators, there are typically hundreds of thousands of legitimate values of A between zero and the installed capacity. For such cases, the assumption of continuous A is reasonable.

Examination of typical $F(A)$ functions reveal a similarity with the error function,

$$erf\ (x) = \frac{2}{\sqrt{\pi}} \int_0^x e^{-\lambda^2}\, d\lambda.$$

(Caution: There are alternative definitions of the error function.) The function $F(A)$ in typical cases is similar to

$$F(A) = erf\ [k(C_{inst} - A)],$$

where C_{inst} is the installed capacity. This model is, effectively, a Gaussian process. It may be useful to use a rational approximation for the error function, for example,

$$erf\ (x) \simeq 1 - (\alpha_1\lambda + \alpha_2\lambda^2 + \alpha_3\lambda^3)e^{-x^2}$$

$$\lambda = \frac{1}{1 + \beta x}$$

$$\alpha_1 = 0.3480242 \quad \alpha_2 = 0.0958798 \quad \alpha_3 = 0.7478556 \quad \beta = 0.47047\ .$$

Production computer codes have been marketed using a variation of the Gaussian assumption; the method of cumulants is used. In this method, a Gram–Charlier series type A is used to model $dF(A)/dA$, and a three to six term truncation is found to give an accurate model. This method may be viewed as being a correction to the Gaussian assumption. The Gram–Charlier series type A consists of the Gaussian probability density function as the first term plus the first, second, and successively higher derivatives of $G(x)$ as the second, third, and higher terms. These derivatives are Hermite polynomials which are orthogonal functions. Thus the series is an orthogonal expansion of $f(x)$. References [5,6,10,11] cover the details of this series, and [32] will lead the interested reader to a growing volume of literature on the application of the cumulant method to COT models.

Use of the COT and LDC to Assess Generation Adequacy

The principal use of capacity outage tables is to assess generation adequacy in generation planning and generation schedule planning. Although it is true that the forced outage rates and load duration data are usually derived from historical data, the use of these characteristics gives a reasonable estimate of generation adequacy. Perhaps it is more important to note that the analysis of generation adequacy using the COT and LDC is done in a *consistent* fashion such that comparisons may be made between alternative generation expansion plans, operating schedules, maintenance schedules, and so on. The COT alone will give useful data on generation reliability; for example, the expected value of available generation, $E(A)$, is given by

$$E(A) = \sum_k A_k f_k$$

(the sum is carried over all outage states), and the expected capacity on outage is

$$E(U) = \sum_k U_k f_k \ .$$

The expected time for which the system demand is greater than the generation is a very useful index for generation adequacy assessment. Let $E(t)$ denote this expectation; some authors refer to this expected time as the *expected load loss*. Let an outage of U_k occur in a system with installed capacity C_{inst}; then available generation is

$$A_k = C_{inst} - U_k \ .$$

The probability of this outage state is $f_k = f(A_k)$. Figure 9.1 illustrates this outage state plotted on an LDC: note that in the time horizon of interest, T, there will be t_k time units of negative generation margin conditions. Thus

$$E(t) = \sum_k t_k \ f_k ,$$

where $\sum_k$ denotes summation over all outage states.

Thus, to obtain $E(t)$, construct the COT, and for each table entry, calculate $t_k f_k$. The t_k parameter is read from the LDC (which is usually a table). The sum of $t_k f_k$ over all outage states is the required expected time of negative generation margin. Typical figures for $E(t)$ are in the order of 0.05 day/year (or 1 day in 20 years).

9.5 STATE ESTIMATION

The term *state estimation* is used in electric power engineering to refer to a broad class of techniques for the calculation and/or approximation of system bus voltage magnitudes, phase angles, and certain other related quantities. These methods differ from power flow studies and other exact (i.e., within prescribed tolerance) algorithms in two important ways: certain input data are either missing or inexact, and the algorithm used for the calculation may entail approximations and other methods designed for high speed processing. In addition to these features of state estimation methods, certain key words often arise in their description: measurement, stochastic models, minimum (but nonzero) error, and faulty data. The state vector of a power system in the sinusoidal steady state is X consisting of:

1. Bus voltage phase angles (at all PQ buses, all PV buses except the swing bus, and at all TCUL buses).
2. Bus voltage magnitudes (at all PQ buses).
3. Tap changing transformer positions (at all TCUL buses).

State estimation methods are used to calculate elements of X for reasons of assessing appropriateness of operating regimes, including contingencies. The system states are not the only parameters that are needed for such assessments. For example, line loads are often the limiting factors of system security assessment. For this reason, state estimation algorithms are frequently more general in scope than strictly *state* calculation procedures; additional quantities which may also be estimated include line loads, bus loads, phase shifter settings, fixed tap transformer settings, the status of circuit breakers and other interrupters, generator settings (both $|V|$ and P), and a variety of information on the operating status of capacitor banks and reactors. A further characteristic of state estimation algorithms relates to the specific systems to which they apply. Commercially available state estimators are commonly "tailor-fit" to specific applications and power systems. The reason for this system dependence is that state estimators are often used for control, alarm, and other operating information functions. The designs of these controls and alarms are system specific. The principal methods and features of state estimators which are common to many applications and systems are the topics of this section. A great deal of literature exists on state estimators. References [25–27] from the journals and texts [28,29] are representative samples. Wood and Wollenberg [30] give a concise summary of the methods in common use. Sterling [33] gives an excellent discussion of bad data supression and dynamic state estimation.

Estimation of Elements of the State Vector

Figure 9.2 shows a pictorial of an electric power system. The state vector of this system is of dimension n_x. This vector is composed as follows:

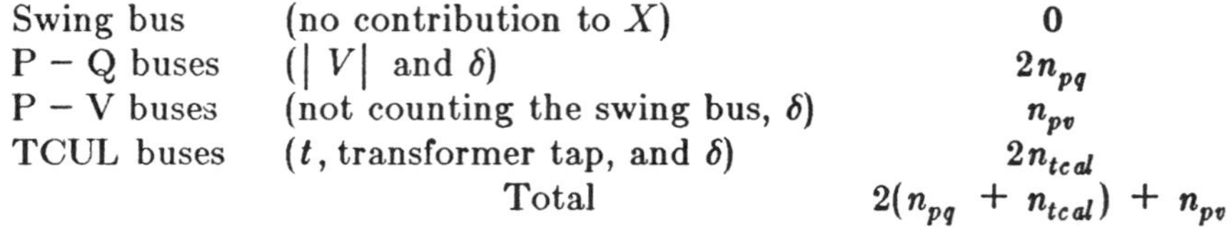

Swing bus	(no contribution to X)	0
P – Q buses	($\|V\|$ and δ)	$2n_{pq}$
P – V buses	(not counting the swing bus, δ)	n_{pv}
TCUL buses	(t, transformer tap, and δ)	$2n_{tcul}$
	Total	$2(n_{pq} + n_{tcul}) + n_{pv}$

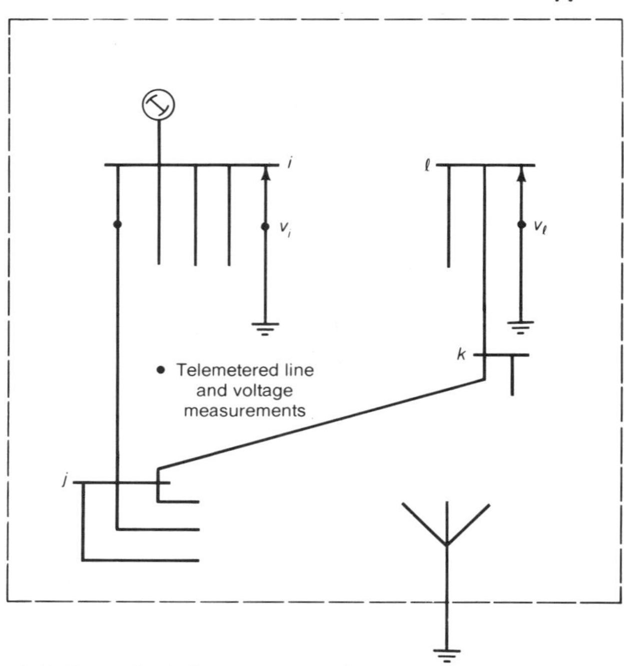

FIGURE 9.2 Pictorial of an electric power system for state estimation.

Therefore,

$$n_x = 2(n_{pq} + n_{tcul}) + n_{pv} .$$

Consider that some measurements are taken in the system (see Figure 9.2). If it is desired to estimate some or all of the n_x states (i.e., elements of X), it is necessary to model the system volt-ampere response. The measurements shown in Figure 9.2 consist of n_1 bus voltage measurements and n_2 line complex voltampere flows. Consider first the dc case. If line losses are ignored, the n_2 line flow measurements may be considered to be line current measurements. For unity bus voltage profile, this assumption is exact. If the bus voltages are not unity, an iterative procedure is required to calculate I, the line current vector. Let V_{bus} be partitioned into a "measured state subvector," $V_{bus\ m}$, and an "unknown state subvector," $V_{bus\ u}$

$$V_{bus} = \left[\begin{array}{c} V_{bus\ u} \\ \hline V_{bus\ m} \end{array}\right] \begin{array}{l} \} \ n_{bus} - n_1 \\ \\ \} \ n_1 \end{array} .$$

Similarly, partition the line current vector $\overline{I}$ into measured and unknown subvectors,

$$\overline{I} = \left[\begin{array}{c} \overline{I}_u \\ \hline \overline{I}_m \end{array}\right] \begin{array}{l} \} \ n_{line} - n_2 \\ \\ \} \ n_2 \end{array} .$$

Then note that

$$\overline{Y} L V_{bus} = \overline{I}, \tag{9.45}$$

which is a simple consequence of writing the line current in any line $\overline{i}_{jk}$ as $(v_j - v_k)\overline{y}_{jk}$. In (9.45), $\overline{Y}$ is a diagonal n_{line} by n_{line} matrix of primitive line admittances, $\overline{y}$, and L is the line–bus incidence matrix, which is of dimension n_{line} by n_{bus}. Substitute the partitioned vectors V_{bus} and $\overline{I}$ into (9.45),

$$\overline{Y} L \begin{bmatrix} V_{bus\ u} \\ V_{bus\ m} \end{bmatrix} = \begin{bmatrix} \overline{I}_u \\ \overline{I}_m \end{bmatrix} . \tag{9.46}$$

and partition $\overline{Y}$ and L as follows:

$$\overline{Y} = \left[\begin{array}{c} \overline{Y}_a \\ \hline \overline{Y}_b \end{array}\right] \begin{array}{l} \textit{first } n_{line} - n_2 \textit{ rows} \\ \\ \textit{last } n_2 \textit{ rows} \end{array}$$

$$L = [\ \underset{\leftarrow \text{ first } n_{bus} - n_1 \text{ columns } \rightarrow}{L_a} \ | \ \underset{\leftarrow \text{ last } n_1 \text{ columns } \rightarrow}{L_b} \ ].$$

Submatrices $\overline{Y}_a$ through L_b are of the following dimension:

$\overline{Y}_a$	$n_{line} - n_2$	by	n_{line}
$\overline{Y}_b$	n_2	by	n_{line}
L_a	n_{line}	by	$n_{bus} - n_1$
L_b	n_{line}	by	n_1.

Substitute the partitioned $\overline{Y}$ and L matrices into (9.46) and expand

$$\overline{Y}_a L_a V_{bus\ u} + \overline{Y}_a L_b V_{bus\ m} = \overline{I}_u \tag{9.47}$$

$$\overline{Y}_b L_a V_{bus\ u} + \overline{Y}_b L_b V_{bus\ m} = \overline{I}_m \ . \tag{9.48}$$

The student should check to see that (9.47) and (9.48) are dimensionally correct. These equations are solved in the mean square sense using the Moore–Penrose pseudoinverse (see Appendix B),

$$V_{bus\ u} = [\overline{Y}_b L_a]^+ \ [\overline{I}_m - \overline{Y}_b L_b V_{bus\ m}] \tag{9.49}$$

$$\overline{I}_u = \overline{Y}_a L_a [\overline{Y}_b L_a]^+ [\overline{I}_m - \overline{Y}_b L_b V_{bus\ m}] + \overline{Y}_a L_b V_{bus\ m} . \tag{9.50}$$

Focus now only on (9.49). This is a state estimation equation since $V_{bus\ u}$ is a subvector of X. If it is desired to correct and update the approximation that $\overline{I}$ is $\overline{S}$, the estimates of $V_{bus\ u}$ and the measured $V_{bus\ m}$ may be used at this time. Wood and Wollenberg [30] show an example. A special case of (9.49) should be highlighted: the case of no bus voltage measurements. For this case (9.49) becomes (9.51),

$$n_1 = 0$$

$$V_{bus} = [\overline{Y}_b L]^+ \overline{I}_m \ . \tag{9.51}$$

Estimation of Additional System Parameters

As stated earlier, state estimation computer programs have evolved beyond the strict *state* estimation phase. From a practical point of view, there are many other system parameters outside the X vector which are of considerable importance. Most of these additional parameters are closely related to line flows, $\overline{S}$. The line flows, of course, are vital to system operation. Overload and system security considerations are often limiting factors in this regard. Also, the status of each system circuit breaker (or other interrupter) may not be properly telemetered. With an estimate of line flow, the status of these devices may also be estimated. Now consider (9.50). Under cases of unity bus voltage,

$$\overline{S} \simeq \overline{Y}_a L_a [\overline{Y}_b L_a]^+ [\overline{I}_m - \overline{Y}_b L_b V_{bus\ m}] + \overline{Y}_a L_b V_{bus\ m} . \tag{9.52}$$

If bus voltage measurements are unavailable, (9.52) becomes (for $n_1 = 0$)

$$\overline{S} \simeq \overline{Y}_a L_a [\overline{Y}_b L_a]^+ \overline{I}_m \ . \tag{9.53}$$

Bad Data Identification

System measurements are made in power networks for a variety of reasons, including protective relaying, satisfying legal and contractual requirements, improvement of operating capabilities, billing and other economic applications, and to obtain data for operational studies. With regard to the latter, state estimation is often an integral part of network analysis procedures since all the required data for such analyses are not directly available. These network measurements usually include bus voltages, line flows, and the status of interruption devices in the system. The transducers that make these measurements for supervisory control and data acquisition are usually termed ***remote terminal units (RTUs)***. An RTU may supply an analog or digital signal to the supervisory control and data acquisition (SCADA) system.

Power system environments are often challenging to the engineer because low signal levels in data circuits must operate at high precision in an environment shared with high voltage, high current, and high power circuits. Additionally, RTUs must be expected to be exposed to lightning transients, switching surges, extremes of weather, and infrequent maintenance and recalibration. For these reasons, it is very important to consider failure of the RTUs themselves, and/or the telemetered channel by which the RTU communicates with the SCADA. Additionally, during start-up and commissioning of SCADA systems, it is not unusual to find RTU's failed or improperly wired. Obviously, bad data, if allowed to be read by a network analysis program, have the potential of creating totally erroneous results or non-convergent cases in power flow studies.

Detection of bad data from RTUs is accomplished in two ways. Some RTUs have the ability to indicate failure modes upon interrogation. This type of failure detection is often effected by failure of the RTU to send back a valid response upon being interrogated. Some RTUs require a "handshaking" procedure before data are read (i.e. establishment that the channel is operational). If the handshaking fails, the RTU is considered to have failed. The alternative to direct reading of failure modes is the identification of bad data by an estimation procedure. By such methods, the parameter in question is first estimated, and the estimated value is compared with the RTU output. If the comparison is unreasonable, the RTU is considered to be failed. As an illustration, suppose that the Oak Street bus voltage magnitude is estimated (by a state estimator) to be 1.05 per unit. If the Oak Street bus voltage RTU shows 2.50 per unit, the unit is considered to have failed. Alternative criteria for the decision to declare an RTU as failed are considered below.

Consider (9.49) and (9.50) in a case in which measured parameters, $\overline{I}_m$ and $V_{bus\ m}$ are potentially contaminated by errors,

$$\hat{\overline{I}}_m = \overline{I}_m + \epsilon_i \tag{9.54}$$

$$\hat{V}_{bus\ m} = V_{bus\ m} + \epsilon_v, \tag{9.55}$$

where $(\hat{\cdot})$ denotes data obtained by telemetering and ϵ_i and ϵ_v are multivariate random vectors of density f_i and f_v. These vectors model channel noise. Then

$$V_{bus\ u} = [\overline{Y}_b L_a]^+[\overline{I}_m - \overline{Y}_b L_b V_{bus\ m}] + [\overline{Y}_b L_a]^+\epsilon_i - [\overline{Y}_b L_a]^+ \overline{Y}_b L_b \epsilon_v$$

$$\overline{I}_u = \overline{Y}_a L_a[\overline{Y}_b L_a]^+[\overline{I}_m - \overline{Y}_b L_b V_{bus\ m}] + \overline{Y}_a L_b V_{bus\ m} \tag{9.56}$$

$$+ \overline{Y}_a L_a[\overline{Y}_b L_a]^+\epsilon_i + (\overline{Y}_a L_b - \overline{Y}_a L_a[\overline{Y}_b L_a]^+ \overline{Y}_b L_b)\epsilon_v. \tag{9.57}$$

If f_v and f_i represent statistical estimates of nominal fluctuations in line and voltage telemetered data, the terms $\hat{\epsilon}_v$ and $\hat{\epsilon}_i$ represent statistical estimates of nominal fluctuations in $V_{bus\ u}$ and $\overline{I}_u$, respectively

$$\hat{\epsilon}_v = [\overline{Y}_a L_a]^+\epsilon_i - [\overline{Y}_b L_a]^+ \overline{Y}_b L_b \epsilon_v \tag{9.58}$$

$$\hat{\epsilon}_i = \overline{Y}_a L_a[\overline{Y}_b L_a]^+\epsilon_i + (\overline{Y}_a L_b - \overline{Y}_a L_a[\overline{Y}_b L_a]^+ \overline{Y}_b L_b)\epsilon_v. \tag{9.59}$$

The "error terms" $\hat{\epsilon}_v$ and $\hat{\epsilon}_i$ are the random processes that contaminate $V_{bus\ m}$ and $\overline{I}_m$ [see Eq. (9.54) and (9.55)]. If it is assumed that these processes have statistics identical to the statistics of the entire V_{bus} vector and the entire $\bar{I}$ vector, respectively, it is possible to identify bad data by checking for measurements far outside the nominal range of these data. This point is illustrated by the following cases:

No Voltage Measurements Taken. If no voltage measurements are made in the state estimation process, (9.58) and (9.59) become

$$\hat{\epsilon}_v = [\bar{Y}_b L_a]^+ \epsilon_i \tag{9.60}$$

$$\hat{\epsilon}_i = \bar{Y}_a L_a [\bar{Y}_b L_a]^+ \epsilon_i \ . \tag{9.61}$$

If the covariance of the measurement error vector, ϵ_i, is $\sum_i$,

$$\sum_i = E(\epsilon_i \epsilon_i^t) \tag{9.62}$$

[it is assumed that ϵ_i is real valued; otherwise, ϵ_i^H must replace ϵ_i^t in (9.62)], the covariance of $\hat{\epsilon}_v$ and $\hat{\epsilon}_i$ are readily calculated;

$$\begin{aligned}\hat{\sum}_v &= E(\hat{\epsilon}_v \hat{\epsilon}_v^t) \\ &= E([\bar{Y}_b L_a]^+ \epsilon_i \epsilon_i^t ([\bar{Y}_b L_a]^+)^t) \\ &= (\bar{Y}_b L_a)^+ \sum_i ((\bar{Y}_b L_a)^+)^t \end{aligned} \tag{9.63}$$

$$\begin{aligned}\hat{\sum}_i &= E(\hat{\epsilon}_i \hat{\epsilon}_i^t) \\ &= \bar{Y}_a L_a [\bar{Y}_b L_a]^+ \sum_i ((\bar{Y}_b L_a)^+)^t L_a^t \bar{Y}_a^t \ . \end{aligned} \tag{9.64}$$

Do not confuse the standard symbolism for covariance, $\sum_i$, and $\hat{\sum}_i$, for summation. Equations (9.63) and (9.64) are more compactly written

$$\hat{\sum}_v = L_{a1}^+ \bar{Y}_{b1}^+ \sum_i \bar{Y}_{b1}^{+t} L_{a1}^{+t} \tag{9.65}$$

$$\hat{\sum}_i = \bar{Y}_a L_a L_{a1}^+ \bar{Y}_{a1}^+ \sum_i \bar{Y}_{b1}^{+t} L_{a1}^{+t} L_a^t \bar{Y}_a^t, \tag{9.66}$$

where

$$L_{a1} \equiv \bar{Y}_b^+ \bar{Y}_b L_a$$
$$\bar{Y}_{b1} \equiv \bar{Y}_b L_{a1} L_{a1}^+ \ .$$

If one assumes that a measurement of a specific line value very far (i.e., many "standard deviations") from the expected value is *most* unlikely, one arrives at a simple criterion for rejection of a data point. Thus if a line flow is estimated at "μ", and the measured value differs from μ by $k\sigma$, where $k\sigma$ is several (e.g., five) times the square root of the diagonal entry in $\hat{\sum}_i$ [Equation (9.66)], it is deduced that the datum is bad (i.e., from a failed RTU).

These imprecise statements may be quantified and placed on a mathematical basis if the probability density function of ϵ_i is known - or assumed. If $\hat{\epsilon}_i$ is multivariate Gaussian σ with as the variance of the measurement error, neglecting correlation, a $\pm 5\sigma$ band around the estimated value will serve as a useful criterion to capture all reasonable measurements. Data outside this band may be taken as "bad" and should be rejected by the estimator. These data should be flagged for output purposes so that operators recognize that the RTU is inoperative.

Both Voltage and Line Data Taken. The more general case of both bus voltage and line data is considered in an analogous way by inspection of (9.65) and (9.66). Again, it is necessary to assume a probability density function for $\hat{\epsilon}_i$, $\hat{\epsilon}_v$. The multivariate Gaussian case is readily handled since the sum of two multivariate Gaussian processes is also multivariate Gaussian. For this case, it is readily shown that

$$\hat{\sum}_i = A\sum_{ii}A^t + B\sum_{vv}B^t + B\sum_{vi}A^t + A\sum_{iv}B^t, \qquad (9.67)$$

where

$$\sum_{ii} = E(\epsilon_i\epsilon_i^t)$$

$$\sum_{vv} = E(\epsilon_v\epsilon_v^t)$$

$$\sum_{vi} = E(\epsilon_v\epsilon_i^t)$$

$$\sum_{iv} = E(\epsilon_i\epsilon_v^t)$$

$$A = \overline{Y}_a L_a(\overline{Y}_b L_a)^+$$

$$B = \overline{Y}_a L_b - A.$$

Note that $\hat{\sum}_i$ defines the Gaussian process $\hat{\epsilon}_i$ since the mean of ϵ_i is assumed to be zero. The same procedure of establishing a band, $\pm k\sigma$, around the estimated value is used to capture reliable data. The reader is cautioned that a $\pm 5\sigma$ band, for example, does not imply that the probability of identifying bad data is calculable using the elementary error function. Although it is true that the error function is readily usable for the univariate case, the probability function of a multivariate process is much more complicated. This point is considered further in [12]. Another point to consider is that the Gaussian approximation is exactly that – *an approximation*. Actual data are distributed in an arbitrary way, and although there are some broad implications and applications of the central limit theorem, the errors ϵ_i and ϵ_v are not Gaussian. Thus the reader is cautioned about expecting high accuracy. The methods described above, on the other hand, are practical. Note that if a measurement is rejected as "bad" and the rejection is in fact a false alarm (i.e., the datum is "good"), the only harm done is that a measurement point is lost. In a well designed estimator, this loss should not have a disastrous effect.

References [9,30,31] contain many more details on bad data identification.

Accuracy of a State Estimator Versus Number of Measurements

The accuracy of a state estimator depends on the algorithm used, the number and site of placement of RTUs, the failure status of those RTUs, the characteristics of the noise contaminating the measurements, and the operating point of the system. It is difficult to make highly conclusive generalizations concerning the accuracy of the estimation, but it is evident that the trend of the accuracy must be such that as measurements increase in number, the estimator quality also increases. When a mean square estimator is used, one can make a few additional remarks about accuracy since the mean square error characteristic versus number of measurements has been studied [18]. The distribution of the singular values (termed the *singular spectrum)* controls the mean square error in general; however, the input data distribution also affects the error. If the singular spectrum has gaps between very large singular values, the mean square error versus number of measurements will usually exhibit similar gaps, as illustrated in Figure 9.3. Generally, the mean square error decreases logarithmically, but due to effects of measurement effectiveness, the decrease is only a qualitative trend. The mean square error is ***not*** usually monotone decreasing.

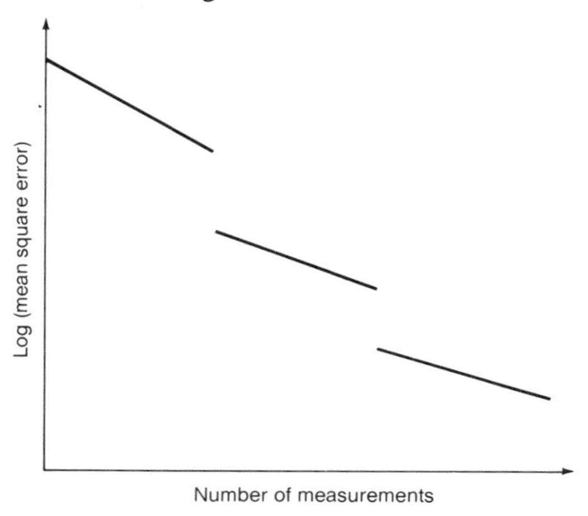

FIGURE 9.3 Mean-square-error characteristic of a state estimator.

9.6

APPLICATIONS OF STATE ESTIMATION IN POWER ENGINEERING

The primary applications of state estimators in power engineering are in real time SCADAs. Figure 9.4 shows a pictorial of a SCADA. The network analysis (i.e., load flow study, interchange study, unit commitment analysis, optimal dispatch) requires system data. These data come from base cases (stored for future use; often 2 to 20 cases of typical interest are stored), stored data, specially prepared data (which may be input from several sources, including the console), and real time measurements. The specifically prepared data may be entered interactively via a cathode ray tube input; production programs have been very inovative in the man/machine interface. The several sources of data are rarely used alone – that is, some data may be obtained from a base case, some from RTUs, and some from the operator via the console. In any case, actual field measurements are often a chief source of network data. Since measurements of *all* required data are impractical, and since

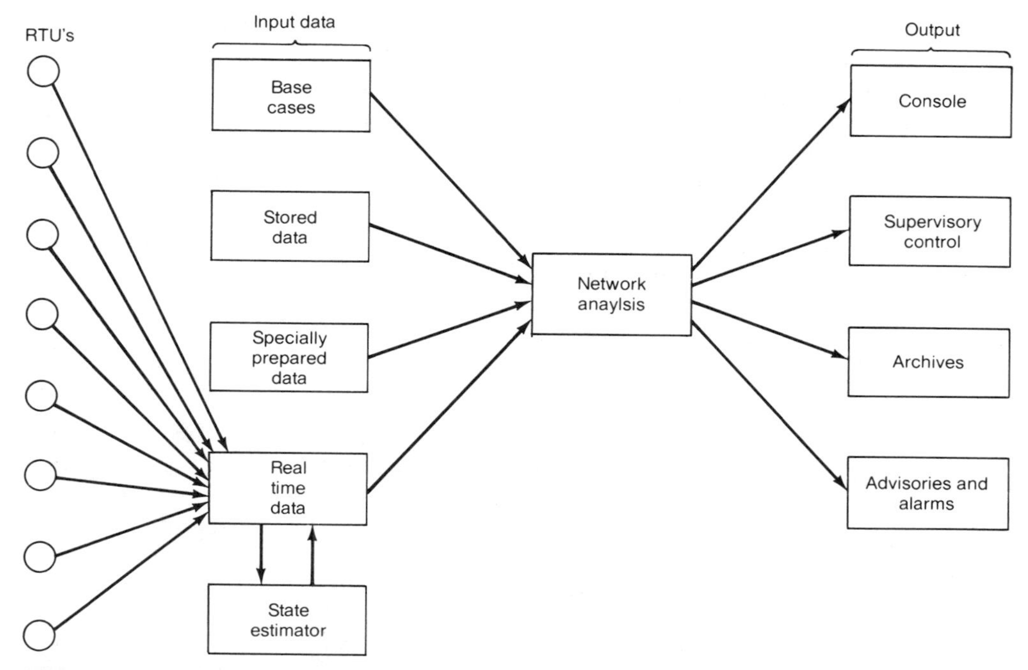

FIGURE 9.4 SCADA system.

some RTUs may be effectively failed, a state estimator is used to obtain the remainder.

Since about 1975, SCADA's have become an important part of most new operating systems. Standardization of all system features has not been made, but some common elements are as follows:

Data Refreshment Rate. Usually, RTUs are time multiplexed into the computer via dedicated telephone-grade lines and/or microwave channels. The channel capacity in bits per second, C, is related to the bandwidth in Hertz, BW, by

$$C = BW[log_2(1 + S/N)] \tag{9.68}$$

where S/N is the signal to noise ratio. In (9.68), a base 2 logarithm is used. Typical audio quality channels have a bandwidth in the 2 kHz class; thus for a signal to noise ratio of 63, the channel capacity is 12,000 bits/s. To convert this to *words* (i.e., numerical data to be telemetered), consider the clustering of 16 RTUs on a common, time multiplexed channel. If 32 bit words are used, the channel has a theoretical capacity of about 23 words per second. Typical practical capacities are lower, but the channel itself is rarely the limiting factor. Usually, the salient speed limitation is the network analysis software. The network analysis routines are specifically designed fast routines (i.e., decoupled load flow, initialization on previously solved cases, use of acceleration factors, and alternative fast methods to calculate the Jacobian matrix). Also consider that console display is usually graphic using a color cathode ray tube. The character generation required and the display may further limit the speed with which the output may be generated. Actual data refreshment rates are system specific, but multiple levels of data priority are often used, with the highest priority data refreshed at the rate of once every 1 to 5 s and the lowest level of data priority at a rate two to five times slower.

1. *State Estimation Error.* The critical factors in assessing the error introduced by state estimation are the number and placement of RTUs, the probability density function of the processes measured, the algorithms used for the estimator, and the singular spectrum. Generally, the number of measurements made is very large compared to that suggested by singular spectrum analysis. Because the mean square error generally falls very rapidly as the number of measurements increases, it is not too difficult to reduce the expected mean square error below that warranted by other sources of error. On a per unit scale, the root mean square error of the estimated and actual state vector will be 10^{-3} or less (due to state estimation error alone).

2. *Size of State Estimators.* The number of states that must be estimated in a large power system is typically in the range 5 to 40% of the total number of system states. Obviously, this figure is highly system dependent. If RTUs telemeter system load and status at many points, the primary need for state estimation is at failed channels. For systems with few telemetered points, state estimation is required to "generate" missing data. For a transmission system of 1000 buses, approximately 2000 states occur and about 960 load buses typically exist. If 75% of these load buses are either telemetered $(P + jQ)$ to the SCADA, or are reliably estimated in one of the base cases, 240 other values of $P + jQ$ must be estimated. Although loads are not states of a power system, there is a close relation between these parameters; the numerical illustration of (240)(2)/2000 or 24% will serve as an illustration of the order of magnitude of the state estimation problem. Failed RTUs typically number on the order of a few percent of the total.

3. *Network Analysis Requirements.* Since state estimation software and SCADA functions in general are on-line applications, the network analysis is essentially a load flow study. Because of the emphasis on speed, the load flow study is usually a decoupled load flow study or distribution factor study. Initialization is a critical procedure in such fast studies; for this reason, initialization is accomplished using a previously solved case or a similar base case. Input/output management is especially important due to the speed requirements. Commercial network analysis packages for use with SCADAs are generally capable of handing 500 to 5000 buses.

Unobservable States. The state estimation problem as described in this section is essentially a mean square error (mse) solution of $Ax = b$. There are other formulations, but the mean square error solution is most widely used. The mean square error problem is, in essence, the solution of x such that

$$||Ax - b||$$

is minimized. Let the vector x be decomposed into two vectors of identical dimension,

$$x = x_o + x_u \,. \tag{9.69}$$

and reconsider the equation $Ax = b$,

$$A(x_o + x_u) = b. \tag{9.70}$$

Further, define Ax_u as being zero, and use the term *unobservable* in connection with x_u,

$$Ax_u = 0.$$

Then, (9.70) has the same mean square error solution as

$$Ax_o = b,$$

namely

$$x_o = A^+ b. \tag{9.71}$$

The term *observable* applies to x_o. The practical significance of the existence of unobservable states in a power system is that given the circuit topology and measurement sites, there may be components of some states [i.e., elements of state vector x in (9.69)] which will not be identified by (9.71). This contributes to the total state estimation error. Obviously, if A was of full rank in (9.70), the entire x vector is calculated by

$$x = A^+ b$$

and the vector x_u is zero. This point is discussed by Johnson [24].

Bibliography

[1] M. Kendall and A. Stuart, *The Advanced Theory of Statistics*, Vol. 1 to 3, Macmillan, New York, 1977.

[2] A. Papoulis, *Probability, Random Variables, and Stochastic Processes*, McGraw–Hill, New York, 1984.

[3] H. Cramer, *Mathematical Methods of Statistics*, Princeton University Press, Princeton, N.J., 1954.

[4] K. S. Miller, *Multidimensional Gaussian Distributions*, Wiley, New York, 1964.

[5] W. Feller, *An Introduction to Probability Theory and its Applications*, Wiley, New York, 1971.

[6] P. J. Bickel, "Edgeworth Expansions in Nonparametric Statistics," *Annals of Statistics*, v. 2, no. 1, 1974, pp. 1–20.

[7] R. Billinton, R. Ringlee, and A. Wood, *Power Systems Reliability Calculations*, MIT Press, Cambridge, Mass., 1973.

[8] R. Billinton, *Power System Reliability Evaluation*, Gordon and Breach, New York, 1970.

[9] F. Broussolle, "State Estimation in Power Systems – Detecting Bad Data through the Sparse Inverse Matrix Method," *IEEE Trans. on Power Apparatus and Systems*, v. PAS-97, no. 3, May–June 1978, pp. 678–682.

[10] P. W. Sauer, "A Generalized Stochastic Power Flow Algorithm," Ph.D. thesis, Purdue University, West Lafayette, Ind., December 1977.

[11] P. W. Sauer and G. T. Heydt, "A Generalized Stochastic Power Flow Algorithm; *1978 IEEE Summer Power Meeting*, Los Angeles, July 1978.

[12] P. W. Sauer and G. T. Heydt, "A Convenient Multivariate Gram–Charlier Type A Series," *IEEE Trans. Communications,* v. CM-27, no. 1, January 1979, pp. 247–248.

[13] G. T. Heydt, "Stochastic Power Flow Calculations," *1975 IEEE Summer Power Meeting,* San Francisco, July 1975.

[14] M. Kendall, "Proof of Relations Connected with the Tetrachoric Series and Its Generalization," *Biometrika,* 1941–1942, pp. 196–198.

[15] J. Dopazo, O. Klitin, and A. Sasson, "Stochastic Load Flows," *IEEE Trans. Power Apparatus and Systems,* v. PAS-94, no. 2, 1975, pp. 299–309.

[16] G. L. Viviani, "Stochastic Optimal Energy Dispatch," Ph.D. thesis, Purdue University, West Lafayette, Ind., May 1980.

[17] G. L. Viviani and G. T. Heydt, "Stochastic Optimal Energy Dispatch," *IEEE Trans. Power Apparatus and Systems,* v. PAS-100, no. 3, July 1981, pp. 3221–3228.

[18] A. Albert, *Regression and the Moore–Penrose Pseudoinverse,* Academic Press, New York, 1972.

[19] F. Galiana, "Short Term Load Forecasting," *EPRI Report SR-31,* Proc., Forecasting Methodology for Time-of-day and Seasonal Electric Utility Loads, Palo Alto, Calif., March 1976.

[20] G. Box and G. Jenkins, *Time Series Analysis and Control,* Holden Day, New York, 1970.

[21] F. Galiana, E. Handschin, and A. Fletcher, "Identification of Stochastic Electric Load Models from Physical Data," *IEEE Trans. Automatic Control,* v. AC-19, December 1974, pp. 887–893.

[22] S. Vemuri, E. Hill, and R. Balasubramanian, "Load Forecasting Using Stochastic Models," *Proc., PICA Conference,* June 1973, pp. 31–37.

[23] N. Uri, "Intermediate Term Forecasting of System Loads Using Box–Jenkins Time Series Analysis," *EPRI Report SR-31,* Proc., on Forecasting Methodology for Time-of-Day and Seasonal Electric Utility Loads, Palo Alto, Calif., March 1976.

[24] B. P. Johnson, "Power Flow Study Solutions using the Pseudoinverse," M.S.E.E. thesis, Purdue University, West Lafayette, Ind., May 1985.

[25] P. Bonanomi and G. Gramberg, "Power System Data Validation and State Calculation by Network Search Techniques," *IEEE Trans. Power Apparatus and Systems,* v. PAS-102, no. 1, January 1983, pp. 238–249.

[26] J. Gu, K. Clements, G. Krumpholtz, and P. Davis, "Topologically Partitioned Power System State Estimation," *IEEE Trans. Power Apparatus and Systems,* v. PAS-102, no. 2, February 1983, pp. 483–491.

[27] R. Bischke, "Power System State Estimation: Practical Considerations," *IEEE Trans. Power Apparatus and Systems,* v. PAS-100, no. 12, December 1981, pp. 5044–5047.

[28] T. Assefi, *Stochastic Processes and Estimation Theory with Applications,* Wiley, New York, 1979.

[29] T. McGarty, *Stochastic Systems and State Estimation,* Wiley, New York, 1974.

[30] A. Wood and B. Wollenberg, *Power Generation Operation and Control,* Wiley, New York, 1984.

[31] A. Monticelli and A. Garcia, "Reliable Bad Data Processing for Real Time State Estimation," *IEEE Trans. Power Apparatus and Systems,* v. PAS-102, no. 5, May 1983, pp. 1126–1139.

[32] N. Rao, C. Necsulescu, K. Schenk, and R. Misra, "A Method to Evaluate Economic Benefits in Interconnected Systems," *IEEE Trans. Power Apparatus and Systems,* vol. PAS-102, no. 2, February 1983, pp. 472–482.

[33] M. Sterling, *Power System Control,* Peregrinus, London, UK, 1978.

[34] R. Sullivan, *Power System Planning,* McGraw Hill, New York, 1977.

Exercises

Exercises that do not involve computer programming

9.1 Equation (9.14) was found by noting that the characteristic function of x is the Fourier transform of the probability density function,

$$\phi_x(\omega) = \mathcal{F}(f_x(x))$$

$$= \int_{-\infty}^{\infty} f_x(x)e^{-j\omega x}\, dx.$$

In this case, $e^{-j\omega x}$ is expanded in a Taylor series about $x=0$ and the result is integrated term by term. Assume that each of these integrals is well defined and the term by term integration is valid. In this exercise, the scalar x will be replaced by a multivariate vector X.

a. Consider the bivariate case, $X^t = (x_1 \;\; x_2)$. The characteristic function is the two dimensional Fourier transform,

$$\phi_X(\Omega) = \mathcal{F}(f_X(X))$$

$$= \int_{-\infty}^{\infty} f_X(X)e^{-j\Omega^t X}\, dX,$$

where the integral indicated is, in fact, two dimensional. Expand the integrand in a Taylor series about $X = 0$ and integrate term-by-term. Find the bivariate form of (9.14).

b. Repeat the process for arbitrary dimension of vector X.

9.2 The second characteristic function of a scalar x is defined as the natural logarithm of the first characteristic function,

$$\psi_x(\omega) = ln(\phi_x(\omega))\ .$$

a. Show that the Taylor series expansion of this expression is of the form (9.15).

b. Assume that the kth order cumulant of x is implicitly defined by (9.15). Find a formula for $\tau_x^{(1)}$, $\tau_x^{(2)}$, and $\tau_x^{(3)}$ in terms of the first three moments of x.

c. Consider the bivariate case, $X^t = (x_1, x_2)$. Now ϕ_x is a function of ω_1 and ω_2 (see Exercise 9.1). Find the bivariate form of Eq. (9.15) for $\psi_x(\Omega)$.

9.3 In the discussion of peak load forecasting using fitting functions, a simple forecast was obtained using

$$\hat{d}(t) = A^t f(t) .$$

In this case, for nondiscounted calculations, the expected mean square error was found to be σ_ξ^2 [see Eq. (9.22)].

Find the expected mean square error when discounting is used ($W \neq I$).

9.4 Consider peak load data are as follows:

Year	Peak Load (MW)
1	450
2	475
3	501
4	531

Using a straight line forecast, find A in the forecaster,

$$\hat{d}(t) = A^t f(t) .$$

Then repeat the process using a quadratic fit. For this purpose you may wish to forecast $d(t) - 450$ rather than $d(t)$ itself. Do not use discounting.

9.5 Repeat Exercise 9.4 using a discounting factor of 0.6. Use the form

$$W = diag(\beta^{(k-1)/2}) .$$

9.6 In some stochastic power flow studies, a Gaussian model of vector $[P\ Q]^t$ is assumed. In this exercise a "test" is proposed to determine the appropriateness of the Gaussian assumption. Let this vector be X.

a. Find a formula for the higher statistical moments of a scalar, x, which is Gaussian and standard measure,

$$f_x(x) = G(0, 1) .$$

b. Examine (9.14) and (9.15), noting that

$$\psi_x(\omega) = ln\ (\phi_x(\omega)) .$$

Expand the natural logarithm in a Taylor series and equate coefficients of like powers of ω. In this way, you will find formulas for the first few cumulants of x in terms of the first few moments. Do *not* look for a general formula, but find a formula for at least the first four cumulants.

c. Find the first four cumulants of Gaussian x in standard measure.

d. In part (c) you should have found that higher cumulants of Gaussian x are zero. This is also the result for multivariate X. With this in view, develop a measure based on the first three cumulants of $[P\ Q]^t$ to determine whether the assumption of Gaussian X is valid in a stochastic power flow study. Propose a tolerance on the third cumulants of X for this purpose.

Exercise that requires programming

9.7 In this problem, it is desired to prepare a computer program to calculate a capacity outage table for a system with the following generating units:

Unit No.	Capacity (MW)	Forced Outage Rate
1	400	0.005
2	400	0.005
3	800	0.004
4	431	0.013
5	60	0.017
6	25	0.019
7	15	0.024

The COT will be used to assess generation adequacy.

a. Print the COT using a table with 100 rows. The total installed capacity is 2131 MW. Prepare the program such that the table resolution is uniform in A (with $\Delta A = 2131/99$).

b. Often, COTs are prepared with the lower entries deleted (i.e., the large values of U deleted). These deletions generally are for states with very low probability. Comment on a threshold value of f_k that might be usable to delete outage states.

c. Consider the system under consideration to have peak demand 2000 MW. Thus the capacity margin is a nominal 131 MW. If the annual load duration curve is linear between 2000 and 1000 (the minimum demand), find the expected days per year for negative generation margin.

APPENDIX **A**

Simplex Method for Linear Programming

A.1

LINEAR PROGRAMMING AND THE SIMPLEX METHOD

Linear programming is a technique by which a linear function of several variables is maximized [1]. Also, linear combinations of the several variables are constrained by inequality constraints. There are numerous variations of the formulation and its solution; one very popular variation is produced here.

Consider the scalar function c of vector-valued argument, which is a linear function

$$c(X) = C^t X, \tag{A.1}$$

where C^t is a row vector. Constrain the region of X of interest,

$$AX \leq B \tag{A.2}$$

$$0 \leq X, \tag{A.3}$$

where the inequalities hold if each row holds, A is m by n, X is an n-vector, and B is an m-vector. The linear programming problem is to find X^* which satisfies (A.2) – (A.3) and maximizes $c(X)$.

The region in X that satisfies (A.2) and (A.3) is depicted in Figure A.1. The several inequalities in each row of (A.2), if plotted as equalities, form hyperplanes. The interior of these hyperplanes satisfies (A.2). It is assumed here that the elements of A are positive and that the intersection of the regions in X that satisfy each row of (A.2) is nonempty. Equation (A.3) forms hyperplanes defined by the coordinate axes. In the three-dimensional depiction in Figure A.1, these hyperplanes are $x_1 = 0$, $x_2 = 0$, and $x_3 = 0$. The point

$$X = 0$$

generally satisfies both (A.2) and (A.3). It is clear that if the elements of vector C are positive, and if the polytope that satisfies (A.2) and (A.3) is convex, the extremal $c(X)$ must occur on an edge of the polytope (or at a vertex). The n-dimensional polytope is termed a *simplex,* and the method of organized search of its vertices is termed the *simplex method.* The essence of the simplex method is the selection of a feasible solution (usually $X=0$), and the updating of that selection by movement along the steepest edge of the simplex until X^* is obtained. Reference [2] contains a theoretical development of the method.

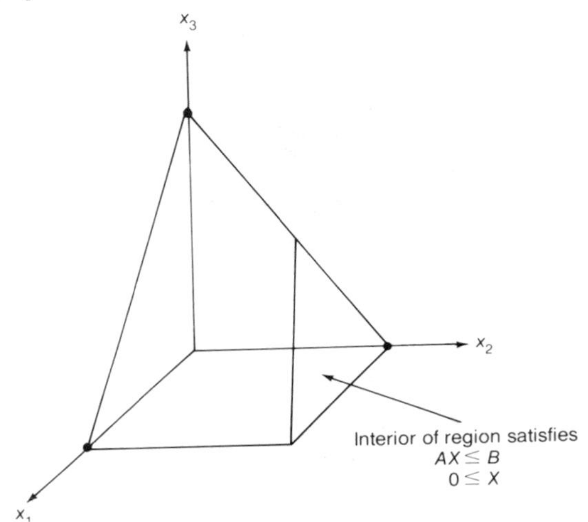

FIGURE A.1 Simplex.

If (8.2) is augmented on the left-hand side by slack variables, equalities are obtained. These slack variables are nonnegative,

$$AX + \begin{bmatrix} x_{n+1} \\ x_{n+2} \\ \vdots \\ x_{n+m} \end{bmatrix} = B \, .$$

If $\hat{X}$ is X augmented with the m slack variables, one obtains

$$\left[A \;\; \begin{matrix} 1\,0\,0 & \cdots & 0 \\ 0\,1\,0 & \cdots & 0 \\ 0\,0\,1 & \cdots & 0 \\ \cdots & & \\ 0\,0\,0 & \cdots & 1 \end{matrix} \right] \hat{X} = B \qquad \text{(A.4)}$$

$$0 \leq \hat{X} \, . \qquad \text{(A.5)}$$

A popular form of the simplex method employs the use of a table known as the *simplex tableau;* Figure A.2 depicts the initial state of such a table. At this initial state, a trial vertex is selected, generally

$$X = 0 \qquad \hat{X} = [0 \ \ B^t] \ .$$

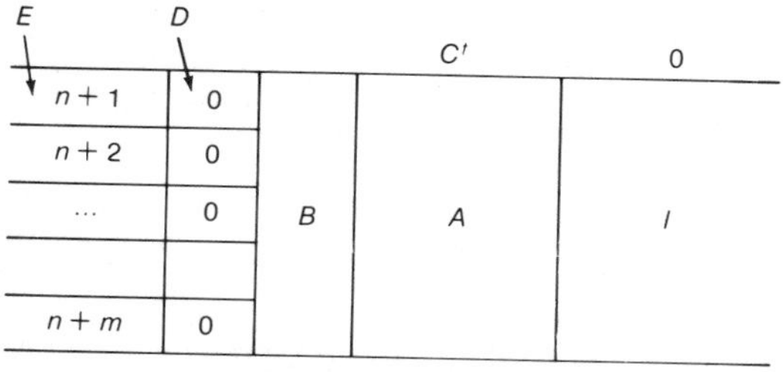

E	D		C^t	0
$n+1$	0			
$n+2$	0			
...	0	B	A	I
$n+m$	0			

FIGURE A.2 Initial state of linear programming simplex tableau.

This trial vertex appears in the simplex tableau in the sections labeled E (these are the subscripts of $\hat{X}$); D (these entries are the coefficients of $\hat{x}_{n+1}, \hat{x}_{n+2}, \ldots, \hat{x}_{n+m}$ in the cost function $c(X)$; initially the D column contains all zeros since the slack variables do not appear in $c(X)$; and B (these entries are the corresponding values of $\hat{x}_{n+1}, \hat{x}_{n+2}, \ldots, \hat{x}_{n+m}$). In Figure A.2, the section [A I] depicts the coefficient of $\hat{X}$ in (A.4). The section $[C^t \ 0]$ depicts the total coefficient of $\hat{X}$ in the cost function,

$$c(\hat{X}) = [C^t \ \ 0]\hat{X} \ .$$

Having established the initial state of the simplex tableau, the initial value of $c(\hat{X})$ is calculated,

$$c\left(\left[\frac{0}{B}\right]\right) = 0.$$

The simplex algorithm is then as follows:

1. Calculate an F row vector (1 by $n+m$) according to

$$F_i = \sum_{j=1}^{m} D_j [A \ \ I]_{ji}$$

where $[A \ \ I]_{ji}$ is the row j, column i entry of the body of the tableau. The F row is entered as shown in Figure A.3.

2. The $C^t - F$ row is calculated according to

$$[C^t - F]_i = [C^t]_i - F_i$$

and entered as shown in Figure A.3. The largest positive value of $[C^t - F]_i$ is identified and this column is the *key column.*

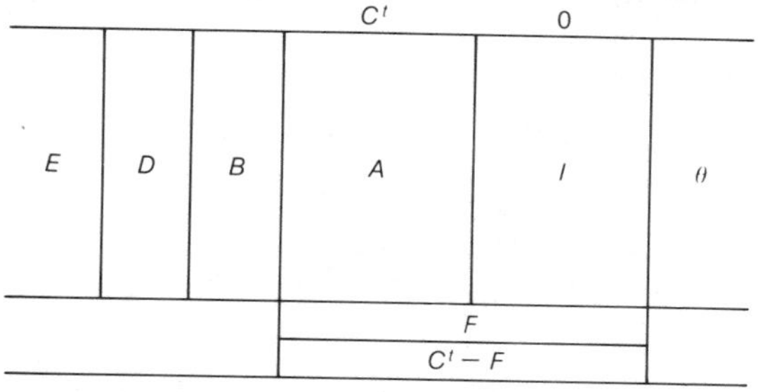

			C^t	0	
E	D	B	A	I	θ
			F		
			$C^t - F$		

FIGURE A.3 Simplex tableau showing intermediate vectors.

3. The θ column is calculated according to

$$\theta_i = \frac{B_i}{[A \quad I]_{ik}} \qquad k = key\ column$$

and this m-vector is entered in the tableau as shown in Figure A.3. The minimum positive value of θ is identified and this row is the ***key row.*** The intersection of the key row and column is the ***pivot element.***

4. Kron reduce the $[B \ A \ I\,]$ array using the pivot identified in step 3, for $i \neq p$, $j \neq p$,

$$[B \quad A \quad I]_{ij}^{new} = [B \quad A \quad I]_{ij}^{old} - \frac{[B \quad A \quad I]_{ip}^{old}[B \quad A \quad I]_{pj}^{old}}{[B \quad A \quad I]_{pp}^{old}}$$

and replace the pivot axis last by all zeros excepting the pivot position itself, which is replaced by unity.

5. Replace the entry in E in the key row by p, where p = the key column. Also, update D in the key row with its corresponding $[C^t \quad 0]$ entry.

6. At this time, the trial vertex is the point where all entries of $\hat{X}$ are zero except those elements that appear in E. The corresponding entries in B are the nonzero values of $\hat{X}$. Calculate $c(\hat{X})$.

This process is repeated until all $C^t - F$ entries are negative. At this time, the trial vertex is $\hat{X}$. This is the solution vector.

Example A.1

At a coal fired operating station, three types of coal, x_1, x_2, and x_3 are used with relative heat content per kilogram of 5, 6, and 7, respectively. The total required heat is 10^6 units. The costs per kilogram are 0.03, 0.04, and 0.07. Combustion of coal type 1 is limited by availability,

$$x_1 \leq 10^4 ,$$

and environmental considerations limit combustion of types 2 and 3,

$$3x_2 + 4x_3 \leq 7 \times 10^5 .$$

Find the most economical fuel mix.

Solution

The constraint equations are

$$x_1 \qquad\qquad \leq 10^4$$

$$3x_2 + 4x_3 \leq 7 \cdot 10^5$$

$$5x_1 + 6x_2 + 7x_3 = 10^6 .$$

Rewritten with two slack variables, x_4 and x_5,

$$\begin{bmatrix} 1\ 0\ 0\ 1\ 0 \\ 0\ 3\ 4\ 0\ 1 \\ 5\ 6\ 7\ 0\ 0 \end{bmatrix} \hat{X} = \begin{bmatrix} 10^4 \\ 7 \times 10^5 \\ 10^6 \end{bmatrix}.$$

The cost is $c(\hat{X})$,

$$c(\hat{X}) = [0.03 \quad 0.04 \quad 0.07 \quad 0 \quad 0]\hat{X}.$$

			0.03	0.04	0.07	0	0	θ	
x_4	0	10^4	1	0	0	1	0	10^4	← Key row
x_5	0	$7 \cdot 10^5$	0	3	4	0	1	∞	
x_3	0.07	$\frac{10^6}{7}$	$\frac{5}{7}$	$\frac{6}{7}$	1	0	0	$2 \cdot 10^5$	
			0.05	0.06	0.07	0	0		
			−0.02	−0.02	0	0	0		
			↑ Key column						

FIGURE A.4 Simplex tableau for Example A.1.

This formulation is a variant of that given above due to the presence of an equality constraint. This equality constraint gives no difficulty and the simplex tableau in Figure A.4 is written. In this tableau, the trial vertex

$$[x_1\ x_2\ x_3] = [0\ 0\ \frac{10^6}{7}]$$

was selected; this is the case since $X = \mathbf{0}$ does not satisfy the equality constraint. The stated initial trial vertex was arbitrarily chosen. If it were found difficult to locate a feasible trial vertex, the equality constraint could be used to eliminate a variable (e.g., x_3), and proceed with the reduced system. Using the initial trial vertex,

$$c(X) = c(\hat{X}) = 10^4 \ .$$

Row 3 of the table must be divided by 7 to cause the A_{33} position to be 1. Note the appearance of $10^6/7$, 5/7, 6/7, and 1 in row 3 of Figure A.4.

			0.03	0.04	0.07	0	0
x_1	0.03	10^4	1	0	0	1	0
x_5	0	10^4	0	3	10	0	1
x_3	0.07	$\frac{95}{7} \cdot 10^4$	0	$\frac{6}{7}$	1	$-\frac{5}{7}$	0

FIGURE A.5 Simplex tableau showing intermediate steps.

The $C^t{-}F$ and θ sections are calculated and indicated in Figure A.4. Both the column 1 and column 2 entries of $C^t{-}F$ qualify as key columns; column 1 is chosen arbitrarily. Row 1 is the key row. After Kron reducing the body of the tableau, Figure A.5 results. Figure A.5 also shows x_1 replacing x_4 in the trial vertex. At this point

$$X = \begin{bmatrix} 10^4 \\ 0 \\ \frac{95}{7} \times 10^4 \end{bmatrix} \qquad c(X) = 0.98 \times 10^4 \ .$$

			0.03	0.04	0.07	0	0
x_1	0.03	10^4	1	0	0	1	0
x_2	0.04	$\frac{1}{3} \cdot 10^4$	0	1	$\frac{10}{3}$	0	$\frac{1}{3}$
x_3	0.07	$\frac{92}{7} \cdot 10^4$	0	0	1	$-\frac{5}{7}$	$-\frac{2}{7}$

FIGURE A.6 Final simplex tableau.

The subsequent and final step reveals that the key column and row are 2, 2. The final tableau is shown in Figure A.6. this is recognized as the final tableau since the $C^t - F$ row is nonnegative. At the solution

$$X = \begin{bmatrix} 10^4 \\ \frac{1}{3} \times 10^4 \\ \frac{92}{7} \times 10^4 \end{bmatrix} \qquad c(X) = 0.96333 \times 10^4 .$$

A.2

APPLICATIONS OF LINEAR PROGRAMMING IN POWER ENGINEERING

Linear programming is a useful procedure in general engineering applications. The algorithm described above is readily modified to accommodate inequalities reversed to the sense in (A.2), namely

$$AX \geq B. \tag{A.6}$$

This is done by defining a new variable, Y, such that

$$Y = D - AX . \tag{A.7}$$

In (A.7), D is chosen such that vector Y has all positive entries. Note that the defining inequality is now

$$Y \leq D - B \tag{A.8}$$

instead of (A.6). Thus (A.8) returns the problem to that described earlier. There are many other variations of the formulation and its solution; a brief overview of selected applications in power engineering concludes this section.

Optimal Fuel Mix Calculations

The energy content and energy conversion efficiencies of alternative fuels may be modeled in a piecewise linear fashion. Under this assumption, the total fuel cost becomes a linear function of quantity of each fuel. Availability and operating constraints are of the form

$$AX \leq B,$$

where X is a vector of fuel types and matrix A and vector B are model parameters.

Linearized Optimal Reactive Power Dispatch

Under assumptions of linear bus voltage magnitude versus bus reactive power dispatch,

$$\Delta V = \Xi \, \Delta Q . \tag{A.9}$$

In (A.9), ΔV and ΔQ denote the incremental change in bus voltage magnitude and bus reactive power dispatch and Ξ is the inverse of the $\partial Q / \partial V$ quadrant of the Jacobian matrix. If the objective of the optimal dispatch is to support bus voltage,

$$\Delta V \geq B , \tag{A.10}$$

from which

$$\Xi \Delta A \geq B \ . \tag{A.11}$$

Equation (A.11) is rearranged using (A.7) to obtain the inequality in (A.2). The objective function in this case is $c(\Delta Q)$,

$$c(\Delta Q) = [1\ 1\ 1\ 1\ \cdots]\Delta Q \tag{A.12}$$

if the shunt capacitor costs are linear with respect to the reactive voltamperes "generated." Other costs models are attainable using proportional entries in the coefficient row vector in (A.12).

Maximum Interchange Capacity

In bulk power transmission system operation, it is sometimes useful to know what total simultaneous interchange power is available via interties. This maximum total is termed the ***simultaneous interchange capacity*** (SIC). The SIC may be a short term figure limited, in part, by the short term generation available from neighboring companies. The SIC may be a long term level based, in part, on the long term power available over tie lines. Note also that the transmission system presents limits in the form of line and component ratings. If the SIC is limited primarily by the transmission system, the term ***transmission limited*** applies, while cases of intertie power limits are termed ***generation limited.***

Let vector X denote the interchange power levels entering a system from several neighbors (see Figure A.7). It is desired to maximize $c(X)$,

$$c(X) = [1\ 1\ 1\ \cdots\ 1]X \ .$$

Generation constraints are of the form

$$X \leq B \ ,$$

where B_1 is an n-vector of limits corresponding to the n interchanges $x_1, x_2, \ldots, x_n$. Note that each interchange may be composed of several intertie buses. Transmission limits are of the form

$$A_2X \leq B_2,$$

where A_2 is a k by n array which describes the assumed linear loading of k lines with respect to the n interchanges (note that elements of matrix A_2 are essentially distribution factors), and B_2 is a k-vector of line ratings. Then

$$\begin{bmatrix} I \\ A_2 \end{bmatrix} X \leq \begin{bmatrix} B_1 \\ B_2 \end{bmatrix}, \tag{A.13}$$

which is of the form (A.2).

Example A.2

The Home Power Company has interties with the Utah Power Company, New Mexico Power Company, and the Bureau of Power Management. It is desired to find the short term simultaneous interchange capacity of the Home Power Company in order to evaluate the possibility of negative generation margin. The three interchanges have generation limits

$$x_{utah} \leq 1000\ x_{nmpc} \leq 2000\ x_{bpm} \leq 300 \ .$$

Each interchange described is composed of several intertie buses. The several interties with Utah, for example, form the Utah interchange. Three key lines, L1, L2, L3 are known by operators to limit the SIC. The sum of the distribution factors for L1 and all intertie buses in the Utah interchange is 0.3000. Similarly, the key lines here incremental loads ΔP_1, ΔP_2, ΔP_3,

$$\begin{bmatrix} 0.300 & 0.100 & 0.000 \\ 0.010 & 0.010 & 0.010 \\ 0.100 & 0.100 & 0.010 \end{bmatrix} \begin{bmatrix} x_{utah} \\ x_{nmpc} \\ x_{bpm} \end{bmatrix} = \Delta P \ .$$

The maximum incremental loads in L1, L2, L3 are each 400 MW.

Find the SIC for the Home Power Company.

Solution

It is desired to maximize $c(X)$,

$$c(X) = (1 \ \ 1 \ \ 1)X$$

subject to

$$\begin{bmatrix} 1 & 0 & 0 \\ 0 & 1 & 0 \\ 0 & 0 & 1 \\ 0.300 & 0.100 & 0.000 \\ 0.010 & 0.010 & 0.010 \\ 0.100 & 0.100 & 0.010 \end{bmatrix} X \le \begin{bmatrix} 1000 \\ 2000 \\ 3000 \\ 450 \\ 450 \\ 450 \end{bmatrix} . \tag{A.14}$$

Inequality (A.14) is of the form (A.2), and linear programming is used to obtained a solution. The details of the linear programming are omitted, but the solution is

$$X = \begin{bmatrix} 1000 \\ 1500 \\ 3000 \end{bmatrix} .$$

Hence, the SIC is 5500 MW. In this case, the pivot rows of the simplex tableau [using the ordering in (A.14)] were 1, 4, and 3, and these rows of (A.14) are at their limit. It is concluded that this system is predominantly *generation limited.*

Optimal Secure Operating Schedule

A *secure* power system is a system which is operated at sufficient line loading margin and generation margin such that a disturbance does not result in negative margins in either regard. Thus there is a tradeoff between system security and system load: as the system load is increased and operated with a high economic return, the loading and generation margins decrease, thereby creating a less secure condition. If the operating point is viewed as a point in multidimensional space, the security constraints may be viewed as hyperplanes that contain secure operating points. If the security constraints are formulated as linear functions of generation power settings, the hyperplanes are flat and the volume of the operating region may be maximized using linear programming. The large operating region translates into high operating flexibility. Fischl and others give additional details and examples [3–5].

Bibliography

[1] G. Dantzig, "Maximization of a Linear Function of Variables Subject to Linear Inequalities," *Activity Analysis of Production and Allocation,* T. Koopmans, ed., Cowles Commission Monograph no. 13, Wiley, New York, 1951.

[2] D. Pierre, *Optimization Theory with Applications,* Wiley, New York, 1969.

[3] R. Fischl, W. Huckins, T. Ho, and S. Hurtak, "Methods in Transmission Network Planning," *1977 IEEE Summer Power Meeting,* Mexico, D.F., Mexico.

[4] R. Adler and R. Fischl, "Secure Allocation of Regulating Reserve in the Presence of Bus-Demand Uncertainty," *IEEE Trans. Power Apparatus and Systems,* v. PAS-97, no. 6, November–December 1978, pp. 1994–2004.

[5] R. Adler and R. Fischl, "Security Constrained Economic Dispatch with Participation Factors Based on Worse Case Bus Load Variations," *IEEE Trans. Power Apparatus and Systems,*" vol. PAS-96, no. 2, March–April 1977, pp. 347–356.

APPENDIX **B**

The Moore-Penrose Pseudoinverse

B.1

INTRODUCTION

In this appendix, the topic of pseudoinverses is discussed. This technique is used in a wide variety of advanced engineering applications, primarily in automatic control theory. In power engineering, pseudoinverses are used in "transportation method" power flow analyses, state estimations, load forecasting, linear programming, and certain other probabilistic applications. The pseudoinverse naturally arises in many instances where singular matrices occur, such is the case were the indefinite bus impedance and admittance matrices are found to be pseudoinverses of each other. Like many other topics in mathematics, it is not necessary to introduce the concept of pseudoinverses, but their use simplifies many developments and they often give a different, useful viewpoint of vector and matrix expressions. The pseudoinverse was developed somewhat independently by Moore in 1920 [1] and later by Penrose. References [2,3] are definitive texts on this topic.

The set of linear, algebraic equations

$$Ax = b \qquad \text{(B.1)}$$

has a single solution for vector x when A is square and nonsingular. It is natural to raise the question as to the nature of the solution when A is not square or when A is square and singular. The first of these possibilities (i.e., rectangular A, may be considered as the problem of solving n_r

simultaneous equations in n_c unknowns. When $n_c > n_r$, there are $n_c - n_r$ more unknowns than equations, and (B.1) may be "solved" only to the extent that the excess unknowns will take on values in an even, reasonable fashion. Admittedly, this is an imprecise statement, but it is this thought that makes the mathematics indicated below appropriate to the kind of problems commonly encountered in practical engineering applications. For example, the simultaneous set of equations

$$x_1 + x_2 + x_3 + x_4 = 1$$

$$x_1 = 2$$

is "reasonably" solved as

$$x = \begin{bmatrix} -1/3 \\ -1/3 \\ -1/3 \\ 2 \end{bmatrix}.$$

The implication of this illustration is that the equation

$$\begin{bmatrix} 1 & 1 & 1 & 1 \\ 1 & 0 & 0 & 0 \end{bmatrix} x = \begin{bmatrix} 1 \\ 2 \end{bmatrix} \tag{B.2}$$

is reasonably solved as

$$x = A^+ \begin{bmatrix} 1 \\ 2 \end{bmatrix},$$

where A is the coefficient of x in (B.2) and A^+ is

$$A^+ = \begin{bmatrix} 0 & 1 \\ \frac{1}{3} & -\frac{1}{3} \\ \frac{1}{3} & -\frac{1}{3} \\ \frac{1}{3} & -\frac{1}{3} \end{bmatrix}.$$

The matrix A^+ is the pseudoinverse of A, a concept which will be formalized below.

In the event $n_r > n_c$, there are "too many" equations for the given number of unknowns. The best that one may expect, from an engineering point of view, is that the unknowns will be selected so that the error in satisfying each equation will be minimum. This minimum error is usually done in the least squares sense. For example, the equations

$$x_1 + x_2 = 1$$

$$x_1 + x_2 = 1.5$$

$$2x_1 + x_2 = 2.8$$

cannot be satisfied exactly for any choice of vector x. The mean square error in satisfying these equalities is *mse*

$$mse = \frac{1}{3}[(x_1 + x_2 - 1)^2 + (x_1 + x_2 - 1.5)^2 + (2x_1 + x_2 - 2.8)^2] .$$

The minimum mean square error is found by setting $\nabla_x(mse) = 0$, with the result

$$x = \begin{bmatrix} 1.55 \\ -0.3 \end{bmatrix}.$$

This is a mean square fit of x_1, x_2 to the given set of algebraic equations. One interpretation of the illustration is that the given set of equations,

$$\begin{bmatrix} 1 & 1 \\ 1 & 1 \\ 2 & 1 \end{bmatrix} x = \begin{bmatrix} 1.0 \\ 1.5 \\ 2.8 \end{bmatrix} \quad \text{(B.3)}$$

has the "solution"

$$x = A^+ \begin{bmatrix} 1.0 \\ 1.5 \\ 2.8 \end{bmatrix}$$

where A^+ is the pseudoinverse of the coefficient of vector x in (B.3):

$$A^+ = \begin{bmatrix} -0.5 & -0.5 & 1.0 \\ 1.0 & 1.0 & -1.0 \end{bmatrix}.$$

Again, this remark will be formalized below.

The case of $n_c = n_r$ with singular A may be viewed as a minimum square error problem with "reasonable apportioning" of unknown values.

B.2

DEFINITION AND PROPERTIES OF THE PSEUDOIVNERSE

The terms *singular values* and *singular vectors* are defined as follows for a general m by n matrix A:

The *singular values* of A are σ_i, which are the square roots of the nonnegative eigenvalues of A^tA. These singular values are also the square roots of the nonnegative eigenvalues of AA^t. For the case of zero rank of A, there is one singular value of A and it is zero.

The *singular vectors* of A are e_i and f_i, where

$$A^tAe_i = \sigma_i^2 e_i \qquad AA^tf_i = \sigma_i^2 f_i \, .$$

Evidently, e_i are the eigenvectors of A^tA corresponding to the nonnegative eigenvalues of A^tA and f_i are eigenvectors of AA^t corresponding to these nonnegative eigenvalues. Usually, the singular vectors are chosen as orthonormal.

The Moore–Penrose pseudoinverse of any m by n matrix A may be defined as A^+,

$$A^+ = E^t \begin{bmatrix} \sum^{-1} & 0 \\ 0 & 0 \end{bmatrix} F, \quad \text{(B.4)}$$

where E is the matrix of singular vectors $[e_1 \; e_2 \; \cdots \; e_\ell]$, F is a similar modal matrix $[f_1 \; f_2 \; \cdots \; f_k]$, and $\sum$ is a diagonal matrix of the nonzero singular values of A,

$$\sum = diag\,(\lambda_i) \qquad i = 1, 2, \ldots, q.$$

If the only singular value of A is zero, $\sum^{-1}$ in (B.4) is omitted and only the lower right rectangle of zeros in

$$\begin{bmatrix} \sum^{-1} & 0 \\ 0 & 0 \end{bmatrix} \tag{B.5}$$

is written; in this case, A^+ is a n by m rectangle of zeros. The "definition" in (B.4) is a practical definition which is proposed here for use in power engineering. The more usual mathematical definition of the pseudoinverse of matrix A is as follows: matrix A^+ is defined as the pseudoinverse of A if all four of the following conditions are satisfied,

1. $AA^+A = A$
2. $A^+AA^+ = A^+$
3. $(AA^+)^t = AA^+$
4. $(A^+A)^t = A^+A$.

These four conditions are known as the Moore–Penrose conditions. The equivalence of the Moore–Penrose conditions to Eq. (B.4) for all cases of practical interest is well known, as is the uniqueness of A^+ [2].

When the m by n matrix A is of rank q, there are q nonzero singular values of A and

E *is* n *by* 1

F *is* m *by* 1

$\sum$ *is* q *by* q

$\begin{bmatrix} \sum^{-1} & 0 \\ 0 & 0 \end{bmatrix}$ *is* n *by* m

and hence A^+ is an n by m matrix. Note that in the matrix (B.5), the upper right rectangle of zeros is q by $m - q$, the lower left rectangle of zeros is $n - q$ by q, and the lower right rectangle of zeros is $n - q$ by $m - q$. The Moore–Penrose pseudoinverse is simply termed the pseudoinverse in this text.

The properties of A^+ are documented in [1-6]. They are briefly summarized as (for any m by n matrices A and for any symmetric n by n matrices S) follows:

1. A^+ is unique.
2. The mean square error solution of $Ax=b$ is $x=A^+b$.
3. When Λ_S is a diagonal matrix of the eigenvalues of S,

   $$S^+ = E\Lambda_S^+E^t,$$

   where E is the modal matrix of S. Also note that Λ_S^+ consists of zeros everywhere except in the upper left diagonal, where the inverses of the nonzeo singular values of S appear. For nonsingular S, $S^+ = S^{-1}$.
4. A^+ is expressible as a function of the pseudoinverse of a symmetric matrix as

   $$A^+ = (A^tA)^+A^t$$

   $$A^+ = A^t(AA^t)^+ .$$

$$\lambda^+ = \begin{cases} \lambda^{-1} & \lambda \neq 0 \\ 0 & \lambda = 0 \,. \end{cases}$$

6. When A is decomposed into lower left trapezoidal L and upper right triangular U,

$$A = LU,$$

where L is m by k, and U is k by n, then

$$A^+ = (A^t A)^+ A^t$$

$$= (A^t A)^{-1} A^t \qquad (\textit{if } A^t A \textit{ is nonsingular})$$

$$= U^{-1}(L^t L)^{-1} L^t \; (\textit{if } U \text{ and } L^t L \textit{ are nonsingular})$$

7. If the first k columns of A are linearly independent and if $A = BC$, then

$$A^+ = C^+ B^+ \,.$$

Example B.1

Find the pseudoinverses of the following matrices:

$$A_1 = \begin{bmatrix} 1 & 2 \\ 0 & 3 \end{bmatrix} \quad A_2 = \begin{bmatrix} 1 & 2 \\ 0 & 0 \end{bmatrix} \quad A_3 = \begin{bmatrix} 1 & 2 & 3 \\ 0 & 0 & 1 \end{bmatrix}$$

$$A_4 = \begin{bmatrix} 1 & 2 \\ 2 & 5 \\ 3 & 10 \end{bmatrix} \quad A_5 = \begin{bmatrix} 1 & 1 & 1 & 1 \\ 1 & 0 & 0 & 0 \end{bmatrix}$$

Solution

In case 1, the given matrix is nonsingular and A_1^+ is readily identified as A_1^{-1}. In this case, the singular values are $\sqrt{7 \pm 2\sqrt{10}}$, and the normalized singular vectors are

$$\sigma_1 = 3.65 \quad e_1 = \begin{bmatrix} 0.158 \\ 0.987 \end{bmatrix} \quad f_1 = \begin{bmatrix} 0.585 \\ 0.811 \end{bmatrix}$$

$$\sigma_2 = 0.82 \quad e_2 = \begin{bmatrix} -0.987 \\ 0.158 \end{bmatrix} \quad f_2 = \begin{bmatrix} -0.811 \\ 0.585 \end{bmatrix} .$$

Therefore, A_1^+ is calculated from (B.4):

$$A_1^+ = \begin{bmatrix} 0.158 & 0.987 \\ -0.987 & 0.158 \end{bmatrix} \cdot \begin{bmatrix} 0.274 & 0 \\ 0 & 1.22 \end{bmatrix} \cdot \begin{bmatrix} 0.585 & -0.811 \\ 0.811 & 0.585 \end{bmatrix} = \begin{bmatrix} 1.00 & 0.67 \\ 0 & 0.33 \end{bmatrix}$$

which agrees with A^{-1}.

In case 2, the singular values are $\sqrt{5}$ and 0.0 and $\sum^{-1}$ is recognized simply $1/\sqrt{5}$. The singular vectors are calculated and substituted into (B.4) to give

$$A_2^+ = \begin{bmatrix} 0.447 & 0.894 \\ 0.894 & -0.447 \end{bmatrix} \cdot \begin{bmatrix} \frac{1}{\sqrt{5}} & 0 \\ 0 & 0 \end{bmatrix} \cdot \begin{bmatrix} 1 & 0 \\ 0 & 1 \end{bmatrix} = \begin{bmatrix} 0.20 \\ 0.40 \end{bmatrix} .$$

Note that

$$AA^+ = \begin{bmatrix} 1 & 0 \\ 0 & 0 \end{bmatrix} \neq I .$$

Case 3 is readily pseudoinverted using

$$A^+ = A^t(AA^t)^+,$$

with the result

$$A_3^+ = \frac{1}{5}\begin{bmatrix} 1 & -3 \\ 2 & -6 \\ 0 & 5 \end{bmatrix} .$$

Case 4 is recognized as a product of lower left trapezoidal L and upper right triangular U,

$$A_4 = \begin{bmatrix} 1 & 0 \\ 2 & 1 \\ 3 & 4 \end{bmatrix} \begin{bmatrix} 1 & 2 \\ 0 & 1 \end{bmatrix}$$

and the formula

$$A^+ = U^{-1}(L^tL)^{-1}L^t$$

is used to obtain

$$A_4 = \frac{1}{42}\begin{bmatrix} 45 & 48 & -33 \\ -14 & -14 & 14 \end{bmatrix} .$$

The last case is the matrix seen in (B.2). The pseudoinverse is conveniently found using

$$A^+ = A^t(AA^t)^+ .$$

In this case,

$$A_5 = \begin{bmatrix} 1 & 1 \\ 1 & 0 \\ 1 & 0 \\ 1 & 0 \end{bmatrix} \begin{bmatrix} 4 & 4 & 1 \\ 1 & 1 & 1 \end{bmatrix}^+ = \begin{bmatrix} 0 & 1 \\ 1/3 & -1/3 \\ 1/3 & -1/3 \\ 1/3 & -1/3 \end{bmatrix} .$$

B.3 APPLICATIONS OF THE PSEUDOINVERSE

The primary applications areas of the pseudoinverse stem primarily from the fact that the mean square error solution of a linear set of equations is also the pseudoinverse solution. Applications in power engineering include:

1. *Transportation power flow studies.* The relationship

$$S_{bus} = L\overline{S}, \tag{B.6}$$

where S_{bus} is the bus injection powers, L is the line–bus incidence matrix, and $\overline{S}$ is the line flow vector. A mean square error solution is

$$\overline{S} = L^+ S_{bus} . \tag{B.7}$$

Variations of the use of (B.7) are discussed in Chapter 5. Methods in which entries of either $\overline{S}$ or S_{bus} are estimated from measurements of other entries are termed ***state estimation methods***

2. *Load forecasting.* Straight line fits to data are often encountered in forecasting problems. These are usually either mean square error or weighted mean square error problems. Also, nonlinear function coefficients may be calculated using pseudoinverses (see Chapter 9).

3. *Linear programming.* The linear programming problem is the extremization of an index, c_q, which is a linear function of vector X. Let

$$c_q = c^t X \, .$$

The vector X is constrained by the component-wise inequality

$$AX \leq b \, .$$

Candidate solutions, X_0, are augmented by ΔX such that ΔX does not cause c_q to increase. Therefore, c is orthogonal to ΔX. It can be shown that ΔX must be of the form

$$\Delta X = (I - A^+ A) y,$$

where y is any vector of appropriate dimension. The orthogonality of c and $(I - A^+ A) y$ results in a solution for X which is equivalent to that obtained by the simplex method (see Appendix A).

Bibliography

[1] M. Nashed, *Generalized Inverses and Applications,* Academic Press, New York, 1976.

[2] G. Golub and C. Van Loan, *Matrix Computations,* John Hopkins University Press, Baltimore, Md., 1983.

[3] A. Albert, *Regression and the Moore–Penrose Pseudoinverse,* Academic Press, New York, 1972.

[4] B. Noble, *Methods for Computing the Moore-Penrose Generalized Inverse and Related Matters,* Prentice–Hall, Englewood Cliffs, N.J., 1977.

[5] J. Wilkinson and C. Reinsch, *Handbook for Automatic Computation, Vol. 2: Linear Algebra,* Springer-Verlag, New York, 1971.

[6] J. Bennett, "Triangular Factors of Modified Matrices," *Numerical Mathematics,* v. 7, 1965, pp. 217–221.

Index

N

O

P

Q

R

S

T